Docklands

Docklands
Cultures in Conflict, Worlds in Collision

Janet Foster

© Janet Foster 1999

Published in the UK in 1999 by UCL Press

UCL Press Limited
1 Gunpowder Square
London
EC4A 3DE

and

325 Chestnut Street
8th Floor
Philadelphia
PA 19106
USA

The name of University College London (UCL) is a registered trade mark used by UCL Press with the consent of the owner.

British Library Cataloguing-in-Publication Data
A catalogue record for this book is available from the British Library.

Library of Congress Cataloging-in-Publication Data are available

ISBN: 1–85728–273–6 HB
       1–85728–274–4 PB

Every effort has been made to contact copyright holders for their permission to reprint material in this book. The publisher would be grateful to hear from any copyright holder who is not here acknowledged and will undertake to rectify any errors or omissions in future editions of this book.

Typeset by Graphicraft Ltd, Hong Kong
Printed and bound by T.J. International Ltd, Padstow, UK

For Mark and Emma

Contents

Acknowledgements

There are many people whose support was invaluable while I was researching and writing this book, during which time I have changed jobs twice and become a mum. The project was only made possible in the first instance by the T.H. Marshall Fellowship funded by the *British Journal of Sociology* at the London School of Economics. I am immensely grateful to those involved in establishing this fellowship and for the opportunity it provided me to investigate the impact of urban change and community transition on the Isle of Dogs, and to Paul Rock in particular, who offered support through the good times and the bad. As ever, his insights during the fieldwork and comments on the manuscript were very helpful indeed.

Many staff in the Department of Sociology at Warwick deserve thanks too. The critical mass of qualitative researchers there provided me with greater confidence that ethnography was a *bona fide* form of social enquiry and colleagues including Ellen Annandale, Jim Beckford, Bob Burgess and Margaret Archer in different ways provided intellectual stimulation or other support for which I am most grateful. I would especially like to thank Carol Wolkowitz for many convivial train journeys from Coventry to London in which progress on this research was often discussed; Nina Cope, my excellent teaching assistant; and the Warwick Research and Innovations fund which helped finance the transcription costs. More recently my colleagues at Cambridge, Ben Bowling and Loraine Gelsthorpe, deserve special thanks for listening as I agonized over the final stages of writing and for commenting on all or parts of drafts.

Eve Hostettler at the Island History Trust deserves thanks for her help during the research, her supportive comments on the manuscript, and facilitated the use not only of the Trust's photographs but others in the collection too. Kutub acted as an interpreter and without him the interviews with many of the Bengalis would not have been possible. PC Phil Hanford compiled the racial harassment statistics and DS Giles at 3 Area provided other crime data. Moira Parkes willingly transcribed some of the hundreds of hours of interviews with Neil Dion as well as chasing up references. Most of all she provided moral support when I very much needed it.

I am indebted to all of those, too numerous to mention by name, who kindly agreed to sit with me and my tape recorder to relate their experiences and made me so welcome. Without them this book would never have been written. I would also like to thank all those individuals and agencies who allowed me to reproduce photographs or tables in the book.

Finally, there are two other people who made this book possible. Anita Starck who looked after Emma so that I could write, and my husband Mark, who has supported me throughout and never lost faith in me, even when I lost it myself. I could not have done it without him.

List of figures

List of tables

Introduction

I first became interested in the Isle of Dogs when I was working on two difficult to let council estates in the East End of London in the late 1980s. During this research I became friendly with a woman whose flat had panoramic views of the locality. Every time I visited, the view changed. I watched in amazement as a post-modern landscape rapidly took shape and I became increasingly curious about how Docklands residents felt about the development, which was massive in scale and dramatically altering the nature of the place where they lived. From these initial thoughts an ethnographic study grew when in 1990 I was awarded the T.H. Marshall Fellowship at the London School of Economics to look at a community in transition on the Isle of Dogs.

I read the available literature on the development (most of which was critical) and found that local people's perceptions were rarely documented in any depth, although their dissatisfaction made newsworthy stories for the media. Furthermore, in one of the most extensively researched and dramatic examples of urban regeneration, the views of those responsible for implementing change had not been documented either. It seemed to me that in order to understand the nature of change on the Isle of Dogs, where the heart of commercial development was located, it was necessary to investigate the different and competing perspectives of all those actors involved, both powerless *and* powerful.

Recognizing the need to understand the development from a variety of different perspectives was not without its problems. Until I embarked on this research I had no experience of interviewing powerful people. Given my sociological background and our well established predisposition for siding with the "under-dog" (see Becker 1965:244), I began, not surprisingly, with an in-built sympathy for those whose lives had been turned upside-down by the development with, I suspected, little benefit to them (Foster 1996:151). I assumed that business people and affluent residents would be a rather homogenous group broadly pro-development and anti-community. I discovered, however, that they frequently presented arguments that not only conflicted with my expectations but, even when they reinforced them, were often understandable from their perspective. In fact all of the different, sometimes deeply conflicting, accounts of the local people, councillors, Development Corporation executives, Board members and employees, affluent residents, business people and developers I interviewed had legitimacy if they were considered within the frame of reference of the individual's experience or the interest group from which they originated.

Although my empathy with local people and their plight remained, because they seemed incidental rather than integral to the development, I discovered that "powerlessness" and "power" are not one-dimensional, and as I began to explore the issues in more depth the complexities became increasingly evident. Conflicts within and between different sections of the "powerless", between "Islanders" and non-Islanders, between white and Bengali, between affluent and established residents, and indeed between and within each of these groupings themselves, were as important, and in some cases, more important than, the conflicts with the LDDC and developers. The "powerful" were also more divided than they might have superficially appeared. It is often assumed that the pursuit of profit is the primary goal of developers and business and that this supersedes all other concerns. While the goal was shared, the *means* of acquiring it was perceived in very different ways and the reflections of the business community, affluent residents and members of the Corporation on the development and community gain were far more varied than I had anticipated.

I quickly discovered that status was essential to accessing more powerful groups and for the first time found myself using my "Doctor" title. I contacted the key players in the Development Corporation both past and present by letter. They seemed to welcome the opportunity to discuss their views and experiences and at the end of each interview (most of which were tape recorded) I asked for the names and contact numbers of others with whom they thought I should talk (a technique called "snowballing"). I adopted a similar approach in the "community", initially contacting a variety of active groups and organizations in the area, by attending and observing meetings, and then broadened out through residents' contacts. I also contacted groups like the Docklands Forum and Docklands Consultative Committee whose remit covered Docklands as a whole, attended local council meetings, and observed a variety of liaison meetings.

Different roles were required for different groups and in each I was viewed in very different ways. A new set of clothes and a professional approach were required for the Corporation and business environment. Here I was perceived as an academic researcher from a well known university who was writing a book. Another, less formal role, but one which retained some status, was required for affluent residents keen to establish my credentials, my approach, and what I intended to do with the material. My preferred and accustomed research role with no emphasis on "status" was often most important for established Island residents; while for the Bengali families, as a white woman unable to speak their first language, I required an interpreter for some interviews where language proved to be a barrier.

Moving between such contrasting worlds required tuning into different frames of reference and constantly shifting between diverse views. It also had an impact on my vacillating feelings towards each of the groups involved. I found myself drawn towards the world of the established residents and their sense of place. It was not difficult to empathize with their feelings of anger about the development. However, their long association with the area and the emotions they had invested in it made them a powerful force, and the "special" nature of what they perceived as *their* Island also included denying a place in it to others. I was repelled by the racism and the hostility to which some outsiders were subjected and shocked by the extent to which a minority were willing to go.

I felt ashamed that I had any sympathy for the white working class when I listened to the harrowing accounts of harassment experienced by some Bengali families forced to move to the Isle of Dogs as a result of a "one offer only" housing policy. Even though I knew that these dreadful acts were perpetrated by a minority, somehow the whole white population seemed tainted with racism — a considerable injustice to those who I knew did not sympathize with racist views in any way. I also found, like many newcomers, the negativity and confrontational approach of some Island people difficult.

Despite my initial scepticism, I was impressed by the sheer pace of development during its early stages and, like many of those I interviewed, I found it exciting and breathtaking on one level while also appreciating the impact that change on this scale had wrought on the residential population. These different emotions frequently left me with feelings of confusion, with empathy and detachment simultaneously. That I was always being exposed to another version of events, another story, reinforced these experiences. What I had rather naively believed would be a "nice" community study providing a respite for a short period from my criminological research in fact turned into a deeply challenging and sometimes disturbing experience.

Much has been written about the power relationships between interviewers and interviewees and the importance of a non-exploitative approach. Far less is said about the feelings of powerlessness that researchers sometime, feel. For those in positions of power, interviews in a variety of forms go with the job. Although the influential people I interviewed were less familiar with the largely unstructured approach I adopted, their experience showed in the adept management of the interview. These encounters left me with a very different set of questions and emotions to ponder — had they told me the truth? Were their emotions spontaneous or engineered? How much had they revealed? I was asked questions and my opinion was sought in a way I had not experienced to any significant degree previously, and their responses were more measured and concise. I found myself in an unaccustomed position, under pressure to discuss issues in more depth, express my views and justify my viewpoint.

Although I was rarely conscious of it during the interviews themselves, as I read through some of the interview data I was struck by the degree of control some assumed over the interview process, so that these were more pointedly "conversations with a purpose" (Burgess 1984:102) (although our purposes may have been different) both for myself and the interviewees. In some cases individuals moved between talking "on" and "off" the record, a sure sign that they were consciously constructing and controlling their accounts. Some of the best data were inevitably "off the record" and I have not used this material in the following account.

Affluent residents were less controlling in their approach but sometimes seemed unwilling to be overtly critical, perhaps fearing that such information might be used against them, as it had been in media coverage where the focus was on highlighting the material and social divide between the two groups. When new residents talked of the contrasts between the poorer sections of the community and the affluent housing, therefore, a great deal was left unspoken. For example, one businessman said:

> The house I had was in Island Gardens . . . I mean nothing wrong with the house. The river was just over the back and the park was there, but within

two minutes walk, God you wondered where you were . . . I won't describe what I might think about it.

In most cases I did not probe further, almost as if I had accepted that these were boundaries beyond which the interviewees did not want to explore and comments were sometimes made after the tape recorder was switched off to indicate their concerns. For example, one newcomer said at the end of an interview that she had been really honest "because it's not like a survey or anything where you feel that you have to be positive because otherwise it will affect your house prices" (research notes).

By contrast, when I spoke to the poorer sections of the white Island community about their attitudes towards the Bengalis, for example, I did probe their initial remarks and encouraged them to open up and discuss their negative views — a feeling I was not entirely comfortable with because in many ways I sensed that they were opening up because they felt I was sympathetic to their views (which I was not) and because I knew that I would use such data to illustrate racism in the area despite their claims that they were not racist. At one level these differences in accounts and my approach to them reflect the relative powerlessness of poorer Island residents and the relative power of affluent residents, the business community and Corporation.

There are other contradictions too. Given that much of my prior research involved work with offenders, I had become accustomed to anonymizing those whom I observed and interviewed and pursued the same course on the Isle of Dogs too. This proved fraught with difficulties as it became apparent that some individuals were so key that they were identifiable anyway, or that they needed to be identified, and in some instances it was desirable for them to be so, because the view expressed seemed more significant because they had said it. I am conscious, though, that it is predominantly those who held positions of status and power, for example, the chief executives and the chairs of the LDDC who are named, rather than the councillors, community representatives and ordinary residents. I recognize that this situation is highly problematic, but it was not feasible to go back and ask permission from each person I interviewed, given the time lag between fieldwork and writing, and my initial commitment on anonymity. All the attributed quotations have been agreed by the individuals involved, except those that were already in the public domain, and some altered or elaborated the original transcript. One person I had hoped to name refused to be identified but allowed me to use the quotations; another requested that all quotations be removed from the manuscript. I hope that this anomaly between named and anonymized individuals does not suggest that one set of voices were more important than another because this is certainly not what I intended.

The plethora of views from a variety of different perspectives made the task of writing this book considerably more difficult than it would have been if I had simply focused on one group, for example the local residents, or the business community, or the affluent residents, because it would represent one particular view of change, or a number of different views but from one particular interest group. As it is, the picture that emerges here is one of conflicting and contradictory views about the nature and type of development that was thought necessary, the impact of the development that did occur and the benefits thought to stem from it. There were also differences in the degree to which people felt that the development had a detrimental impact on the established Island community and, even if it had, whether this was an altogether negative outcome.

This diversity, while rich in itself, presented a multitude of difficulties. Cohen (1989:29) summed up the dilemma well in a study of the Whalsay community in the Shetland Isles:

> I found it impossible to generalise in a way which did not merely invite a host of exceptions. Whilst one could abstract a set of principles . . . it was clear that these were not applied uniformly within the community. Rather, the principles were interpreted and applied somewhat differently by and to different people.

It was not simply the diversity of attitudes that was problematic in terms of try-ing to make sense of the data and come to a view about the development and its impact. There was also a prolonged period between the fieldwork for this book (conducted between 1990 and 1992) and the writing up (end of 1994–97). Such a distance can sometimes be beneficial, especially if it heightens critical reflections, but it also has drawbacks. It is often the case that the farther away from the field experience one may be at the point of writing the less sympathetic the account will be, as the closeness of relationships nurtured in the field is less immediate. Without doubt this book is different from the one that it would have been if I had written it immediately after the fieldwork, but I hope that I have managed to capture both the sympathy and confusion that I actually experienced during my two years of research on the Isle of Dogs.

Like many of the approximately one hundred and thirty people I interviewed (ranging in age from 14 to 89), I came to see the Island as a special place. Through attending a variety of community meetings as an observer on a regular basis and talking to those involved in community politics as councillors or activists, I empath-ized with their anger and their belief that regeneration should have included them. I appreciated the slightly more detached perspective of the professionals (teachers, doctors, dentists, community workers and clergy) I interviewed who tried to place the "rawness" of "local" emotions and the development into a broader perspective. I identified with the feelings of exclusion, which even white working-class residents, some of whom had lived on the Island for 20 years, and others more recently arrived, felt in the face of the long-established Island community. I came to understand too, through my conversations with past and present chief executives and chairmen of the Development Corporation, members of the LDDC Board, and its employees, some of whom had been with the Corporation since its inception, that regeneration is not a straightforward task and that the Corpora-tion were themselves limited by wider political and structural factors. The busi-ness men and women with whom I spoke from companies with a few employees to large multinational corporations challenged my rather simplistic perspective. The professional and affluent, often referred to as "yuppies", were more diverse than I had ever imagined and some were committed to the locality and the wider com-munity. Sadness dominated my contacts with the Bengali families because they were victimized simply for being there, and the "venom" unleashed upon them was difficult to accept or fathom. Their treatment also generated concerns that com-petition, in part sparked by the development and by other policies locally, resulted in an Island that had become contested territory — an experience that was all too real to the teenage pupils I interviewed in a local school. The potential consequences, as the discussion in this book demonstrates, were very worrying indeed.

During the course of researching this book I read a wide variety of literature across a number of specialized fields from urban sociology to geography and planning because in different ways they all helped to explain events in Docklands. Much of this work was very impressive (see, for example, Fainstein 1994, Massey 1994, Logan & Molotch 1987, Harvey 1989) and posed a real dilemma about the approach to adopt here because, as the account clearly demonstrates, Docklands' fortunes were integrally linked with wider structural forces emanating from well beyond the confines of the Isle of Dogs and beyond the control of the Docklands Development Corporation.

However, this book is not intended to provide detailed discussion of complex subject matter like globalization, the role of international finance and the extensive theoretical debates about urban planning, or cultural geography. That is not to deny the importance of these broader structural influences on localities or the similarities that globalization has brought between cities as far apart as London and New York (Fainstein 1994). I wanted to explore the regeneration in a different way and try to make sense of urban change through the experiences of those involved in implementing it, and those who were forced to accept it, an approach that largely precluded any extensive discussion of these other issues. However, the obvious pertinence of broader structural factors posed a problem not dissimilar to that described by Harvey (1996:7) following a radio series on urban issues in which he commented:

> There's a tremendous diversity of opinion there, and I think one of the most interesting problems I'm going to have is figuring out how to take all that diversity and do something with it, other than just simply saying: "Well, on the one hand this, on the other hand that, on the third hand this, and on the fifth hand that."

At one level this book seeks to tell a story of the dramatic urban, economic and social change that occurred in one locality, the Isle of Dogs. It describes the experiences, attitudes and perceptions of "local people" who were largely powerless to alter the shape and pace of development and the impact this and other factors had on their sense of "community" and feelings about the area. Their experiences are contrasted with those who were responsible for generating and implementing change who describe in their own words how they perceived their role, and the rationale behind the regeneration strategies they employed. These narratives are presented without much or any substantive discussion of the academic literature. However, all that I read shaped my perceptions and thinking about the issues, and a bibliography is included at the end of the book to acknowledge those influences and for those who might want to explore the literature itself in more detail.

My own strong preferences for ethnographic research and a search for understanding the social world from the actors' perspective made me comfortable with this approach at one level. I am heartened by the recent revival in community studies (see Bulmer 1986, Crow & Allen 1994, Payne 1996) and agree with the need, as Payne (1996:31) argued, "to balance the elegant simplification of sociological theory" with detailed local studies that demonstrate "the complex messiness of human existence" and lead to an appreciation of the "fuller interplay of factors in human experience".

Nevertheless, no matter how small-scale the orientation of this study, macro processes impinged in very overt and important ways that I could not ignore, which explains the particular structure I adopted in writing this book. The first chapter sets the scene describing the remarkable continuity in the area's failures and fortunes over time and provides the framework around which the narratives should be considered. Chapters 2, 3 and 4 focus on those vested with developing Docklands, how they set about altering the image of an almost unknown part of the East End of London and the response to their "dreams and schemes" from locals who wanted regeneration but not of the kind the LDDC enacted. Chapter 5 focuses on the affluent residents: their motivations for moving to the area, their experiences of living there and the nature of their interactions with the more established "local" population. Chapter 6 describes the process of accommodation and conciliation that replaced conflict and opposition among many local people with a change in the LDDC's approach in which, for a short while at least, the community took a higher profile. Chapter 7 focuses on the experiences of Bengali families, many of whom did not want to be on the Isle of Dogs, and the "battle" for public housing in which the anger and resentment once directed at the Corporation, and for a brief period on the "yuppies", was now turned on the Bengalis. Chapter 8, the last descriptive chapter, outlines the changing sense of place and space which Island residents experienced in the wake of the development. The detailed ethnography in these chapters is then placed in a wider context in Chapter 9 drawing in the academic literature and linking the micro aspects of this study with wider structural factors that impinged on the development process and influenced its ability to deliver the jobs that local people so desperately wanted.

This approach, the narratives wedged between two more substantive chapters, may seem a little awkward. I am conscious too that by presenting it as a narrative there is a danger that it may be viewed as "the way it was" rather than my *interpretation* of the accounts filtered through the thoughts of those involved.

Furthermore, given the time lapse between the research and writing, those asked to comment and reflect on their observations of the changing face of the Isle of Dogs when the draft manuscript was completed were considering comments made some five or more years before. In view of this I was cautious about including their comments into the main body of the book, except where the account was significantly enhanced by doing so, or to correct or develop areas already mentioned. However, I also felt it was important to record their thoughts, comments and interpretations so that the reader would have an opportunity to reflect upon them. Therefore, I have included a short postscript that updates events since the completion of the fieldwork and directly quotes from the conversations I had with, and the correspondence I received from, some of those who read the manuscript.

Rather like the development itself, writing this book has sometimes been difficult and there were times when I thought it would never be finished! But I also enjoyed it and it seems to me that the diverse accounts outlined here highlight not only the importance of understanding urban change and conflict from a variety of different perspectives, and that all social processes are multivocal, but also the imperative of a model of development that seeks to *include* people rather than work against them. I wanted to give all of those involved in what was a complex, conflicting and contested development project a voice. I hope that if they read this book they will feel that I did justice to their particular perspective.

Chapter One

"Echoes of the past"

> . . . for all the changes, the overriding conclusion is that there has been
> a remarkable consistency in the underlying processes . . . In pursuit of pro-
> fits, investors have come and gone in the past, and now they are coming
> back again. Historically, what is happening now is not so new . . . Each new
> application and new agency to steer development looks unique but they
> are themselves part of deep-flowing processes. In Docklands everything
> changes, yet everything stays the same. (Hardy 1983a:22)

The vast Canary Wharf tower, which now dominates the skyline to the East of the
City of London and beyond, is situated close to the Isle of Dogs, an area which for
most of its existence was isolated and little known (see Figure 1.1). Almost a century before the controversy surrounding the redevelopment of London's decaying dock areas, Booth, in his study of the London poor, described it as "strangely
remote from the stir of London", where it appeared residents were "neither in it
nor of it" (Booth 1888–91: Vol. 3, Sec. 1:20); a situation that, despite "its international industrial and trading connections" (Hostettler undated a:11), barely changed
in more than a century. "Nobody knew where the Isle of Dogs was" (Foster 1996: 48),
a woman who moved there in the 1970s recalled:

> I called a mini cab . . . and asked him to take me to the Island. He thought
> I meant Sheppey! . . . He was going [in that direction] and I said, "Where
> are you going?" and he said, "You said the Island." I said "The Isle of
> Dogs!" He said "Where on earth's that?" . . . Even now to try and say to
> somebody that you live on the Isle of Dogs they go "Where's that?" Now
> it's recognised as Docklands so we're still not known in that respect.
> (Christine)[1]

Another woman whose daughter went for a job interview in St Katherine's dock,
"not three miles away" summed up the Island's isolation and anonymity

> She started to say where she lived and that and the bloke said "Hold on
> before we start I think it's a hell of a long way for you to travel everyday."
> So she said: "What, the Isle of Dogs?" He said . . . "Where is the Isle of
> Dogs? Is it near the Isle of Wight?" and she said "No. It's just down the
> road!" For years people had never heard of it. I just used to tell people
> that I lived just across the river to Greenwich cos they hadn't heard of the
> Isle of Dogs and it was so much easier. (Sally)

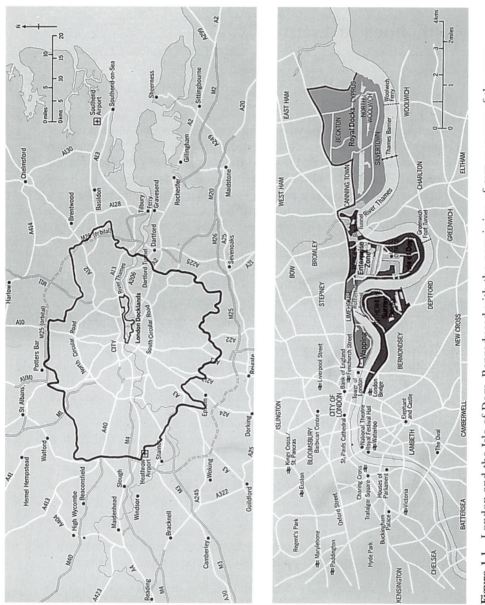

Figure 1.1 London and the Isle of Dogs. Reproduced with the permission of CNT as owner of the LDDC archives. © Commission for the New Towns

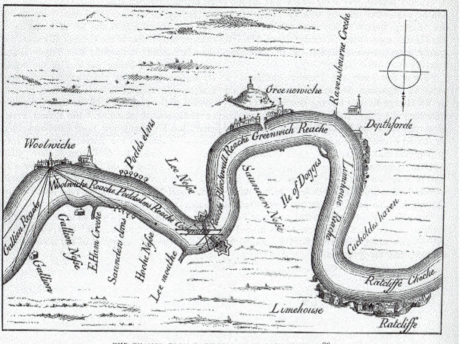

THE THAMES, FROM RATCLIFFE TO WOOLWICH, IN 1588.

Figure 1.2 The Isle of Dogs 1588 (*Source*: Walford, E. c.1900:534)
The Isle of Dogges is marked as a little islet to the south-west of what is now known as the Island. The theory is that as dead dogs floated down the Thames which they did in those days they got lodged by the tide in that part of the river which runs between the islet and the shore. So it became known as the Isle of Dogs, which — as the islet was washed away by the river — got transferred to the present Isle of Dogs. (Hostettler in *The Islander* July 1992:6)

How the Isle of Dogs acquired its name "has never been satisfactorily explained" (Weinreb & Hibbert 1983:411–12). The most popular explanation for its origins was described by an elderly Islander who had lived there all his life: "I was taught . . . that King Henry VIII used to hunt deer in Greenwich Park and used to keep his hunting dogs" here which "usually making a great noise, the seaman and others thereupon called the place the Isle of Dogs" (Walford c.1900:534). This theory has been challenged, however, not least because there were Royal kennels at Deptford on the south side of the river (Hostettler 1992:6), a more practical option for housing hunting dogs used in Greenwich Park. Furthermore, a map (see Figure 1.2) dated 1588 reveals the area of land now called the Isle of Dogs was "not an Isle, indeed scarce a peninsula" (Walford c.1900:534). It is possible, there-fore, that over time the land surrounding the small "Islet" also became known as the Isle of Dogs "even though" this surrounding land was recorded "officially" (as) "Stebunheath (Stepney) Marsh" (Hostettler 1992:6).

Whatever the derivatives of its name, and the land it referred to, the area that is now known as the Isle of Dogs, was not technically an island, as it is surrounded by water on three sides. It only became an "island" during the nineteenth century after the docks were built because, when the bridges that formed the entrances to the docks were raised to let shipping in and out, the area was cut off and inaccessible by road (Hostettler undated a:11, Hostettler undated b:33). The theory about the small islet would, however, explain why an area that is not surrounded by water acquired this title (Hostettler 1992:6). For its inhabitants, whatever its strict geographical characteristics, it was an island, and the territorial boundaries, marked out by the bridges, remained very important to those who had lived there all their lives even after the 1980s redevelopment was well underway. "When I was young I was taught that the Isle of Dogs was definitely an Island", a local in his eighties explained:

> ... The boundaries are still there ... where the City Arms pub is, the City Pride they call it ... there used to be a bridge there used to run to the docks ... that was the start of the Island there. Then you come right the way round the Island ... the first big bridge, the one that lifts (the blue bridge), that was the second part of the Island. You had to cross water and ... Island Gardens ... you know the foot tunnel that goes under the River? That was the three parts. It didn't matter where you went, living in this area, you had to cross water to get off it, that is ... the Isle of Dogs. ... [Now] they're calling parts like Canary Wharf and all round there ... right the way round to Poplar High Street, ... they're still sayin' that's the Isle of Dogs — it's not the Isle of Dogs. (Henry, 89)

The problems that beset this little-known and isolated area of London, its history and fortunes (and those of Docklands generally) were inextricably linked with the economic booms and slumps that characterized the British economy over two centuries which, as Hardy (1983a) eloquently described, left an impression of history repeating itself. Furthermore, as he also pointed out, these communities were not alone. "In many ways the story of Docklands is but a microcosm of Britain's changing economic fortunes — one place amongst many where the impact of these wider changes are experienced and vividly portrayed on the ground" (Hardy 1983a:3).

This chapter briefly describes the history of the Isle of Dogs and the ways in which its geography, and the social and economic factors that influenced its development, shaped the experiences of the people who settled there and the communities in which they lived.

Boom to bust and back again: 1800–1945

Until the nineteenth century the Isle of Dogs was largely marshland used to graze cattle. However, this changed dramatically during Britain's rapid industrialization and "imperial expansion" (Hardy 1983a:4). In a few decades the Island became a densely populated and industrialized area. Its transformation began with the construction of the West India Dock, opened in 1802, surrounded by a similar kind of hyperbole to that of the regenerated Docklands a hundred and eighty years later:

Whoever has enjoyed the satisfaction of visiting [the West India Dock] and viewing the work in its present state must be astonished by the stupendousness of its scale and the extent of human wisdom, skill and industry, which has begun, carried on, and so far completed an "Imperial Work", the proof of past and pledge of future prosperity. (*European Magazine* 1802, quoted in Hollamby, 1990:5)

The West India and later Millwall docks (opened in the 1860s) formed an integral part of London's dock network and contributed to a trade that in the middle decades of the nineteenth century made London the "busiest port not only in Britain but also . . . in the world" (Hardy 1983a:6). It was the area's dependency on the docks and its associated industries that linked the Isle of Dogs' (and the rest of London's Docklands') fortunes with the vagaries of the national and international economy. Nevertheless, during "its Victorian heyday the port must have seemed like the sun, with trade routes radiating like rays to all parts of the world. And it all seemed so simple and enduring — with raw materials and cheap food coming in, and finished goods going out" (Hardy 1983a:11).

The realities, however, for those working in the docks were far less romantic. Dock work was based on casual labour and workers were required to *call-on* twice a day at the dock gates (Hardy 1983b:11, Hill 1976:16–29, Hostettler undated a:2). Mayhew (1860) described this practice as "a sight to sadden the most callous" in which "to see thousands of men struggling for only one day's hire" was "made the fiercer by the knowledge that hundreds out of the number there assembled must be left to idle the day out in want. To look in the faces of that hungry crowd is to see a sight that must be ever remembered" (quoted in Ellmers & Werner 1991:116). The "exploitative and degrading" nature of casualization was the subject of much criticism by social commentators like Booth and later social reformers like Beatrice and Sidney Webb as well as the newly emerging trade unions (Hill 1976:14), and in the early part of the twentieth century the 1908 Port of London Authority Act included provision for "permanent employment whenever possible in London" (Hill 1976:14/15). However, almost thirty years later working practices in the docks had barely changed at all "apart from some voluntary schemes to register those eligible for work in the docks" (Hill 1976:15) (see Figure 1.3). The casual system suited employers because of its flexibility and opportunities to reduce costs, and also because it "placed minimal obligations on" them as they "virtually hired for each job by the piece, paying either for hours worked or tons removed" and "took no responsibility for the welfare of men who sought a living from dock labour" even though their work was essential during peaks in demand (Wilson 1972:17, quoted in Hill 1976:15).

"The pattern of dock life", therefore, "was shaped by common themes" Hardy explained:

The vagaries of trade and uncertainty of work; the sheer physical exertion involved in most of the tasks; and conflicts surrounding constant attempts by employers to minimise labour costs. Inevitably what happened at work was reflected in Docklands life generally and, long after the docks have closed, these historical relationships still have a bearing on what is happening in the area now. (Hardy 1983b:7)

13

Figure 1.3 The "Call" c.1946, © Topham Picture Point. The stevedores began to assemble at six in the morning, in readiness for the "call" at a quarter to eight. There was a second "call" each day at a quarter to one. Stevedores loaded ships; dockers, who unloaded, were "called on" inside the dock area.

Indeed they did. A hundred years after Mayhew's observations, competition for jobs, among men who now had trade union recognition and more job security, was still a point of conflict (see Hobbs 1988:109–110) as this former docker recalled:

> It was a 'ard job . . . even in the late sort of sixties and sometimes even to seventies . . . there would be fights break out on the stones . . . I mean you know pushing and shoving, a right 'ander, broken nose 'ere and there . . . you would get sort of an idea, a good idea there was a good job going and you would all try and shape for it you know and all try to get on there and you'd try to get in front of someone and someone would take offence at that and put you back in your place. . . . You know there's desperation as well. I mean it's born out of working hard for nothing then a good job comes along and someone else is trying to nick your place. (Jim)

"The . . . thing I couldn't come to terms with", this former shop steward continued, "was the physical aspect of no respect for the workers . . . like round where the ship was, there was no washing facilities, no heating facilities, nothing, no toilets . . . I mean 1960 we're talking about. The conditions was like 200 years ago." It was "like stepping back in time" (Jim). Sir Christopher Benson, who became the second Chairman of the London Docklands Development Corporation during the 1980s, had himself sailed from the London docks in his teenage years and recalled his impressions:

> I saw the old method of choosing men to work each shift; they gathered round the dock gates early in the morning were picked off [and] told to go and do duties at various ships and if there wasn't a job for you you went home. . . . I can think of nothing more awful than this sort of cattle wagon treatment . . . the whole method was awful and it must have had an effect on them.

Hill (1976), in an excellent study of London dockers in the early 1970s, suggested that even after the abolition of casual labour in 1967 "the institutional and cultural structure of one distinct and unique system of employment relationships" embodied in casual labour was "transposed in another". (Hill 1976:14). Furthermore, he suggested that the casual system had become so ingrained in people's thinking about dock work that it was not simply the employers who allowed it to continue for so long but the dockers themselves who had "a fatalistic acceptance of the system", believing that the nature of the work required, as employers' argued, "the need for a margin of surplus labour to be hired or fired according to fluctuations in trade" (Hill 1976:15).

Hardy (1983b:4) wrote: "Since the construction of the first docks in the early nineteenth century through to their present redevelopment, the interests of capital have been in clear and open conflict with the interests of labour and local communities." Certainly the nature of the work, exploitation and poor labour relations contributed to an inheritance of struggle and a negativity that remained a hallmark of established Islanders' lives even in the 1990s, as this relative newcomer observed: "All their lives they've been done down, all their lives they've had to live in uncomfortable conditions and [they] can't believe that anything good will ever happen to them." Another said:

> There's always a negative aspect to things . . . I couldn't believe how devastatingly suspicious, how naturally the instinct of Island people is to see the worst and to see a threat in everything. Now that is desperately sad and I can only assume that that comes from having been at the lower end where horrid things have always happened and big people up there have always taken decisions which affected their lives which they've been powerless [to influence] and are always in a defence position, to have to react and never initiate. (Part quoted in Foster 1996:152)

With the docks came a plethora of other industries from "timber wharves . . . cement works, potteries . . . and heavy engineering" (Hardy 1983b:21). In the boom years of the mid-nineteenth century, shipbuilding was a major industry and Brunel's *Great Eastern* was built on the Island (Hostettler a:5–6). A man whose family had been in the area since the 1830s recalled how his great great grandfather, a master shipwright and Huguenot refugee, moved to the Isle of Dogs from Portsmouth

Figure 1.4 Women "Dockies" unloading uniforms from the front, Millwall docks, 1914.
© Island History Trust

because it was "the ship building boom area in those days . . . [and] he came up for work". In fact people from all parts of Britain, Ireland and Europe (see Hostettler undated a:4) migrated to the Island to work in the docks, warehouses and factories, which produced a diverse and fragmented "melting pot" (Hostettler undated a:4). The residential population, which was just 5,000 in 1850 (Hardy 1983b:21), swelled to 21,000 by the end of the century (Hostettler undated b:33) and, for a brief period before the 1880s, the area even attracted an "upper- and middle-class elite" (Cole 1984:347).

Unfortunately, the booms were all too often followed by slumps. By the late 1860s the shipbuilding industry collapsed as the work moved to yards, often in the North of England, deemed more competitive and closer to raw materials (see Anderson 1982:35/6, in Hardy 1983a:13), with devastating consequences for the local population. Journalists who witnessed the "scenes of desolation and despair" on the Island described "people starving, shipyards turned into grass-grown wastes, and no hope of revival" (Hostettler, undated a:6).

At the beginning of the twentieth century, 41 per cent of the population (approximately 20,000) were classified as in poverty on Booth's 1902–3 poverty maps (Eyles 1976:4), and on certain parts of the Island this figure was 60 per cent (Bedarida 1975, quoted in Eyles 1976:4). Booth wrote: "there is in fact no large district in London which is so entirely inhabited by the labouring classes and the labouring classes alone as is the Island" (quoted in Cole 1984:151).

By the beginning of the twentieth century the "industrial" character of the Isle of Dogs and its solidly working-class population was firmly established, with the docks in the centre, and factories around the perimeter, providing employment for both men and women (see Figure 1.4):

The nature of the Island's economy was such that everyone in the family had to start work as soon as they could leave school; many women continued to work after marriage. Island girls and women were employed in the local factories: they sewed tarpaulins, wound ropes, processed and packed a variety of foodstuffs, made sacks, boxes and packing cases, tested cables and shovelled chemicals into vats; they also served in bars, canteens and coffee shops and worked in offices. (Hostettler, undated c:2)

"This was a very big industrial area", an established Islander explained: "Apart from the actual people who worked in the dock there was big industry in heavy engineering, warehousing, manufacturing of all sorts. Moreton's processed food, Pan Yam's was another one . . . You could go through the book and say well if you could buy it we've made it on the Isle of Dogs" (Jack). "Most of us have seen more of the docks" since the development in the 1980s, an elderly Islander explained, "than we ever did when we was living here for 40 years or so before hand because they was guarded by the dock police and anyone who wasn't supposed to be in there was dealt with severely in most cases" (Jack). Hill's study of London dockers highlighted the existence of "a wide mix of mainly manual occupations". Asked to identify the employment of their immediate neighbours, only 12 per cent of men and 3 per cent of foremen in the sample had neighbours working in the docks (Hill 1976:166). However, even in 1970, almost half of those working in the Millwall and West India docks lived within a two-mile radius of them and 20 per cent "within one mile" (Hill 1976:167).

Heavy industrial activity inevitably caused environmental problems and at the turn of the century the area "was likened to the Black country . . . cos of the smoke" (Elizabeth). In 1926 the Port of London Authority suggested that "many of the industries" like "oil refining, colour and chemical manufacture, timber yards, lead smelting, iron and brass foundries and other noisy odour-giving processes were of a type best carried on outside a town" and "the nearness of industry to dwellings gave rise to many environmental hazards", which included toxic chemical dust and lead pollution (Eyles 1976:6).

In the opening decades of the twentieth century, despite more stable economic activity and "brief intervals of relative prosperity for the entire community" (Cole 1984:198) the area "was very very poor", an Islander in her seventies recalled; "there seemed to be crowds of people living in . . . these big houses . . . and you could see young children going to school with nothing on their feet" (Betty) (see Figure 1.5). "In most houses there must have been a minimum of ten people in a house", another recalled: "There were five of us upstairs five downstairs so that was ten . . . There was three rooms up and three rooms down . . . But two doors along the lady had 16 children".

Despite the overcrowding "people were much more helpful to one another" many Islanders said. "You were all on the same level really. You didn't have a lot" (Edith). Bulmer (1986:92) suggested that such assistance was "a realistic response to low incomes, economic adversity and unpredictable domestic crisis" in which no other "safety net" existed. "Neighbourliness and mutual support in times of need" (Hostettler undated a:12) were common experiences for those who could recall the Island in pre-war years. "People were much closer to one another . . . There was

Figure 1.5 Providence Cottage, Emmett Street, pulled down in 1930's, replaced by Providence House, which was demolished in 1980's to make way for road improvements and the Limehouse Link Tunnel. © Island History Trust

more friendliness on the Island years ago", one woman said. "Everybody was on the same level", another commented:

> You just lived . . . and nobody was better than anyone else and people just used to help one another. If anybody was ill you did what you could for them and things like that, you know. We used to keep chickens out in the back garden. [My parents] used to send 'em new laid eggs and things like that if anybody was ill and people used to appreciate it you know. And you could trust people. There was one or two that you knew you couldn't trust. You knew just one bad egg just now and again but then you knew them and so you treated 'em that way. You weren't rude to them but you were just wary of them, you know what I mean. (Georgina, born in 1912 whose grandparents had settled on the Island)

"There was a great neighbourliness, everybody helped everybody else," another said:

> I mean I can remember when my mother had her babies; they were all born at home but as soon as they knew she was in labour there'd be two

ladies coming along, they had white aprons on. You know they were only local women but they delivered the baby . . . If anybody died one of the neighbours would always come and lay the body out. If you had a wedding there was always neighbours there would help out . . . wait at table and do the washing up. Or if there was a funeral the neighbours were there to look after everything while you went off to the funeral . . . It was like that right until the war started. (Elizabeth)

Before the Welfare State, "such a system of neighbourhood interdependence was a practical response to a world in which there were few alternative forms of support to going into the work house in times of need" and among the semi-skilled and unskilled "mutual self help remained a significant source of short-term and emergency social care" well into the twentieth century the bulk of which "was on the basis of reciprocity" (Bulmer 1986:93).

Although the area was predominantly working class and had the appearance that "everyone was the same", there were differences *within* the population, as Hostettler notes,

between the very poor and the better-off, between skilled and unskilled workers, between temperance society members and hard drinkers. Life depended on wages, charity or poor relief. Wages varied according to skill, personal fitness and the state of the economy. Living standards . . . varied too. Some people could afford to dress up in best clothes, have photo-graphs taken, buy a piano to put in the parlour and go on holiday to Yarmouth. At the same time a man could die of starvation in Samuda Street, and when school photographs were taken, children who had no boots or shoes to wear had to sit at the back of the class so that their bare feet would be hidden from the camera. Some families had a house to themselves, or shared between three generations; others managed in one room and a scullery. (Hostettler, undated a:14, 12)

These differences were reflected in older Islanders accounts too, as this woman explained: "It was a very mixed community . . . before the war. There were quite well to do people . . . Teachers . . . the managers of the different factories . . . doctors all lived on the Island you know" (Betty). "The housing on the main road were largely three-storey with an area basement . . . and you regard[ed] them as the best houses" said another.

But all the middle of the Island, which was not very much cos most of it was docks, . . . All the streets . . . were all sort of houses with front door, smack on the pavement, three up (three down with two families in them). [That's] where I was born. Those that had a little bit of area in the front was the, not the upper crust, but it was regarded as something all right, you know. But the rough part, that we regarded as the rough part, not that they were any different from us, there was a very sort of sameness . . . much of a muchness all over. Within that sort of even context of wage earners that was regarded as the really rough was the dockside areas, the very areas that (are) now regarded as something special. (Jack)

Figure 1.6 West Ferry Road 1910, The corn chandler's shop at 87 West Ferry Road in 1910. This shop was owned and run by the Line family at this time, and Albert, their eldest son, is standing in the doorway wearing a white coat. West Ferry Road which dates from the early 19th century, was built up with houses and shops along both sides. The shops included several butchers, bakers, grocers, hardware stores, clothing, boot and shoe shops, dining rooms, pie shops, wet and dry fish shops, greengrocers, hairdressers and drapers, a post office, and there was also Squires, the pawn brokers, to supply ready cash in times of urgent necessity, which were frequent for some. © Island History Trust

"We were only saying this the other day," another elderly Islander recalled:

> A woman came in about a photograph and I said to her "What was your mum's name then?" She said "Frederick" so I said "Eta Frederick?" So she said "Did you know her?" So I said "Yes, I remember your mum." Talking about her mum like I said: "Well I think when we were kids we sort of felt she was a little bit posh." I said "they seemed to be a little bit cut above the others." . . . I mentioned it to a niece of mine . . . and said did she remember her and she said "Yeah. I always used to think they were posh didn't you?" We did have those sort of people. The father was the foreman up at one of the local firms. I said "Yeah well if he was a foreman they probably were more comfortable than other people, when you've got a lot of casual work around, so a man who's a foreman and he only had two children they probably were a little bit [posh] . . . she probably dressed nice and all the rest of it." (See Figure 1.6)

Although the mutual aid and supportive aspects of pre-war community life often came more readily to older Islanders minds, they also remembered the less positive aspects of living in difficult and overcrowded conditions. "There used to be a few quarrels in this street", one recalled:

There was a few here that'd have dust-ups now and again. . . . There was a couple of . . . families — I don't know what used to spark it off. My mum . . . she liked to have a little bit of privacy. She used to say "It's all these in and out of others houses all the time. In the end it always finishes up like this you know (quarrels)." They'd sort of carry it on rather publicly in the street . . . Since then, that was pre-war days, people don't seem to do that now. (Beatrice)

"There was fighting, but not sort of violent fighting. . . . it was more fist fighting," said another and painted a vivid picture of the hardship and violence that had characterized her grandmother's life:

There was a street where my grandma lived . . . the other side of the Island, Millwall side . . . and it was called "do as you like street". My Nan had a very small house. She had two bedrooms and two rooms down and a great big scullery and an outside toilet. . . . My Nan had eight children in that house . . . At Christmas . . . we'd only be there a couple of hours and a couple of the brothers would start fighting . . . Now down that street was absolutely deadly . . . There was one family down there, . . . the husband and wife would literally fight in the street and no one would interfere they would just let them get on with it. Women I think were very very crushed. I mean my grandma was. [My grandfather] used to like a drink and my Nan used to do ship washing. The ships used to dock at the bottom [of the street] and she used to do a lot of washing. I mean it must have been awful . . . and if my grandfather came in in a bad mood or if he came home from work and she happened to be standing at the street, which a lot of them did in those days — they would get their chairs out and sit just outside their doors — and if my Nan happened to be doing anything like that he would think nothing of coming in and throwing all that washing out in the dirt and she'd have to do it again. She used to put up with that. He would go out on a Saturday night. He'd probably come home worse for drink and he would just [smash the china in the house] . . . My Nan would stand up to him as well. If he hit her she would hit him. Mum always used to tell us the tale when my grandfather was going out and my grandma happened to say something and he just swiped her round the mouth and her mouth was bleeding and she just spat all this blood all over his clean shirt. (Betty)

In 1851 this particular street had 461 occupants from 98 families crammed into just 63 houses (Hostettler, undated a:4).

Widgery (1991:73), in his compelling account of the East End he came to know as a GP, suggested that "because of the long history of men working in short-term and in seasonal employment, woman are more assertive, responsible and self-confident". Certainly as a local man commented: despite the masculine image of the East End, "you know with all the jack the lads and blokes sort of swanning around the street" and the hard labour associated with the docks and heavy industry, "it's a matriarchy" where life "revolves around the mum in the family" (see Young & Willmott 1972). Nevertheless, from the account above, and no doubt numerous others like it, matriarchy did not protect some Island women from the additional oppression associated with their gender or the traumas of domestic violence. Indeed,

Massey, drawing on research in the mining towns of the north of England, observed that: "miners, themselves oppressed at work, were often tyrants in their homes, dominating their wives in an often oppressive and bullying fashion" (Massey 1994:194). The structure of neighbourhood life then was, as Rose (1990:428) explained, "gendered, . . . not only by family ties" but by "affective networks among women . . . strengthened by their common experiences as women".

The matriarchal family structure was central in the creation and maintenance of kinship and neighbour networks, especially in the first half of the twentieth century as the Island's population stabilized at around 21,000 (Cole 1984:152). "If anybody came down the street you always knew where they were going, what house they were going in", a woman who had lived on the Island all her life recalled. "It was very very close knit, very" (Georgina). "Everything was done in a family sort of way", her son continued, "some streets were virtually . . . more or less related, banks of 'em, you know, like a dozen houses or more and they would all be the same family or . . . certainly related you know." "Everybody knew everybody and people very seldom moved away", another explained: "They was just Island, what we call Island people you know but we didn't term it then. We just knew 'em as neighbours" (Jack) (see Figures 1.7 and 1.8).

A woman whose family were of Irish descent and moved to the Island in 1900 described her perceptions of the area before the war:

> At one time you knew everybody's faces round here. If you didn't know them personally you knew their relatives. Everybody who comes to the house with the kids, when they were in primary school, and you say "What's your name?" You all sort of knew something about 'em, or they would be some vague relation, or somebody you knew had married their cousin. You always knew a little bit about their background. Very few of them came in that was a complete stranger. (Beatrice) [See Figure 1.9]

"The intensity and strength of this pattern of neighbouring," according to Cole (1984:167), resulted from the Island's "small size . . . and demographic stability" because both "made it possible for neighbours to know one another quite well", a situation further facilitated by "the district's geographic isolation [and] social uniformity" (Cole 1984:153). Massey (1994:163) suggests that Docklands, along with coal mining and cotton towns, were among the few areas where place and community were intertwined, unlike most other locations where "'places' have for centuries been more complex locations where numerous, different, and frequently conflicting communities intersected". The Isle of Dogs would become contested territory with a diverse range of conflicting interests following the Docklands redevelopment in the 1990s, but at this stage in its history the picture was a more homogenous one. Nevertheless, Damer's (1989:102) description of the Glaswegian community of Govan, which emphasized the way in which the area was defined by the work rather than the place, had some resonance with the Isle of Dogs:

> Govan . . . had all the characteristics of a "community" . . . [but] this feeling is not derived from Govan as a *space*, it is a reflection of the long struggle of the working people of Govan for survival . . . Their attachment to their neighbourhood . . . was not some simple kind of *Gemeinschaft*, it was one of the expressions of developing solidarity, between workers of

Figure 1.7 Street Scene c.1930s. A Union Castle liner, berthed in the West India Docks dominates this picture of children playing in Alpha Road, on the Isle of Dogs, about 1930. The cobbled streets between the rows of terraced housing were used for play and neighbourly meetings. Back yards and gardens were too small for games, and the overcrowded homes became cramped and stuffy in the warm weather; there were few parks and open spaces for recreation, but in the quiet side streets of Millwall, children could join in group games and mothers could watch over them. One of the people who lived in the houses was Jane Mooney, who had a family of 13 or 14 girls. She lived to be 90. © Island History Trust

Figure 1.8 St Hubert's House 1935. The flats were financed by the Isle of Dogs Housing Society, after a fund-raising campaign, led by the women who worked at St Mildred's Settlement. The "Settlement ladies" as they were known, who came from well-to-do backgrounds, were horrified by the conditions in which some Island families were housed, and publicized these "slum conditions." As a result, many families were re-housed from dilapidated early-19th century cottages into new flats, which had bathrooms, toilets, hot water and electricity. St Huberts House was damaged by a land-mine during the Blitz, but was repaired, and remains home to many Island families in the 1990s. © Island History Trust

Figure 1.9 An Island family. A 1920s studio portrait of Ethel May and Alec Thomson, and a young relative, Edith Mary. This is an example of a long-standing family, for Alec's father had been brought to the Island as a child in the 1850s and his descendants still live locally in the 1990s. © Island History Trust

different background, religion and status, gained in a series of bitter class struggles. (Damer 1989:102, emphasis in the original)

However, communities are more differentiated than Damer's account suggests. Certainly in the case of the Isle of Dogs it is difficult to separate the Island as a space, especially its geography and isolation, from the industry and labour that fundamentally influenced those who worked there and the communities in which they lived. Furthermore, as Dickens (1988:125, quoted in Crow & Allan 1995) argues, "we need to be sensitive to how 'community' and associational life tie in to other kinds of social relations (especially those of class, race, property and gender)". "Communities" are not singular entities (see Crow & Allan 1994) in which all have equal investment and identical experiences.

On the Isle of Dogs before the Second World War, for example, divisions existed between different parts of the Island. "You ... never used to call it the Island", a local explained "because the Isle of Dogs wasn't used very often then you know." Instead residents identified with the particular part of the Island where they lived. It was a "very close community, very close" an Island woman explained:

> Although there were two, well, I was going to say almost three communities really ... Because there was where I live, which is more towards the bridges up that end and then there was a kind of gap ... and there was those that lived sort of round here by the Island Gardens ... and then there was another gap and then there was those that lived at Millwall. It was almost three communities. (Gladys) (See Figure 1.10)

Cole defined these three areas more precisely: "The northern half of Millwall lying between the west entrance to the West India Docks and the entrance to the Millwall Docks ... the most distinct and self contained of the three. The southern half of Millwall running from the Millwall dock bridge to East Ferry road" and "the reminder of the Island, stretching from East Ferry Road around the eastern entrance of West India Docks" which formed "Cubitt Town proper" (Cole 1984:154).

For many locals the divide was essentially between Millwall and Cubitt Town (so named because William Cubitt let the land that was developed for a mixture of industrial and residential uses from the 1840s onwards; see Hostettler undated a:9). "It was a very, very close [community], more like a village", another Islander explained "... everybody knew everybody. Well I say everybody. The Island was split into two halves ... Where we live here is Cubitt Town. If you wanted to talk to somebody past the catholic church we would say she comes from over the bridge and they had their little community there and we had ours this side" (Georgina).

"The Island has always been in distinct parts", said another "and they very seldom impinged on one another ... When I was born and lived over that side of the Island [Millwall] that was the elite side of the Island and the other side ... Well I don't have nothing to do with 'em and [then] Cubitt town became the elite side and the other side was the working side. Nonsense really but I mean that was the interpretation of it" (Jim). In fact a woman born in Millwall, who later moved to Cubitt Town said: "I'm a bit of a stranger over here and that's how it was. When I was a kid I was never allowed round here. Couldn't go round this side and if someone was a bit simple they'd say, 'Oh yeah but they're from the other side of the Island. That's what they're like'. But it is *so* territorial" (Elizabeth). These differences were also reflected in the childhood of a woman born in the 1960s:

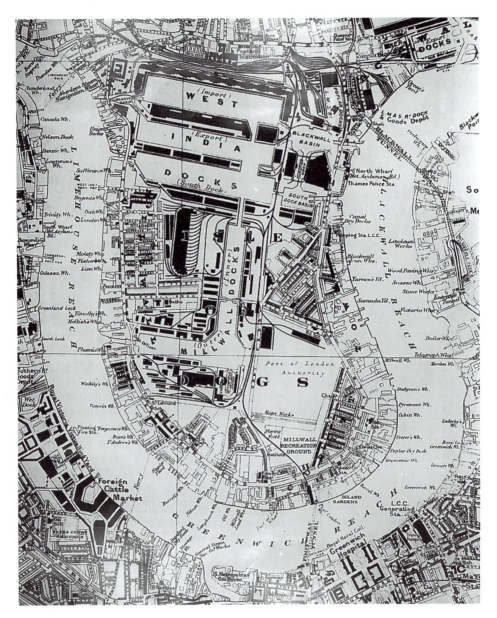

Figure 1.10   The Isle of Dogs 1930. Reproduced from the 1930 Ordinance Survey map

[The Island was] desolate really. There was just one bus. Just, you know, like a village — not like a country village but a very small community. We lived round the other side of the Island [Millwall] where me mum lives but I never ever came round here [Cubitt Town] 'til I was about 11 or 12. I'd come past on the bus [but] to actually get off the bus you didn't, but it's only ten minutes away.

The stability and networks of communal association established in the first half of the twentieth century were devastatingly disrupted by the onset of the Second World War. The Island's geography and strategic importance made it highly vulnerable to attack. "There was an awful lot of bombing", an Islander explained: "Lots of people . . . literally walked out and left their homes intact and just went" (Edith). "It really devastated the Island", another woman recalled:

> They reckon the Island was one of the worst bombed areas. On the first day of the Blitz, on September 7th in the afternoon they dropped quite a few bombs round the Island, round all the dock area so they came with a vengeance . . . We went away the Monday morning [after the start of the Blitz] [we couldn't] keep the children here. We didn't really have anywhere to go . . . [so we] took lodgings in — . . . Three families in this street all got a lorry and they all went down there and they all got put up in the village hall. I used to go and visit 'em there. (Beatrice)

Hostettler (undated d:45) described "the Island at War":

> The overall impression from records made at the time and from what was remembered years later, is of people struggling to behave normally in a world where houses, shops and factories had collapsed in heaps of burning rubble, a world of fire, of dust, of falling debris and black stinking smoke; and when this had cleared away, a world of desolation, where once-familiar streets were scarcely recognisable and neighbours and kin had vanished without trace. In September 1940 an official observer wrote: 'The Isle of Dogs is like a district of the dead, and nearly everyone has gone.' (Hostettler undated, d:45) (See Figure 1.11)

"The Isle of Dogs was a truly crippled community after the Blitz in 1940" and its history from this stage on was characterized by "almost constant crises and disturbances" (Cole 1984:349). By 1945 a pre-war residential population of more than 21,000 had been whittled down to approximately 9,000 (see Hostettler undated d:45). "One third of all warehousing was destroyed, as were many public buildings . . . In 1945 in Cubitt Town 75 per cent of homes were found to be unfit for habitation" (Hostettler undated, d:45).

This was our place

The relative stability of the first four decades of the twentieth century that contributed to close-knit and neighbourly relations on the Island were altered not only by the devastation of the war but by the redevelopment that followed it. As in many other inner city areas in Britain, it was highly controversial. Furthermore, as in so many other aspects of change on the Isle of Dogs in subsequent decades, it was not that renovation, redevelopment and improvement in living conditions were not required, and desirable, but the form the redevelopment took that was problematic.

27

Figure 1.11   Bomb Damage. © London Metropolitan Archive

Most of the housing that had survived the bombing was approximately a hundred years old by the 1950s, and had been in a very poor state for more than half a century owing to "the neglect of private landlords" (Hostettler undated a:12), and the problems associated with close proximity to the river, where "mud and water under the floor boards was common" and exacerbated by "occasional floods", all of which, contributed to the "general decay" (Hostettler undated a:12). Given the poor state of the housing, and extensive bomb damage, the Island to which evacuees returned was in desperate need of reconstruction.

"We felt we were Londoners, we felt this was our place", an Islander who returned after the war said:

> I used to think that the kids would hate to be dumped in a country town when they were really Londoners . . . I was thinking I was denying their birthright not to bring them back to London. So we went to the council to see if there were any houses in this area and we got offered this one which was the one I was brought up in but had been stood empty for five months after it got damaged in the bombing. So we had this one and I'd been bought up in it — it was funny really, it was a coincidence. Yet —'s a lovely town and had we had any sense we probably would have stayed there. (Beatrice)

Figure 1.12 "Pre-fab" Housing. © Island History Trust

Some, whose former homes were uninhabitable, moved in to pre-fab housing (see Figure 1.12) erected as a temporary measure to deal with the housing shortage. "Most of the houses had gone from that end where we lived", one woman explained:

> There were very few houses . . . we moved from our house, well, what was left of it, into a pre-fab. They were marvellous. We thought we were on holiday when we moved into that . . . we had running hot water . . . we had a bathroom which we'd never had before and a really enormous garden and we grew vegetables. . . . We had a wonderful sense of community when we were in the pre-fabs because most of the people where I was were old Islanders and old neighbours, you know, so everybody in that little block knew each other. (Gladys)

This became Gladys' home for 25 years!

Not everyone was so fortunate in their housing . "I get a little bit uptight when I read about people" complaining about the housing situation today, another Islander told, me. "Young people today should sort of know how we were when we got married. The housing people just would not entertain us at all because we had no children. So we had to come in here [privately rented], where we had no bathroom, [and] . . . only outside toilet" (Betty). It was thirty years before this couple were finally offered council housing.

Some were forced to look elsewhere for accommodation: "We got two rooms at the Elephant and Castle when we first got married [in 1948]," one Island woman said. "There were three couples and we had two rooms each. [It was] very rough, oh extremely rough. It really shocked me I didn't realize there were places like it then . . . We did that until my mother got us two rooms locally . . . That's where I lived from [19]52 up until we moved here [one of the council estates] about 1969."

Despite the damage and the reduction in the Island's residential population, many of the Islanders who either stayed during the war, or returned after it, quickly re-established networks. "The community itself was still strong", one recalled "it always has been a very close community." "After the war . . . it was more like a little village" another said; "it was lovely . . . really nice . . . If you were in trouble your neighbours would be there. I mean everybody knew everybody . . . and you would always sort of walk along with somebody cos you knew everybody" (Sheila). Even in the late 1940s and early 1950s it seemed that "most of the people that lived here were old established families for donkeys' years sort of thing" a man from an old Island family, who grew up on the Island at this time said:

> It was a bit cliquey . . . There was always a saying you daren't talk about someone at a bus stop because it was guaranteed that the person standing behind you . . . was related to the person you was talking about. So it was a very friendly place . . . It was mainly sort of families intermarrying. And the same as the docks; I mean the gangs [were] made up in lots of instances of say about two or three families . . . because it was father and son as well, you couldn't get in there other than that . . . It was a fairly close knit community then. (Jim)

As recently as 1970, at the time of Hill's study, he found significant overlaps between kin ties both inside and outside the workplace where "direct father–son inheritance accounted for 67 per cent of the men" in his sample and where "other catagories of relatives, such as brothers, fathers, and brothers-in-law, also included large numbers of dock workers" (Hill 1976:174–5).

It was during the period when "they knocked down all the old streets" and replaced them with new council housing that the Isle of Dogs became contested territory. Given the damaged and decrepit state of housing on the Island it is not surprising that redevelopment was initially welcomed. "There was no opposition at all to it", one Islander said; "they was all eager to get out and get a bathroom and a garden." "I think most of the older people weren't very happy about them", said another, "but they was just programmed I suppose" and "we didn't realize I don't think . . . quite what was happening." As concrete high-rise, and frequently poorly constructed, flats replaced the damaged and decaying streets of Victorian terraced houses, local people were frequently dispersed (a familiar pattern in slum clearance and post-war housing in many parts of Britain; see Power 1987, Reynolds 1986, Young & Willmott 1972), further damaging kinship networks and breaking with the traditions of living and working in the immediate locality. "The unfortunate part about it was", an Islander explained,

> when they were going to start building these new flats the people that was in the old houses, and mind they were old and they did need demolishing,

they all had to come out first, and they moved those all to Poplar and they didn't want to go to Poplar. They were Islanders. They wanted to stay on the Island. So a lot of the Islanders were all moved to Poplar and then when these places were rebuilt they were bringing people from Poplar and Stepney down to the Island. (Gladys)

Although there were new opportunities for mobility in the post-war years of which some Islanders happily took advantage, housing policies were not sensitive to many Islanders' attachment to the locality. While the difference between Poplar and the Isle of Dogs might have seemed unimportant to an administrator and only amounted to being re-housed less than a handful of miles away, it might as well have been in a "foreign" land to many Islanders and there was a good deal of resentment, as this local explained:

> As they started building the flats so we got people from Poplar, Bow and I think as far as 'ackney. They just came down here and resented it — just moaned about the Island . . . They hated it. The Island had nothin' for them. It didn't mean a thing . . . and there was a lot of bad feeling then between the actual Islanders you know. We always said "Well we've always lived here. We like it" and I think there was a feeling too that their families were coming onto the Island — different families from outside coming on to the Island — and some of the Island people couldn't get houses in the end and they had to move out and of course this caused bad feeling.

"I didn't want to come to the Island", a woman who was moved to the Island at this time said. "People that lived in Poplar and other parts of East London thought ooh I don't want to live right down there. That's a long way away [and] you can't get on and off" (Shirley).

The close-knit networks of "Islanders" and their strong sense of attachment to their locality did not make them receptive to these "outsiders" as a woman who moved to the Island at this time described:

> When I first came down here . . . when the estate was going up . . . we were foreigners and I was born just at the bottom of the East India Dock Road [about a mile from the top of the Island], you know, but we were foreigners . . . The likes of me were treated dreadfully . . . you'd sort of go into a shop . . . and you'd hear 'em say "Bloody hell, not more of 'em" kind of thing "not down 'ere". It was just the general attitude especially from the older people. (Kay) (Foster 1996:149)

There is no exaggeration in the use of the word "foreigner" as Gladys, an old Islander, illustrated by a spontaneous comment about the redevelopment: "That's when you sort of got a real influx of foreigners. When I say foreigners I mean, you know, people from other parts of London." However, Gladys denied that there was any resentment towards those who moved on to the Island as a result of slum clearance:

I wouldn't say there was resentment, no. I don't think we really thought about it. No you'd say, "Oh it's not like the old Island is it ... Not like Islanders." And I still often think they're not like Islanders. I think you can tell the differences between Islanders and the people who have moved on to the Island. We always liked to think we were a breed apart. ... But you know when people used to come down here they used to say "Have you got your passport? You have to change your money." Silly things like that you know. ... On one occasion they was talking and said something about the Island and they said "Oh yes, the Islanders, they're the aristocrats of the East End" and I think we thought we were. I honestly do think we thought we were a cut above Poplar or Stepney ... They were rougher, ... and that's [why] when they moved down here ... and we thought they were rougher people ... I think we kept up certain standards perhaps ... Really in every community, I know lots of people who came from Poplar and Stepney who are very nice people indeed ... [but] maybe they had more of the rougher than the nicer ones, I just don't know.

"We didn't know their roots did we", another Islander commented:

It's just that, I mean, our neighbours we knew their parents and grand-parents and things like that, but when the people came to the Island (from other areas) we only just knew them. You see so, that's the sort of feeling it is because when you go out, when girls go out for a walk with the babies in the pram and you say "Ooh yeah, I can see you're so and so and that", you know, and it can be about four generations away. You see it's a strange sort of feeling I don't know why it is. (Georgina)

"It was like making new friends all over again" one Islander said of the period following the redevelopment, a process that was not always easy for newcomers as "Islanders" formed a powerful and formidable "clique". Some even referred to them as "The Mafia". Islanders recognized what an Anglican priest had characterized as their "affectionate clannishness" (see Free 1904:91). As one explained: "Some didn't like it 'cos they said we was too cliquey. Well we didn't notice it because we were the clique." The differences drawn between "Islanders" and "foreigners", insiders and outsiders, is a common feature of many urban areas and villages across the world that once had relatively closed residential populations (see Bassett et al. 1989, Cybriwsky 1978, Payne 1996:24, Crow & Allen 1994 for summary of studies). However, it was especially ironic that an area that was a "melting pot" of diversity a century before with worldwide trade on which their livelihoods depended were resistant to people who lived just a few miles down the road. As this woman who moved to the area in the 1970s explained: "People that have not been here for generations ... they're still treated as outcasts to a large degree, which is horrible cos there are very few people that are true Islanders. ... There should be a community spirit ... not division."

Even the different housing policies of the local borough and the former Greater London Council produced divisions: "This side of the road is different from that side of the road, between two sides of the same street" an Island woman explained, because this side of the road was Tower Hamlets and they was mostly re-housing local people. But over there was the old GLC and that's when they came from everywhere all different directions. So that side is more mixed up than this side. See I can point to those houses there and the first house over that side of the road they've always lived on the Island. Not the next house . . . but the house next door [but one] they've always lived on the Island, and that one there they have, and that one there they have, you see, but you don't very often [get that] over there; you won't get it, just one or two of them.

As we will see in later chapters, the sensitivities about Islanders' access to local public housing and hostility to "outsiders" continued to be an issue far beyond the immediate post-war decades.

Elias' (1976:xvi–xix) observations on "the established and the outsiders" are most apposite for the descriptions above:

> As soon as one talked to people . . . one came up against the fact that residents of one area where the "old families" lived regarded themselves as "better", as superior in human terms to those who lived in the neighbour-ing newer part of the community. They refused to have any social contact with them apart from that demanded by their occupations; they lumped them all together as people less well bred. In short, they treated all new-comers as people who did not belong, as "outsiders". (Elias 1976:xvi)

Furthermore, Elias argued that "these newcomers themselves, after a while, seemed to accept with a kind of puzzled resignation that they belonged to a group of lesser virtue and respectability, which in terms of their actual conduct was found to be justified only in the case of a small minority" (Elias 1976:xvi). Elias found it initially "surprising that the inhabitants of one area felt the need and were able to treat those of the other as inferior to themselves and, to some extent, could make them *feel* inferior" (Elias 1976:xvii), especially as they, like established Island resid-ents, differed only in the length of time they had lived in the neighbourhood, not by their class, ethnicity or other key distinguishing features. However, he suggests, and as was to become manifestly evident on the Isle of Dogs, "exclusion and stigmatization of the outsiders by the established group" are "powerful weapons" that allow the maintenance of "their identity", enable them to "assert their super-iority", and serve to keep "others firmly in their place" (Elias 1976:xviii).

It was not until "outsiders" settled on the Isle of Dogs during the redevelop-ment from the 1950s onwards that the identity and concept of being an "Islander" became important because, until this time, those with long family associations had never previously questioned their right to be in the area. Now they found them-selves having to defend their territory and, as Suttles (1972:13) observed, "it is in their 'foreign relations' that communities come into existence and have to settle on an identity and a set of boundaries which oversimplify their reality".

Islanders sought exclusivity and this status was only conferred on those born on the Isle of Dogs, as this woman now in her seventies explained: "There were eleven of us . . . Most of them were born on the Island so they are actually Islanders, but then I came as a child so I think I am an Islander, I consider myself an Islander. I mean 62 years you've gotta be aint yer really!" (Rose). Some, however, would deny her this status, as a *real* Islander explained: "That one (my grand-daughter) her father, her other grandmother . . . they came from Bethnal Green, they weren't Islanders but they've been here . . . about forty years now haven't they? But they'll never be classed as an Islander for some reason or the other, strange innit. My husband's been here . . . 57 years on the Island . . . But he's never classed as an Islander and yet he's only been here 57 years!"

If this appears extreme, Strathern's (1982:257–9) study of the Essex village of Elmdon found that the only people who could be categorized as "real Elmdoners" were a very small group associated with just four surnames, two of which dated back to the seventeenth century and the other two to the beginning of the nineteenth. Those who moved in before the First World War were perceived as "belonging to Elmdon" but were not "real" Elmdoners. As on the Island, spouses of "real Elmdoners" could not assume that status.

Certainly many of those who moved to the Island in the 1960s and 1970s got the message that they were outsiders. "I'm not going to say I'm an Islander because no matter how long I've been here", a prominent community activist born in Bow said, "if you're not actually born here (you cannot be an Islander)." "I've got six grandchildren born on the Island," he continued, "but they're not termed as *true* Islanders because they had to go to hospital to be born and that's not on the Island. That's the extreme some people go to."

A woman who moved to the Island in the early 1970s felt the hostility too, not only initially but a decade later when she moved onto another estate. "I had something said to me when I moved in here by the lady next door. She was saying to me like 'You're not an Islander' and I looked at her and I said 'Why do I have to be?'"

> There is still an element of back stabbing with the old Islanders' fraternity which is a shame really. They live in their own little worlds. I mean they must have known that when they got the new build in the fifties and sixties when the high rise blocks . . . started . . . in the sixties and seventies, they must have known they were going to get an influx of people, right? So when those people started moving in I s'pose they was ignored and caned the same way I was when I came to the Island. The same way as the Asian people, the same as the so called yuppies. (Christine)

Many others identified with this. "People would simply say 'Well you're not an Islander' you know", another prominent community activist said, "and . . . on one occasion I was raising . . . [an] issue, . . . and someone made a point, in fact they said: 'Well you know blimey he's only lived here for twelve years' . . . If they wanted to make a point about a particular thing they'd say 'yeah but you're not an Islander' . . . In a way . . . you're identified down a bit, you know, and it restricts you, it restricts the things you can do [and] . . . the things you can say."

When I asked how long you had to live on the Island to be an Islander, he replied: "Well I suppose after about 20 odd years. . . . upper of 20 odd years . . . the people would begin to talk to you and they'd say 'Well you're not an Islander you know, blah blah blah blah' but even so . . . and even now you're still getting people say 'Oh yeah but you wasn't born on 'ere was yer, you wasn't born 'ere was yer'." His experiences were aptly reinforced in an exchange between two Island women. One described him as a "genuine man, he's always worked for the Island". Her friend's immediate response was "He's not an Islander." "I don't know whether he's an Islander or not", came the reply, "but I mean he has always sort of been a voice." Another man said "I hate this business of not being an Islander. I am an Islander because I've been here for ten years. Even if I'd only been here for one year it was my choice to live here. I am an Islander and I hate the thought of being rejected" (Bill).

Although they would always have a different status, "those that came in the beginning" (1960s) had by the 1980s been "absorbed into the community" and some of those who had reluctantly accepted offers of housing on the Island in the 1960s and 1970s changed their view of the Island, as this woman explained: "When I came here I really liked it cos it was sort of a little bit suburban because it was quiet . . . and you came away from the hustle and bustle of the main A13 road and it was kind of quiet and nice you know" (Shirley). Others too, who had taken their flats saying "As soon as I can I'm going", twenty years down the line came to defend their right to remain on the Island too. "You try to get these people off [the Island now], they're still over there" an Islander explained "and . . . they'll fight as much as the original Islanders will to retain what they think is theirs and to retain really your own identity." On a small estate of council houses an Island couple felt their estate had eventually become "more like the old Island used to be because everybody knows each other again at the moment". "It's taken a long time", they said, but "this little part" is "more or less back to where we used to be." The community did "settle down", said another, "but not anything, not anywhere like it did before . . . It's not so nice now."

"There is just not that closeness that there was", an Islander lamented; "there are people who have lived [in the road opposite] for the last 20 odd years that I wouldn't even know." While a woman, who had lived in the same home all her life (over 60 years) said: "In this street . . . I know quite a few [people] but there's still a lot of people that I don't actually know. You think to yerself I've lived here all my life and people have come in and they've not sort of mixed you know" (Sheila).

Before the developers

An image of close-knit, working class communities — an image well founded though exaggerated by folklore — has faded over the years as people moved away and work patterns changed. What remains exhibits little of its traditional strength, but there is still a sufficient sense of belonging to form a basis for resistance to current changes that threaten to remove the last of yesterday's Docklands. (Hardy 1983b:11)

Despite the intensity of its economic activity during the nineteenth and much of the twentieth centuries and the changes that war and post-war reconstruction brought, prior to the 1980s development the Isle of Dogs was still a relatively homogenous white working-class area, just as it was in Booth's account almost a hundred years earlier. Over 80 per cent of the Island's 12,000 population lived in council accommodation (Hardy 1983b:21). Neither had its geography altered. It was, in the words of one local, "a massive cul-de-sac" that, according to Cole (1984:296), did "much to create, reinforce and preserve a strong sense of community among its residents".

Kinship networks, though diluted from their very concentrated pre-war form, were still in evidence both among Islanders who stayed in their pre-war housing and among some of those who moved to new housing estates, as Beatrice's account demonstrates. "There's a couple next door", she explained:

> She's lived in this street since she was about 12. She's lived on the Island all her life. Next door here . . . now they came to live here [recently] and I didn't know 'em they came from round Millwall . . . the daughter [of the couple who moved into the house] knocked on the door. She knew my name . . . and she said: "Oh Mrs . . . do you happen to be the mother of . . . So I said "Yes, it's me daughter" . . . It turned out that my daughter worked with their niece and their niece and her husband was god parents to our grandson — me daughter's first boy. So that was one connection with them although I never knew 'em. They was just coming from the other side of the Island. Then it turned out that their nephew was married to my niece, my sister's daughter and that's how it is on the Island. (Beatrice)

Even in 1980, 40 per cent of residents had always lived on the Isle of Dogs (AIC 1990). Nevertheless the dilution of the Island's population continued apace as the new estates and slum clearance progressed and by the 1970s the Isle of Dogs was so unpopular that many of the poorest council properties were classified as "hard to let" and it often proved difficult to persuade people to move there.

"I'm here because nobody else wants to live here", a man who moved in to the area and on to one of these estates said, "that's how I got my flat because they couldn't give them away in 1981 when I came to the Island" (Bill). The Island's lack of popularity was due in part at least to its poor transportation links and facilities. It did not even have a supermarket until the mid-1980s. Bill described the estate on which he lived:

> It was like a microcosm of the Island if you like, a mixture of me, new, and older people there who'd lived [here a long time] and resented my being there until they got to know me and then found out that I was all right . . . The neighbours were wonderful, I mean they're really nice people . . . Next door to me . . . the flat above . . . there was a couple . . . He'd done time for GBH and her father had murdered her mother so they were a really salubrious couple and they had two children but they would take my milk in off the doorsteps so that nobody nicked it, you know . . . wonderful . . . I had a car and . . . very few people in our block had a car and . . . if ever I wanted anything doing to the car, you know, just mention it and it was done, and in return every now and then they would ask me could I drive them South of river . . . and just drop them off and

say "We'll be back in half an hour" and I'd sit there and in half an hour they'd come back . . . in return for that my car was mended, you know, if anything went wrong. There's this wonderful loyalty and, I mean I actually now, although I don't consider myself an East Ender, I mean I consider myself to be part of the community and I like it very, very much here. I think, the people have got a wit . . . maybe not quite so biting or as sharp as the Liverpool wit but they have and . . . I've got to know a lot of people, I get on very well with them. (Bill)

The Island was still frequently likened to a "village" not only by established Island residents but by more recent residents too as the comments of this black woman who moved into a council flat on the Island in the late 1980s revealed:

It was very very friendly . . . my neighbours were brilliant . . . you know, back door open, all the stuff that East Enders says, it was like that three years ago. It's like that [now] . . . The people . . . are incredibly lovely people . . . often if I'm driving up to Mile End and I see a woman waiting for the bus I'll offer them a lift . . . I don't think there's many places in London where you'd feel safe doing that kind of thing.

Even after the development was well underway a newcomer said: "You'll find people on the Island who believe that this is the only real genuine East End left and it's probably true." But as many recognized, the positive features of Island life were also accompanied by more negative ones including exclusivity and racism and, as the accounts contained in this chapter and in the rest of the book demonstrate, there were many who discovered this less attractive side. One woman, for example, who moved to the area under the hard-to-let scheme in the 1980s said she never came across the sense of community "but I was an 'outsider' wasn't I".

The very worst treatment was meted out to Bengali families who moved to the Isle of Dogs during the late 1980s and early 1990s. The account below could not have contrasted more dramatically with images of a cosy community:

The people of this Island . . . they don't like to see any strangers, they don't [want] to mix . . . When we came . . . here . . . people had terrible looks . . . and say "Why these bloody strangers come from?" We tried . . . to introduce ourselves with neighbours but some peoples will not answer me even if I say "Hello, good morning" . . . First of all it was very hard for us . . . there was lots of racial harassment, racial abusement and racial attacks. (Quoted in Foster 1992:178)

"All that is solid melts into air"

The underlying economic problems associated with the docks and related industries that would eventually lead to their demise were obscured for some time after the Second World War by national economic prosperity in the middle decades of

Figure 1.13 Shipping and grain discharge West India Docks 1961, Courtesy of Museum in Docklands, PLA Collection

this century (Cole 1984:274). For the first time there was relatively full employment and higher wages (see Nossiter 1978). Absolute poverty declined and the post-war boom brought greater affluence and rising living standards. "Before the war . . . there was a terrific amount of unemployment", an Islander recalled. After it, "things settled down . . . it was sort of boom years, wasn't it. Everyone was in work . . . we were doing very well . . . I mean things did improve as far as money and that sort of thing were concerned. I mean people started going on holiday perhaps that you hadn't done before . . . even going abroad, which was practically unheard of" (Gladys).

"There was plenty of work", a former lighterman recalled, "the Docks were beginning to pick up . . . through the late forties and fifties and sixties [and] . . . . in the mid-sixties handled their largest cargo in a year that it had ever handled in its existence" (see Figure 1.13). "It was very busy . . . I suppose at the time it was almost impossible to be out of work . . . It was really booming." "People complain about the traffic today, but it was far worse in those days", a businessman who worked in the area recalled: "You'd have a job to get in the docks . . . between all the lorries parked there. The congestion was incredible."

Despite all this activity the underlying trends were more disturbing. Since the 1920s the port and manufacturing industries concentrated in London's Docklands' had "been in a state of relative and absolute decline" (Cole 1984:275) and jobs were disappearing as firms moved out of London. In East London alone 75,000 jobs were lost between 1971 and 1981 (Census of Employment, 1981, quoted in Church 1991:2), part of the 1.8 million British manufacturing jobs lost in that decade (Martin & Rowthorn 1986).

However, on the Isle of Dogs, as elsewhere in London, a corresponding decrease in the local population (see Church 1988a:188), which by 1951 was just over 10,000 (Cole 1984:279), meant that work was plentiful. Indeed in the 1950s and 1960s there was a "shortage of unskilled labour" (Cole 1984:279). Sadly, this boom and the growing prosperity that accompanied it could not be sustained and by the 1970s the Island was once more experiencing devastating job losses and a dying dock industry.

The speculative nature of dock construction in the nineteenth century was in part responsible for their demise (Hardy 1983a:9). Hardy argues that the "historical over-provision" of both warehouse and dock space resulted in "dock companies struggl[ing] constantly to achieve even a modest level of profitability" from the outset. Church suggests that although the docks were "never particularly profitable" the reasons for their decline were complex:

> International forces — such as, a slump in world trade, changing trade patterns, technical advances in vessel size and containerization — combined both with national factors, most notably competition from other UK ports, and also with local factors, namely restrictive union practices and mistaken management strategies, to precipitate the decline of the docks. (Church 1988a:188)

By the mid-1960s it was clear that the dock industry was in terminal difficulty. In just twenty years the numbers employed in London's docks fell from 31,000 people in 1955 to 9,800 in 1975 (Spearing 1978, in Cole 1984:275) and in a 15-year period barge traffic declined from 13 million tons a year in 1963 to only 3 million in 1978 (Spearing 1978:231–7 in Cole 1984:275–6). London could no longer compete as an international port. When oil became the primary "sea-borne commodity" and containerization was introduced, the type and size of dock facilities needed (see Hardy 1983a:11) changed, making it cheaper, easier and more efficient for ships to go to Tilbury. In the late 1960s "seven container births" in Tilbury could "handle seven eighths of the entire non-bulk traffic of the Port of London" (Hebbert 1992:116, citing Rees 1967:121).

After years of exploitation, casualization was finally abolished and "for a brief period . . . the docker enjoyed a measure of autonomy rare in industrial society" (Hobbs 1988:128). During the sixties and seventies the dockers also became in-famous for their militancy, "a well deserved notoriety", Hill (1976:105) argued, "if militancy is defined as the willingness to withdraw labour in furtherance of an industrial claim or interest." The dockers supported not only their own claims but, because of their pivotal role, those of other workers' causes too (see Hill 1976:149). A legacy of mistrust born of poor labour relations, exploitation and insecurity made many dockers inflexible and resistant to change (Hostettler undated a:2, Hardy 1983b:13), as this shop steward explained:

> Quite a few people in the union side . . . was frightened . . . They was always being put on, so they thought anything that was changing was gonna be

> detrimental to them so they was always suspicious of any change ... And they was always concerned about their livelihood and the manpower situation. In other words if they had 13 men for a gang and they'd always had 13 men for a gang they didn't see any reason why they should have 10 in a gang ... Particularly the older ones, because don't forget they had gone through it all ... and any change was always looked on it's not changing for us, sort of business. (Jim)

However, by any measure the work was still physically strenuous and sometimes dangerous. One docker, for example, recalled seeing a man lose an arm when a case, three times the declared weight, broke the winch. The wires snapped, "whipped round and chopped his arm off completely" (Jim). Others, lured by the prospect of extra money, worked with dangerous substances like asbestos or in especially dangerous conditions (see Hill 1976 for a detailed discussion of dockers, their work and its structures).

Nevertheless, militancy plagued the image of the docks and was certainly a factor in preventing the Port of London Authority (PLA) considering London's docks for "short-haul" traffic or "roll-on, roll-off ferries" (Roger Tym & Partners 1983, quoted in Hebbert 1992:116). Massey (1994:83) however points out that focusing on militancy is part of a process of regions being "blamed for their own decline" when responsibility lies with broader social and economic factors. Furthermore, on the Isle of Dogs militancy did not prevent the Fred Olsen Line from creating a new £1million ferry terminal from the West India Dock during the early 1970s (Hebbert 1992:116) that operated until its closure in 1980.

The fate of the West India and Millwall docks on the Isle of Dogs was characterized by almost ten years of uncertainty because throughout the 1970s there were "lingering hopes" of retaining the docks (Hardy 1983b:21). "From the middle 1970s ... the Port of London Authority was ... issuing contradictory statements", one local explained:

> They were sayin' in public "We're gonna close the docks next year" and then (say) "Oh no we're not gonna close the dock we're gonna keep it open ... we might keep it for another five years." And then they'd [PLA] say "Well the shipping companies were very reluctant to sort of come into the docks." ... It's no wonder the shipping companies stopped coming in ... you wouldn't put your bloody car in a parking space if the bloody traffic warden says to you: "Well you're not gonna be able to park 'ere [long]. Five minutes time we're gonna move the parking meter." ... So they (the PLA) created the uncertainty.

A member of the Docklands Joint Committee saw the situation differently:

> I remember in the mid-seventies going along to North America with PLA people trying to sell the Port of London to overseas container companies, shipping companies. But it was quite obvious that there was no way the port could compete in international markets by retaining itself in that locality. The water was too shallow, there was no back-up land for containers ... and it was quite obvious to those in commercial business that the port

would have to move downstream to Tilbury, to deeper water and recreate a new port there to compete with container ports on the continent.

Despite this broader economic picture that made the future of the docks look hopeless, some did not believe their closure was inevitable, as this campaigner argued:

> I believe that the docks closed not because . . . the ships stopped coming, because the types of ships that you see in the Millwall and West India could still use 'em. In fact the Olsen Line, in particular, who'd actually developed a passenger cargo terminal in the Canary Islands didn't wanna leave the Docks. . . . they were keen to go on using the dock but the Port of London Authority had made their mind up that they were gonna close it. I think the Port of London Authority closed these docks down as part of a longer term strategy . . . If they really wanted to they could have kept the Docks open.

Another committed campaigner told me that the belief that the dock could have stayed open was not simply nostalgia. "Even on the day Millwall dock closed", he said, "there were five ships fully loaded." He characterized the Port of London Authority as an "inept, inefficient organization" that "lacked vision" and saw the closure of the docks as a way of selling the land and therefore recouping some of the money they owed. The union's role in closing the docks was grossly overestimated (Research notes).

Whatever the causes of dock decline, the consequences were clear. By the 1970s the problems that had been bubbling beneath the surface finally became visible. An industry that had supported 25,000 London dock workers in 1960 was reduced to just 4,100 by 1981 (Church 1988a:188) and the losses did not end there. As the docks declined so did the related industries, a process that was accelerated by unfavourable business rates. It was simply cheaper and easier for companies to move out, often to green field sites or areas of new-town development. One job lost in the docks left a further four vulnerable in related industries (GLC 1985:7). Another recession, this time prompted by the rise in oil prices, led to serious economic decline and in just six years unemployment trebled from 5.2 per cent in 1971 to 16 per cent in 1977 (see Cole 1984:281–2). On the Island, "by the mid seventies things . . . weren't desperate, but I mean things were on the slide . . . Firms were still 'ere but we were getting down really to the sort of, the bedrock with firms", a local recalled. But by the end of the decade decline was very serious indeed. In just seven years between 1975 and 1982, employment on the Island was devastated (AIC 1982): 8,000 jobs in 1975 were reduced to just 600 by 1982 (AIC 1982). As Hostettler (undated a:18) remarked, "Once again, the Island, so isolated and yet so much a part of the world economy, was feeling the effects of changes taking place far beyond its borders."

The final closure of the docks was devastating for some. "It was the lowest I've ever felt in me life", one docker explained, "I just felt dead, numb and that went on for weeks. I was depressed and miserable . . . I felt more anger and bitterness than any other sort of emotion" (Jim). A community that had grown up around the work and whose lives had been fundamentally shaped by it were left individually

Figure 1.14   Dock Desolation. © Mike Seaborne

and collectively, like so many others affected by deindustrialization and globalization, "to adapt to the void in what was their working world, their identity, their
community and their social life" (Taylor et al. 1996:7) (see Figure 1.14).

Massey (1994:146–8) argues that globalization brings into question the meaning of "place" and encourages simplistic comparisons between "then" and "now".
The Isle of Dogs, however, had, since the beginning of the nineteenth century,
supported a dock industry intimately related to global markets and those who
worked in them had observed the peaks and troughs in international trade. They
also had a strong sense of place and pride. "You were protective of that trade," one
worker said, "even though it gave yer some bloody hard times." However, as I
describe in Chapter 2, the new plans for London's dock areas did not seek to
embrace their industrial past and made the transition from old to new Docklands
one fraught with conflict. Furthermore, the process of trying to find a new sense of

place and self-worth in a rapidly changing world was highly problematic, not least because the cyclical booms and slumps that had marked the Isle of Dogs' history did not end with the closure of the docks. The echoes of past experience continued to resonate in the present and future.

Communities past

memory and identity

The vivid memories contained in the preceding accounts paint a picture of an Island that was close-knit and supportive until it was devastated and disrupted by war and the insensitive redevelopment that followed it. One where the work gave shape to people's lives and the communities in which they lived. But was it that simple?

Some have argued that treasured images of close-knit past communities were themselves more "imagined" than real (see Anderson 1991, P. Cohen et al. 1994). Although nostalgia and a tendency for reflecting on the past through rose-tinted spectacles certainly exist, as Crow & Allan (1995:152) observe, "it is hard to think of empirical examples of communities which are purely symbolic, despite all the talk of 'imagined' communities." However, "communities" are "symbolic" expressions of individuals' experiences and the meanings they attach to them (see A. Cohen 1989; 12–13) and, as Cohen highlights, we should not take people's descriptions too literally because "in the public face internal variety disappears or coalesces into a simple statement"; for example, the community was characterized by "togetherness", "everyone knew each other" and so on. However, "In its private mode, differentiation, variety and complexity proliferate" (A. Cohen 1989:74). A sentiment aptly illustrated by the comments of this Island woman: "It was a great sense of community. It really was. People would lend you anythink", but "they'd also talk about you if you didn't conform. I mean, the flats that my mother lived in, you washed the landing . . . If you didn't clean the stairs everyone would talk about you. My friend's mum didn't clean the stairs and she was sort of ostracized. It was terrible" (Sally). Similarly, the Island that was referred to in such glowing terms was also the Island that some had wanted to leave in their youth. "I hated the Island when I was young", an Islander told me:

> I didn't like the Island . . . It seemed to me a bit rundown I suppose, scruffy and I would have liked to have gone p'raps more into suburbia where there was more green grass. I didn't like the Island one bit. And I always said to myself there's only one thing, when I get married I shall leave this place. I shall go somewhere else and live. So what do I do? I marry a man who loves the Island, who loved it. He came from Bow, mind you, and he thought the Island was lovely . . . He used to say to me . . . when I used to nag about moving, he said: "When you go out shopping", he said: "you meet people. 'Hello Lily how are you this morning?' You'd get a bit further up the road, 'Oh hello how are things?'." He said: "If you moved away", he said: "you'd go from here to the shops and back and you wouldn't speak to a soul." He said: "You know everybody, or not everybody, but a

whole lot of people around here." In that way he was right, you know. He saw that sense of community. (Lily)

It is certainly the case that memories play tricks on perceptions, as Pearson (1983) highlighted. The "golden age", he discovered, has always been just out of reach, in the past, some twenty or thirty or fifty years ago. It is also understandable that Island people focused on changes within their own locality symbolized by particular events (the war, slum clearance and later the redevelopment) and their impact upon the changing nature of their community. However, wider and fundamental changes in British society influenced patterns of social life on the Isle of Dogs just as they did in other towns, villages and cities across the nation, although their impact was sometimes less immediately tangible. The introduction of the Welfare State, changes in leisure patterns, technological change, growing individualism, greater mobility and more privatized lifestyles (see Bulmer 1986, Crow & Allan 1994) all played their part in changing "traditional" patterns of neighbourhood life. Furthermore, it is possible that some changes may have been for the better, others for the worse. For example, the levels of recorded crime would certainly indicate that some aspects of life were better two or three decades ago.

P. Cohen in his analysis of Island narratives writes:

> We are presented with an image of an "island paradise", a state of perfect self contained social harmony which has been suddenly, almost overnight, invaded and destroyed by aliens. This island exists in people's minds as occupying a homogenous time and space: it is a place where the population is supposed to have reproduced itself identically from generation to generation . . . and where its cultures of kinship and community have somehow survived despite the closure of the docks, and remained immune, or at least highly resistant, to wider changes in society. (P. Cohen 1996:187)

So why did Islanders cling to a world where change was at least, if not more, marked as continuity? A. Cohen (1982:22), in a study of a Scottish Island (Whalsay), suggested that Islanders were able to neutralize the changes they observed around them by remaining attached to tradition, leaving their "mind set" unaltered: "The reconciliation of the two themes of change and continuity is the essential feature of the management of social identity in Whalsay", he wrote. "Clearly, the structural basis of the society *has* changed", he continued, and: "The relative importance attached to its various elements alters over time and with use. Yet the change is masked by the rhetoric of continuity which Whalsay people employ in their own accounts of their social organization."

In a world of increasing fragmentation and uncertainty (Giddens 1984, 1991) many have suggested that reflecting on the past rather than the present or future provides a refuge. Massey (1994:151), assessing the merit of some of these arguments, wrote: "The search after the 'real' meanings of places, the unearthing of heritages and so forth, is interpreted as being, in part, a response to desire for fixity and for security of identity in the middle of all the movement and change." A "sense of place", which Islanders clearly had,

> of rootedness, can provide — in this form and on this interpretation — stability and a source of unproblematic identity. In that guise, however,

place and the spatially local are then rejected by many people as almost necessarily reactionary. They are interpreted as an evasion; as a retreat from the (actually unavoidable) dynamic and change of "real life", which is what we must seize if we are to change things for the better. On this reading, place and locality are the foci for a form of romanticised escapism from the real business of the world. (Massey 1994:151)

As the following chapters demonstrate, the conflicting perceptions outlined by Massey above were central themes in the conflicts surrounding the development of London's Docklands in the 1980s and 1990s but the "real business" of the world is itself differentially interpreted too.

Conclusion

This chapter has briefly described something of the history of the Isle of Dogs and the experiences of those who lived there prior to the 1980s development. What happened between 1800 and 1980 was as important as the events surrounding the development itself. History in this instance had an unpleasant habit of repeating itself, leaving the area and its people victims of national and international economic trends that they were powerless to influence and from which any benefits often proved to be transient or illusory.

The echoes of the past were reflected in the present. The speculative ventures of the East and West India Dock Companies on which so much pride was bestowed ended in bankruptcy (Hardy 1983a:9). A hundred years later, Olympia and York's famed Canary Wharf development built on the West India Dock went into receivership. Sixty years before this the Wall Street crash had a marked impact on the Island (see Cole 1984:198–202) as did "Black Monday", the financial crash of the late 1980s. The Island's history has without doubt been one of boom and bust and back again.

Hardy wrote at the end of his survey of Docklands history:

Writing these . . . papers . . . has felt rather like sketching a view through a train window; no sooner do you think you have grasped the essence of what you see than it is replaced by something else. . . . events which have shaped Docklands in the past . . . can help not only to locate what is happening now but also possible directions for the future . . . [the use of historical evidence is] not to escape to yesterday's world but rather to help find a route through the complexity and transience of current developments–hopefully . . . to make sense of the London Docklands. (Hardy 1983b:30)

The task of this book is to "make sense" of one small area of Docklands, where the most dramatic social and economic change of the 1980s development took place. The journey has already begun because as this chapter, and Hardy, have ably demonstrated, the processes involved in understanding and "making sense" of what happened on the Isle of Dogs are rooted as much in its past as its present. But the rest of this book does not focus on history, except in as much as by the time of writing the events described here are already in the past. What I set out to capture was the experience of *transition* from old to new in which the pride and industry of the once-thriving docks and factories were now just powerful memories. The question was what would replace them and what place would those who had invested so much in the area have in them. The words of a redundant steelworker quoted by Rause aptly described their concerns:

> There are two Pittsburghs . . . There's the upscale Pittsburgh of the renaissance downtown, and there's this Pittsburgh that most people don't see. I'm not bitter about all this promotion of the new corporate image — that's good, I think they should do it. But they shouldn't turn their backs on the people that built the town. (Rause 1989:59, quoted in Holcomb 1993:142–3)

Note

[1] Pseudonyms have been used for all interview extracts in this chapter with the exception of Sir Christopher Benson.

Chapter Two

Dreams and schemes

The redevelopment of London's Docklands was not a painless affair and was controversial long before the London Docklands Development Corporation (LDDC) was established. The creation of the LDDC in 1981 was the culmination of a decade of competing plans for the future of London's dock areas. That something needed to be done to rejuvenate the 5,100 acres of land in the heart of the capital was never at issue. The problem lay in determining what kind of regeneration should occur; to what extent Docklands communities should be consulted; the balance of public and private investment; and whether redeveloping the area was a "national" concern, the preserve of Government, or a local matter for the five London boroughs that had dock areas within their boundaries. The debate went on for more than a decade as Docklands' fate became embroiled in conflicting political visions, with the production of one detailed plan after another. This chapter briefly describes these competing plans and their impact, the circumstances that led to the creation of the LDDC and the experiences of those who worked in the Corporation in its early phases.

The Docklands problem

The political debate surrounding Docklands development in the 1970s involved a conflict between the Conservative Party's contention that the area was a national resource that required public and private sector involvement to introduce new industry, commerce and private housing, and the Labour Party's approach, which favoured public sector investment in social housing and private industry suited to the skills of local people (Brownill 1990). These sharp political differences punctuated plans formulated for Docklands' future as the two parties moved in and out of government over the decade (see Table 2.1 for key events).

The Secretary of State, Peter Walker, made the Conservative position clear when he determined in 1971 that the task of redeveloping Docklands was beyond the experience and skills of local government. He saw the "opportunity" to develop Docklands as "totally unique" requiring "very imaginative work" (House of Commons Expenditure Committee 1974–5:113, quoted in Hebbert 1992:117). Travers Morgan, a firm of consultants, were commissioned "to make an urgent study of the

Table 2.1   Significant events in Docklands development 1967–89

1967–1970	Closure of East India, St Katherine's, London and Surrey Docks
1970	Conservatives come to power
1971	Secretary of State Peter Walker announces the Travers Morgan study of Docklands
1973	GLC becomes Labour controlled Joint Docklands Action Group (JDAG) formed Travers Morgan study published
1974	Docklands Joint Committee (DJC) established Labour government comes into power
1976	London Docklands Strategic Plan (LDSP) published
1978	Docklands Development Organization established GLC goes Conservative
1979	Thatcher Conservative government elected
1980	West India and Millwall docks close
1981	LDDC established
1982	Enterprise Zone created on the Isle of Dogs
1984	Docklands Light Railway (DLR) Bill receives go ahead
1985	Canary Wharf development proposed
1986	Docklands Highway plans outlined
1987	Olympia and York begin building Canary Wharf Docklands Light Railway opens Reg Ward resigns
1988	Social Accord with Tower Hamlets Community Services Division established
1989	Approvals for Jubilee Line and Docklands Light Rail extension to Beckton Work starts on Docklands Highway

Source: Based on tables in Brownill (1990:17, 57).

possibilities for comprehensive redevelopment" (London Docklands Study Team 1973:1) with "central government and the private sector" interests uppermost in their minds (Brownill 1990:23).

Highlighting the problems that would beset Docklands regeneration from then on, the newly commissioned study "made an inauspicious start" alienating both the Greater London Council (GLC) (the elected authority for London at that time) and the boroughs that had dock areas within their boundaries (Hebbert 1992:119). Although the GLC gave "their formal co-operation", as it was Conservative led, there was "frustration at officer level", particularly among "strategic planners" who felt their role had been usurped (Hebbert 1992:119). The boroughs, on the other hand, "detected . . . an attempt to deny them access to the windfall of surplus land . . . [and] also, with some reason, resented the fact that in the interests of the 'total approach' and to prevent 'fragmented' development, they suffered a ministerial embargo on private development permissions at the height of the property boom" (Hebbert 1992:119).

The Travers Morgan report was published in 1973. It suggested five possible scenarios (from an original list of eighteen) for change in Docklands including a water park, office development and mixed-tenure housing with plans to attract between 50,000 and 100,000 new people into the Docklands areas (London Docklands Study Team 1973). These proposals were deeply unpopular with those living in the different Dockland communities, and Hebbert (1992:119) suggests that insufficient consideration of, and consultation with, local people (who could only comment on the five options that professionals had identified) left a "lasting legacy . . . of grassroots radicalism". The plan also

> happened to coincide with the arrival of a new style of inner city politics grounded in tenants' organisations and local service and advice projects. The Travers Morgan study helped them mobilise an appeal of localism and community defence against outsiders and "gentrification" . . . activists pilloried both the technocratic style of the report, and the substance of proposals for golf courses, luxury flats and convention centres in workaday E14 and SE1. (Hebbert 1992:119)

The Joint Docklands Action Group (JDAG), formed at this time, "allied . . . with the unions and the Labour Party", to oppose the Travers Morgan report and pushed for a different approach to redevelopment. In the same year, when the Tower Hotel was opened in St Katherine's dock (the nearest to the Tower of London, and developed by the GLC and private investment from Taylor Woodrow), a flavour of the protest to follow was ably demonstrated. "While guests inside sipped champagne they watched outside as men and women paraded posters with the slogans 'Homes before Hotels' and 'People before Profits'" (*East London Advertiser*, 21 September 1973, in Brownill 1990:23). The contrasts between private sector development in Docklands and the continued existence of deprivation and poverty in the very same areas of the East End were ongoing and persistent themes throughout the redevelopment of the area.

The formation of the Joint Docklands Action Group also marked the beginning of another schism. Although their sentiments about the proposed development were very similar, and despite an agenda that seemed to be founded "on a clear political analysis of events" (Brownill 1990:23), the Joint Docklands Action Group failed to gain widespread support and were regarded by "other local organisations . . . as politically motivated outsiders, and not 'real' Docklanders (Darby 1973 in Brownill 1990:23)". This theme, and the strains it produced, were to repeat themselves throughout the two decades or more of Docklands redevelopment (Brownill 1990:23), especially on the Isle of Dogs; so too were the competing political visions for Docklands and its development that made conflict inevitable.

Much to the relief of local people and the boroughs at the time, the Travers Morgan proposals were discarded as the political and economic climate changed. Before the Conservative government lost power they, unlike their successors in 1979, were "not prepared to set up an authority to implement development over the heads and wishes of local councils", and therefore, "accepted borough and GLC arguments for a new planning body" (Brownill 1990:23) which "came into operation on 1st January 1974" (House of Lords Select Committee 1981:5). So the inner London boroughs and the now Labour-controlled Greater London Council took over the responsibility for London's declining dock areas and created the Docklands Joint Committee (DJC) with representatives from the GLC, the five London boroughs with dock areas (Lewisham, Southwark, Greenwich, Tower Hamlets and Newham), government nominees, and "after a fight" (Brownill 1990:24), two representatives from the Docklands Forum, an umbrella organisation representing community groups across Docklands.

This time, the needs of the local communities and consultation were at the forefront of the team's approach (DJC 1976:93) because, as one of those involved with the Docklands Joint Committee said, the Travers Morgan "options" had failed to "really respond, or relate, to the deficiencies of the area". The consultative process was lengthy:

> A series of thematic Consultation Papers were published during 1975, each of them drafted, circulated, discussed, redrafted and adopted before publication. All policies were referred for consultation to the Docklands Forum. To further entrench grass roots' participation, the committee funded accommodation and three full-time officers for the Joint Docklands Action Group, and it set up a Consultation Working Party to monitor opinion through a longitudinal survey of a panel of 1000 local families. (Hebbert 1992:120)

In 1975, while these lengthy consultation exercises were taking place, Docklands came under the scrutiny of the House of Commons Expenditure Committee, who found "the politics, professional demarcations and territorial rivalries of local government" contrasted unfavourably with the "single minded development corporations" set up for the new towns (Hebbert 1992:120). The Joint Docklands Action Group and local activists gave critical evidence suggesting that the Docklands Joint Committee was "top-heavy and hierarchical", while the Greater London Council's chief strategic planner suggested that the difficulty lay with "an unholy alliance of local agitators and borough bosses". Much of the evidence heard by the Committee "was pessimistic in tone" and Hebbert (1992:120) suggests that without the intervention of Bob Mellish, Labour MP for Bermondsey, pressing the case for the Docklands Joint Committee, that a development corporation similar to those in the new towns would have been endorsed. Their thinking was certainly in line with a government-appointed independent advisory body, the South East Economic Planning Council, which also recommended the need for a New Town-style development corporation "free from political interference" in order to "give the private investor the necessary confidence to come in" (Hall 1988:354).

Emerging from the House of Commons Expenditure Committee's review with their task still intact, the Docklands Joint Committee continued its work and a year later the London Docklands Strategic Plan (LDSP) (DJC 1976) was published. This outlined a 20-year regeneration strategy (House of Lords 1981:6), £1,138 million of public funding, with another £600 million of private investment (DJC 1976:81). The ambitious aim was for "development to redress the housing, social, environmental, employment, economic and communications deficiencies of the Docklands area" and to have some impact on surrounding inner-city areas and other parts of East London too (DJC 1976:2).

Unlike the Travers Morgan report, which saw future employment in office-based skills, the strategic plan was based on manufacturing and industrial employment suited to the skills of people living in the area. All docks that were no longer working would be reclaimed for building. Despite having laudable aims, the plan's efficacy was highly questionable, as one of the Committee's participants explained:

> They [the boroughs] realised that things were going to change and they were against that change because they knew in economic and social terms there would be a catastrophe . . . Its [The DJC's] first remit was retain the status quo and it just wasn't possible and it gave false hope to all the people here. So there was a lot of play acting and a lot of those plans were unreal in terms of the economic financial context in which they were devised.

The plan "reflected the then political realities" (Hall 1988:352), was "inward looking" and "utilitarian" with an emphasis on "the primacy of borough interests" (Hebbert 1992:120). Even "its own studies identifying the monopoly of council landlordism as a key factor behind the out-migration of firms and families" did not encourage an increased emphasis on owner occupation to promote redevelopment and less than a fifth of the 23,000 new homes proposed in the plan were set aside for owner occupation (Brownill 1990:26). *The Times'* verdict summed up the problems that dogged the DJC's approach to Docklands in which there was a "belief that flair and grand vision are not quite compatible with democracy" (in Ledgerwood 1985:123).

It was not just the foundations of the plan that were problematic but its execution too. There was little political will at the time to invest in London (because other areas in the North were regarded as more deserving) and the Docklands Joint Committee had no independent finance to buy land, the bulk of which was owned by the Port of London Authority (PLA) and other publicly owned bodies like British Rail and British Gas. "The whole plan was unreal", said one of the Committee's participants:

> The DJC had no real powers, it had no means to actually acquire land from the various bodies. It was a sort of consensus committee . . . they had no money apart from the money they had as part of their normal funding. There was no budget, no box marked "Docklands" They had no special additional powers. The other thing of course was that those authorities still had their various jobs to do outside the Docklands part . . . and to everybody Docklands became a problem. It wasn't an opportunity. They weren't happy to let the private sector put money in . . . because . . . the private sector would be putting up offices . . . and people here didn't have office type skills. So the whole basis was to bring back manufacturing industry, port related activities and of course all over London . . . industry was leaving London and one of the reasons why there was so much vacant land in Docklands was that many companies who had a very close relationship with the port decided they couldn't survive on their own account and

they wanted to sell their sites and start again perhaps in the new towns and assisted areas. (Part quoted in Foster 1992:171)

The Committee's strategic plan also required sizeable financial investment at a time when economic crisis reduced local authority spending (see Hall 1988:352) so that "the . . . likelihood of either no development or piecemeal development" (Eyles 1976:26) seemed the most likely outcome. And this is what happened, as another Docklands Joint Committee representative described:

The great difficulty was the Joint Committee was a planning body . . . so the authorities would come together in a committee and take a view of what was needed for Docklands and agree a programme but then the individual elements of that programme had to be implemented by the individual constituent authorities . . . Once it got back to the Borough then it got into the melee of priorities within the Borough and some had given more priority to the Docklands part of their Borough than others so the whole thing was a bit lacking in cutting edge. I can remember about [19]79/80 the DJC ran a very large advertising campaign on TV . . . quite good as advertising goes, quite successful and they had got a lot of enquiries from people saying "We're interested in Docklands. What do we do next?" But they (DJC) couldn't sell them any land cos they didn't own any. They had to send them away and say, "No. You have to go and talk to the Boroughs" and it all sort of disappeared into that morass.

The committee structure was very complex indeed (see Figure 2.1). The Docklands Development Organisation (DDO) serviced the Docklands Joint Committee and between these two groups were "the Officer Steering Group consisting of senior staff from the GLC and the five boroughs" as well as "the Docklands Executive Committee consisting of the Leaders of the GLC and the five London boroughs, and the Chairman of the DJC" (House of Lords 1981:6).

It was not simply the structure that created difficulty but the rivalry between the different boroughs, and opposing interests as each, not unnaturally, wanted to secure the best for their own areas. Consequently, although the boroughs were supposed to act in concert, they sometimes acted independently (Rupert Murdoch's News International Plant in Wapping was one such example, Hebbert 1992:121). Lord Marshall, briefed to look at London government in the late 1970s, highlighted the "conflict and friction" that could "easily lead to indecision, compromise, or simply no action at all" (The Marshall Inquiry on Greater London, Marshall 1978:37, 107, quoted in Hebbert 1992:121). This analysis was reflected in the comments of a third Docklands Joint Committee member:

I think the DJC was doing quite a good job. The problem was that it was locked in local politics. There were five boroughs working together, but sometimes working against each other. . . . There was also the fact that there was at that time a resistance to people coming in and making "a profit", they couldn't conceive the thought of the public sector and the private sector working together.

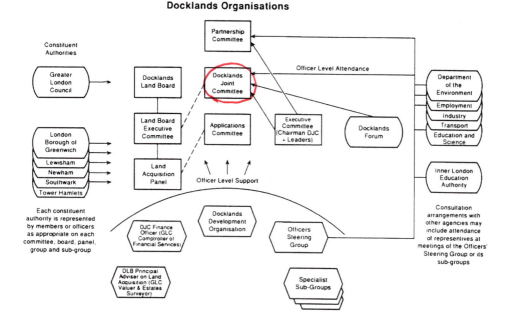

Figure 2.1 Docklands Organisations. *Source*: House of Commons Expenditure Committee Report HC99. © Crown Copyright

Although significant sums of private sector funding had been identified in the plan, "nowhere in the consultation process had anyone bothered to talk to the managers of pension funds, insurance companies and banks" (Hall 1988:354). The plan "existed in a financial vacuum" (Hall 1988:352). An improvement in private sector links did occur in the later stages when consultants were brought in to look at the role that private enterprise might play. "But it was too late by then", one participant said, "because, although we had the right skills, it was a partnership of private sector and public sector skills, there was no additional money."

Some of the action groups including, for example, The Joint Docklands Action Group (JDAG), were opposed to mixing public and private investment, believing that it was impossible to "balance" commercial and social interests and that directed policies aimed at the needs of the local population should be the priority (JDAG 1976, Brownill 1990:29). These ideas continued to be promulgated long after the London Docklands Development Corporation was established, and led to the creation of a *People's Plan* for the Royal Dock area, a joint initiative between the Docklands Forum, the local authority (Newham) and the GLC, in which the proposed development was consistent with the needs of the local population (GLC 1985:57).

Tenant groups and activists were not only unhappy with the London Docklands Strategic Plan because of private sector involvement. They also felt "local views were inadequately represented by the DJC" even though consultation had been a key determinant of their approach. Brownill (1990:28) suggested that while "participation increased and groups were given funding, central power and decision-making was deliberately safe guarded by politicians and planners". "In effect the community could discuss the fringes, after the major decisions relating to jobs and housing had been made."

The consultation process itself also raised questions about what to do in the event of strong opposition to proposals. On the Isle of Dogs, for example, there was vehement opposition to plans for a road that would divide the Island in two. "When they proposed a motorway right the way across the middle of the Island . . . we had [a] . . . big demonstration about that" one of the Islanders said. "The community itself . . . always did sort of stick up for yerselves if there was something you felt wasn't right then you would get groups together that would do something about it." The strength of feeling about this scheme was such that the road, though mentioned, was not a substantive proposal in the published plan (Ledgerwood 1985:122).

Given the financial difficulties, the problems of co-operation between the boroughs, and the lengthy consultation process, it was impossible for the Docklands Joint Committee to meet the targets outlined in their Strategic Plan. So the problems got worse, more jobs were lost and more people moved out. In the five Dockland boroughs alone (Tower Hamlets, Newham, Southwark, Greenwich and Lewisham) over 75,000 jobs in the manufacturing sector disappeared in one decade (Church 1991:2, drawing on Census of Employment 1971 and 1981 data), and in the dock areas within these boroughs problems were even worse. "In just three years between 1978 and 1981 the area lost over a quarter of its 37,000 jobs" (Church 1991:2). "In 1976 the population of Docklands when we published the plan was some 55,000", a Docklands Joint Committee member explained. "By 1980/81 it was just over 39,000, so as the jobs went the people left as well."

"So the tragedy of a negative planning disaster unfolded. After all the studies, and the documents, and the consultations, came the disappointments of 'implementation'. Economic conditions of the late 1970s exposed each of the numerical output targets in the plan as a hostage to fortune" (Hebbert 1992:122). The job creation and industrial strategy failed to meet their targets. So did the housing programme, where only a third of the 6,000 houses originally planned were completed or under construction by the beginning of the 1980s (House of Lords 1981:10).

However, evidence provided for the House of Lords Select Committee revealed that the Docklands Joint Committee had made "a substantial amount of progress . . . in areas other than industry and housing", especially the dock filling programme in Southwark (paragraph 5.2), as one of those working with the Committee explained:

> The DJC clearly wasn't going to . . . realise the potential of the area, that's not to say that . . . some of the local authorities hadn't done some jolly good work because they had, particularly in areas like land reclamation. . . . Southwark had done a lot of work and Tower Hamlets too . . . I disagreed with a lot of the decisions that had been taken, particularly the Dock filling programme, that seemed to me to be terribly misguided, but nevertheless the important thing was people were rolling their sleeves up and getting on with it.

Other initiatives included the Billingsgate fish market (spearheaded by Paul Beasley, leader of Tower Hamlets council and the Corporation of London), located just off the Isle of Dogs, the News International plant in Wapping, Canon Workshops (small business units) and an industrial park in Beckton. But these successes could not hope to stem the tide of firms leaving the area or the mounting unemployment. Brownill (1990:29), whose account of the Docklands Joint Committee's efforts is sympathetic, recognized that "the problems facing the DJC were immense and were not solvable at a local level".

While the political debates went on, the lack of impact on the ground was keenly felt by the various dock communities and the Isle of Dogs was no exception. "In the seventies we'd had five major consultation exercises about what was gonna happen to the Island, to Docklands, whichever," a community activist said:

> and on top of that there was the five PLA (Port of London) proposals about the docks as well and in between that time there was all sorts of various little surveys and studies being done. So people were just fed up to the teeth with bloody consultation exercises, you know. People said "For Christ's sake, you know we keep on talking about these bloody things, you must have produced enough paper . . . [to] have denuded half the forests of Canada by now . . ." I think they were in the business of talking because no one really had a clear idea . . . the strategic plan was set out but no one had the confidence I don't think, certainly not in government circles to say, "Okay let's put some money in and make this work." (Ted Johns)

Residents became tired of the debate and longed for action: "There was a kind of resentment, a growing sense of anger about the things that had not been done. [People could] see their area sort of gradually going down and people kept talking about it [but] nothin' was happening", an activist said, and these feelings led to protests throughout the 1970s. The residents' frustrations were reflected among business people too: "For years", a local businessmen in the area at the time said, "the five boroughs involved had this sort of group of people investigating what they should do and for . . . years all they did was produce bits of paper."

"We always used to say this was a forgotten Island", an elderly woman born there said. "I mean we got nothink down here. Everything we wanted we had to fight for more or less." Despite post-war reconstruction, and a growing population (from approximately 9,000 in 1961 to 12,000 in 1971), the social and transport facilities remained very poor indeed (Eyles 1976:8). Endless promises from those in positions of authority that better amenities and facilities would come once there were enough people to sustain them (Cole 1984:269–70) left "the peninsula seething with discontent" when they failed to materialize (Cole 1984:266) and eventually the anger overflowed with "public demonstrations involving hundreds of Islanders" (Cole 1984:171–2). In March 1970, a Unilateral Declaration of Independence (UDI) was announced by a group of Isle of Dogs residents to draw attention to the need for more facilities, in which the bridges at both ends of the Island were lifted. No traffic could get off or on. "The slow fuse of discontent finally caught light", Eyles wrote, "with Islanders closing the access roads, because they thought that Tower Hamlets Council, the GLC (Greater London Council) and ILEA (Inner London Education Authority) were treating them as poor relations in the London money-stakes" (Eyles 1976:8), action that finally persuaded the authorities to build a secondary school/community facility on the Island.

The response to UDI was typical of the complexities of local people's attitudes. "There were quite differing opinions on it", a local journalist said:

> There were those who felt that it was important and useful. But there were a very large number, I don't know how many, who rather resented it and thought it was just [one individual] trying to get a bit of a reputation. I don't think that was true but that's how a lot of people felt. Of course, accidentally the press took it up . . . and it made quite an impression outside the Isle of Dogs, but it didn't really make any in the Isle of Dogs . . . A lot of people who counted for something in the community thought that it was a lot of nonsense.

Here we confront the problem that while communities may appear externally cohesive they are in fact characterized by internal conflicts and differences (see A. Cohen 1989:74). As Suttles observed: "In most instances local communities seemed impossible to find so long as one thought of them as neatly bounded areas whose residents were of a single mind . . . Those who claimed to represent 'the community' were often badly divided and could not settle on the same constituencies and boundaries" (Suttles 1972:44) and, as the discussion in later chapters demonstrates, this continued to be the case during the redevelopment of the 1980s and 1990s.

When The House of Commons Expenditure Committee returned to the topic of Docklands in 1979 Hebbert (1992:122) suggested that "by the measure of the DJC's own targets wretchedly little had been achieved since its inception in 1974." Although some work had been done on preparing the land for development, the DJC was able to demonstrate little success. Sir Hugh Wilson, Chair of an Executive Committee established in 1979 to try and promote more effective implementation of the plan, explained to MPs that the situation was "highly complex" and that the DJC had passed through several stages from the compilation of the plan itself "to a transitional period when that strategic plan gave rise to local plans" that in themselves "are vital to the actual detailed development of particular projects" and finally had arrived at "the period of implementation" (House of Commons Expenditure Committee, 1978–9 col 47). While some maintained their belief in the public sector's ability to develop the area, not surprisingly others felt that more directed and dramatic intervention was required free from the fetters of bureaucracy and planning.

The creation of the London Docklands Development Corporation

After a decade of political conflict and opposing plans, Docklands' fate was sealed when a Conservative government was elected in May 1979 with a manifesto committed to reducing state intervention and public provision, and encouraging private enterprise. The years of talking and relative inaction were over, replaced by a development momentum the speed and veracity of which took many by surprise.

Just a few months after the general election, Michael Heseltine, Secretary of State for the Environment, set out his agenda for tackling the blighted areas of

Docklands in London and Liverpool with the creation of Urban Development
Corporations (UDCs). Like Peter Walker almost a decade before, development in
Docklands was once again linked to the "national interest" (Local Government
Land and Planning Act 1980) and the agenda was to put "land and buildings into
effective use, encouraging the development of existing and new industry and commerce, creating an attractive environment and ensuring that housing and social
facilities are available to encourage people to live and work in the area" (LGLPA,
s.136:120). Their role was to "generally do anything necessary or expedient" to
promote "regeneration" (LGLPA, s.136:120). The Act also allowed for the creation
of Enterprise Zones to attract inward investment with attractive tax incentives, a
ten-year rate-free period, with minimal planning or other controls, something that
proved to be a major factor in developing the Isle of Dogs (in Brownill 1990:31–2).

   Although the London Docklands Development Corporation came to be seen
as "an all-powerful instrument for change within the eight and a half square miles"
(LDDC 1998:6) the Corporation's powers were limited and strongly reflected the
prevailing Conservative government's political ideology. Reg Ward, the first chief
executive of the London Docklands Development Corporation said: "Whilst we
were supposed to be based on the New Town Model we were not by the time other
ministers and civil servants finished . . . The New Town Model to be used was in
fact stripped of all the critical powers of a New Town Corporation." He continued:

> (The LDDC was) not the planning authority, simply the "Development
> Control" Authority . . . within the UDA . . . The councils remained *the* Plan-
> ning Authority. We were not meant to be the prime owner of development
> land . . . We were to be endowed, with effecting compulsory purchase of
> Local Authority owned land that was available for redevelopment. Expli-
> citly, we were to have no responsibility for, no direct involvement in, social
> housing, community development, community support; health, education,
> training, etc., etc. All of these were to remain the complete responsibility
> of the local authorities and public agencies. We were not to be involved in
> major infrastructure or transport developments . . . outside that required
> to service the development sites. (Reg Ward)

   Once the decision to create the Urban Development Corporations was made,
speed was of the essence. There would be no more time-consuming consultation
or impediments to development caused by bureaucratic processes. Urban Development Corporations would be "single minded agencies" not subject to local accountability and the government introduced legislation to establish them via the Local
Government Planning and Land Bill in order to avoid a public inquiry. However,
"In its haste to push through the legislation the government issued a 'hybrid' Bill,
which allowed for petitions against it" (Brownill 1990:33). Consequently, a House
of Lords Select Committee sat for 46 days hearing evidence about the proposed
London Docklands Development Corporation (see Hall 1980, Colenutt & Lowe
1981). Hebbert (1992:123) noted:

> The . . . hearings showed once again the gulf separating the localists, who
> defined the issue in terms of community defence, from the wider view of
> surplus dock land as a resource for London and the nation. In between,

the boroughs uncomfortably tried to distance themselves from the activists, while claiming that the DJC planning arrangements were working well.

The remit of the Committee was "extremely limited" because "it could discuss only whether Docklands was a suitable area for a UDC and whether the precise area proposed was adequate" (Brownill, 1990:33). Drawing on problems recognized privately by many members of the Docklands Joint Committee about the inability "to acquire or dispose of land" (House of Lords 1981:para 6.3), the Government wanted a directed agency dedicated to regenerating Docklands because "each borough is necessarily as much concerned for parts of its area which lie outside Docklands as for the parts which lie inside it and cannot possibly have a single minded concern for the regeneration of 'docklands' as a whole" (House of Lords 1981:para 6.3). Their Lordships, reflecting the criticisms made of the Docklands Joint Committee and the London Docklands Strategic Plan they produced, believed that "the boroughs tend to look too much to the past and too exclusively to the aspirations of the existing population and too little to the possibility of regenerating Docklands by the introduction of new types of industry and new types of housing" (House of Lords 1981:para 6.4).

Brownill (1990:35) suggests that with the exception of the provision for public housing (which was amended in the light of their opposition) the boroughs did not challenge "any of the policies or aims of the UDCs", leaving "local groups . . . abandoned . . . to defend the LDSP (London Docklands Strategic Plan) and local accountability". Docklands residents, whose expectations had been raised by the Strategic Plan, were faced with the prospect of almost total exclusion as Urban Development Corporations were formally accountable only to the secretary of state. Clearly their objections to this situation formed a pivotal part of their opposition but it was not one by which the Lords were entirely convinced:

> The Committee have no doubt that the opposition to the proposed transfer of development control from the boroughs to the UDC, voiced by the local organisations which have petitioned against the Order, is perfectly genuine. These bodies were consulted when the LDSP (London Docklands Strategic Plan) was prepared; they realise that the influence which they can bring to bear on elected councillors is greater than any influence which they are likely to be able to bring to bear on a UDC; and they fear that if a UDC is established, developments of which they disapprove may be carried out without regard to their views. But while the Committee understand — and indeed sympathise with — their attitude, they think that the approach of some of the witnesses to the problem of the regeneration of Docklands was somewhat parochial. (House of Lords 1981:para 8.3)

The strength of local opposition did, however, have an influence on the recommendation that "the UDC must win the confidence of local organizations". The vehicle for this was to be the Docklands Forum, who must "know what the Corporation is thinking of doing, so that they can express their views on any proposal and it must make it clear that such views, even if not accepted, are always seriously considered" (House of Lords 1981:para 8.7). Nigel Broackes, Chair designate of the proposed Development Corporation, confidently predicted that their "success" depended on "harmony with the boroughs and . . . due consideration to the views of the local population as expressed through the Docklands Forum. Once it is seen that the UDC is operating in this spirit, opposition to it will gradually die away", he said (House of Lords 1981:para 6.8).

It was not only the local authorities and interest groups who were concerned about the loss of accountability. An Isle of Dogs businessman, still in the area when the LDDC was established said:

> I opposed it purely on the grounds that it was not an elected body. From a democratic point of view I thought it was wrong. But once it was created one went along with it and I made it my business to get to know the people running it . . . And I think I realized then that if we'd have had a sort of quasi-elected body nothing would have ever happened. You know there are times when you need a war or something to change things and you needed somebody with absolute power to get on with it . . . and I think in the end . . . people will be glad that that did happen, it's like Napoleon laying out Paris isn't it really.

It took nearly two years for the Lords to sanction the creation of the London Docklands Development Corporation in June 1981. In their report they concluded that the demise of London's docks and the loss of related industries was a "calamity" and those now sanctioned with its redevelopment were faced with "a vast task" (House of Lords 1981:para 8.1). The Lords recognized that the transfer of accountability from local mechanisms to the secretary of state was "not easily to be justified" especially in the light of the opposition that had been mounted to it by local groups, but ruled "the Government have made out their case" and it should be accepted (House of Lords 1981:para 8.8).

Getting started

By the time the Lords had officially sanctioned the creation of the London Docklands Development Corporation, a "shadow" corporation, under the Chairmanship of Nigel Broackes, had been "hard at work for the previous year and a half". Broackes, head of the Trafalgar House empire, said: "From my first visit [to Docklands] . . . I was convinced on countless scores that it could work and that the machinery up to then had been wrong . . . My overwhelming impression was that the Island itself was a beautiful place and that alone would guarantee success." What had once been viewed as a problem was now seen as an opportunity, and with his deputy Bob Mellish (Labour MP for Bermondsey, and former minister and chief whip), it was agreed that there would be no more dock filling and that stimulating the private housing market would be a crucial element of the development strategy. "Bob Mellish certainly had a very open mind about the housing", one of the original team said, and "the kind of variety of jobs. I mean he swung it completely round and said 'Why shouldn't kids here come in offices, in the posh shops, why should they just continue working down the coal mine.' He was breaking years and years of tradition."

During the select committee hearings, Nigel Broackes suggested that although the public face of the boroughs was critical, they were privately more accommodating: "There was a lot of doctrinaire opposition to us on the general point of the withdrawal of sovereignty from the Socialist boroughs", Broackes said, but "I was surprised by the underlying reasonableness of most of the participants when you got to talk to them." As a result, the leaders of the three boroughs (Tower Hamlets, Southwark and Newham) were asked by Michael Heseltine to meet with Broackes:

> They came along one morning and they each had something to say. [One] said "You must understand that the reason we opposed you with so much vigour is that we thought if you were appointed you would win and put us in disgrace". [Another] said to me the last words of Mary Queen of Scots when she was executed . . . which were addressed to the executioner to the effect that if I have to lose my head please send me an expert axe man (and the last) said "I've looked you up in *Who's Who* and decided with my colleagues that you can't be entirely bad if you're on the council of the Victoria and Albert Museum! So on that basis we got off to really a rather good start and they all agreed to join." (Nigel Broackes)

A later Board member said: "In my view it was a fairly definitive argument,"

> I don't think even with hindsight that the local authority would have been in a position to [cope] with serious regeneration on their own without the Corporation having been involved. On political lines . . . irrespective of the fact that we'd rather be in power ourselves, the Corporation was there and to be in on the inside is worth something . . . we ought to be in there participating rather than on the outside waving a fist.

So despite their opposition to the creation of the Corporation there was a pragmatic stance from the three local authorities initially: "The people who came on our Board in that first year, who accepted Heseltine's invitation . . . were the old style Labour politicians, very paternalistic . . . but they wanted things for their people", an LDDC executive said. "They actually wanted opportunities . . . [so they] took up the secretary of state's invitation . . . because they realized that this was reality and let's work with it. We failed . . . let's now represent our people by being members of the Board and let's get a slice of the action, not that there was much cake to go about then."

Many of those recruited to work for the newly formed Corporation came from "inner London local government" (Hebbert 1992:124) and they too believed that public opposition and private sentiments differed, as this LDDC executive explained:

> People from the outside sometimes think that [it] must have been very tense, you know, when you told them you were going to join the enemy but it wasn't like that at all. It was all very friendly and supportive . . . Southwark council at that stage was quite a pragmatic authority so that although they . . . publicly opposed the setting up of the LDDC they didn't do it with . . . any venom and when it was set up [they] wanted to work very closely with the Corporation to help them, not out of altruism, but

because they wanted the best for their part of Docklands . . . So the fact that I was moving from (a council) . . . to LDDC was actually all very friendly.

Another reported much the same feeling. "There was a lot of overt opposition which actually masked as I perceived it" different sentiments "particularly on the part of the Boroughs. A lot of people would say to you 'actually this is the best thing that can happen but . . . we've got to oppose' and there was a lot of half-hearted opposition. . . . behind the scenes that seemed to be the position."

Reg Ward, the first LDDC chief executive said, of local authority representation on the Board: "One had a conversation that was running all the time (with the boroughs) despite . . . the public difference of view . . . It was terribly important that, from the local authority's point of view, the leaders were on the board, notwithstanding the strains that it created . . . It was also terribly important that I was actually operating constantly with the other three chief executives on a very personal level." He continued:

> That initial board was actually very broadly based, a nice combination of independent professionals along with people with a track record related both to development and to regeneration along with the very secure base of the local authority leaders . . . We were carrying, as it were, a blue flag into a fairly red dominated area on the face of it . . . [But] never once in any one of the Board meetings did I ever hear, and I really mean this, even with Jonathan Matthews (the Liberal councillor invited onto the board) much later, did I ever hear anybody talking party politics. This was significant when you think that John O'Grady, Jack Hart, and Paul Beasley were all staunch Labour leaders of very traditional Labour Councils, . . . Never once did party politics play a part in decision making round a Board table . . . Beasley, O'Grady and Hart came to do a job within the Board. They addressed the problems on an absolutely even handed basis. The initial Board was very good to work with.

This was how, at least in terms of the statutes, the LDDC was intended to operate: "to work alongside, not usurp local and other public authorities" because although they had "assumed development control powers and could approve or turn down planning applications" they had "with the exception of the Enterprise Zone, to work within the framework of the three borough local plans" (LDDC 1998:6). In fact Reg Ward argued that "we could have been stopped in our tracks if the Districts had realized their ability to bring us to a standstill by pushing through their Local Plans" (although when the councils did put forward alternative proposals for sites within the LDDC's area that were consistent with their local plans, for example, in opposition to the proposed airport for Docklands, and in a scheme in Greenland Dock in Southwark, both of which went to public inquiry, the councils lost (DCC 1985:24)).

The initial cordial relations between the Corporation and local authorities were short lived and in the council elections in 1982, Tower Hamlets, Newham and Southwark moved further left and refused to nominate anyone to sit on the Board perceiving an altogether more sinister agenda in the LDDC with which they did not wish to be associated. This is when "the real opposition" started, an LDDC executive said:

There were some significant changes in Southwark and in the other authorities . . . Although it remained in Labour control the Labour members who'd run it for many years who were . . . typical old style Labour author-ity, lots of things wrong with them but basically decent and reasonable, and pragmatic . . . they were sort of ousted in an internal coup and . . . [the councils] swung violently to the left and the new members . . . [were] very zealously opposed to everything the LDDC was trying to do . . . Relations completely broke down to the extent that they [Southwark] took a resolu-tion refusing to recognize the LDDC [and] . . . issued an edict for their officers not to speak to our officers. Wouldn't answer our letters, wouldn't speak to us on the phone . . . It was quite painful particularly because the people . . . were old chums of mine and to have our two organizations at war was actually quite painful.

The policy of non-co-operation had much to do with the prevailing political climate generally, as well as the history of conflict and dissent about the nature of development in Docklands. In this respect local and national politics could not be separated and the LDDC "was seen in some quarters as a Tory manoeuvre to unseat deeply entrenched Labour housing and economic policies and even per-haps, if owner occupation swamped the area, long-term Labour control" (LDDC 1998:6). The climate was hostile both nationally and locally and many on the left were demoralized and angry. A Conservative government dismantled, with cavalier indifference, local electoral accountability for all of those who lived in the areas covered by the LDDC — most of them Labour supporters, and later abolished the Labour-controlled Greater London Council, the elected body for London, taking out two key democratic tiers, and the legitimacy and powers that went with them.

Furthermore, the Conservative emphasis on private sector development was deeply uncomfortable for the Left. As Brownill (1990:107) notes, "The LDDC was one of the most extreme examples of central government attempting to impose its own market-oriented logic onto a local area" without even seeking to address the social needs that were manifestly evident; in fact these were explicitly ruled out of the LDDC's remit. In their zeal to tackle what they perceived to be profligate "loony left councils" championing minority causes, the Conservatives imposed increasingly stringent financial constraints on local government, and pursued policies through-out the 1980s that undermined the position of local authorities. "The Conservatives believed in home ownership, the right to buy for council tenants, the encourage-ment of the private sector and the development of housing associations as a coun-terbalance to the local monopoly of council provision for people who could not afford alternatives" (LDDC 1998:5), all threats to Labour local authorities.

The refusal of the local Labour politicians in Docklands itself to compromise and negotiate with the LDDC (which did nothing to help Docklands residents, as I discuss later) was equally matched by the doctrinaire approach of the Conservat-ive government, which for almost a decade, refused to be swayed from its socially divisive path. The stage was set for conflict and many on the Left, especially local Labour politicians, fervently hoped that the "Docklands experiment" would fail and that they would gain political advantage from it. Meanwhile the more moder-ate former leaders of Tower Hamlets, Southwark and Newham, appointed as indi-viduals, remained on the LDDC Board until 1986.

The LDDC's "manor"

Within the Development Corporation the mood, by contrast with that in the local authorities, was ebullient. Broackes and Mellish, optimistic about the opportunities for developing the designated areas of Docklands, set about their most crucial task of finding a chief executive and in July 1980 appointed Reg Ward, a charismatic and energetic figure, formerly chief executive of London Borough of Hammersmith and Hereford and Worcester County Council. He passionately believed that Docklands provided a unique opportunity and in his first week in the job was invited to lunch in the Nat West Tower:

> They held the lunch in one of their dining rooms pointing east and so for the very first time one saw one's manor . . . laid out before one. Eight miles of river opening up into the Thames Estuary all the docks and so on. I was totally taken aback and I could not help saying "But where's the bloody problem?" It was the most magnificent waterscape you could ever have hoped to see and yet it was seen as a problem. I came in really . . . perhaps romantically . . . actually not perceiving the problem at all. The problem, in my view, was one of perception, of actually seeing the area differently. (Part quoted in Foster 1992:173)

The LDDC's "manor" contained as Ward explained, several "separate, independent [and] totally deprived areas" (see Figure 2.2):

> *Within Tower Hamlets*; Wapping, Limehouse, Poplar and the Isle of Dogs
> *Within Southwark*; Bermondsey/Rotherhithe/Surrey Docks and Greenland Docks
> *Within Newham*; Silvertown (Beckton)

With the exception of dock filling in the Surrey and London Docks, Beckton marshes (which had been drained) and St Katherine's dock, the area had barely been developed at all. The LDDC's job was without doubt "a vast task".

The discussion that follows in this and subsequent chapters about the development focuses entirely on the Isle of Dogs, that part of the LDDC's eight square mile area where the most dramatic and intense commercial development took place, fundamentally altering the landscape and putting this previously little known area firmly on the map.

"Brimming with opportunities": the view from inside

The LDDC's approach to developing Docklands differed fundamentally from that which had shaped the strategies for the area during the 1970s, and, in so doing, challenged many of the cherished views of planners and local government. "I have

Figure 2.2 The LDDC's "Manor". Reproduced with the permission of CNT as owner of the LDDC archives. © Commission for the New Towns

to acknowledge", Reg Ward said, "that to a large extent dreams, images of what might be, replaced conventional planning." "My experience had led me to have a healthy disregard of conventional planning systems", he continued:

> because if you have an area with major problems, an area where nobody wants to go and where nobody's prepared to invest you can prepare all the plans that you like but nothing will happen . . . there was no point really in pursuing a conventional master planning approach. Equally you had to avoid a problem focus. You had to say, this is not a problem area, it's one that's just brimming with opportunities, it's simply that people have not perceived them.

Reg Ward was a powerful persuader and sought to sell a vision of Docklands with an enthusiasm and verve that was infectious. As a woman who ran her own catering firm and was in search of premises in the early 1980s, shortly after Reg Ward took up his post, described:

> There was this wonderful day when I was . . . doing my directors' lunches . . . There were about twelve men and there was this one person who was holding the entire company spellbound by what he was saying. Not a word was spoken. They were all listening to him and he was describing London's Docklands . . . The more he talked about it the more enchanted I became cos I thought this is it, but where is the Isle of Dogs, must be a million miles away, cos one didn't know about Docklands . . . I had not left the dining room while this conversation was going on and I went round and round the table offering more potatoes, more brussels sprouts and they kept saying "No", but I still went round and round because I wasn't going to leave the room while this was happening. . . . His name was Reg Ward. . . . When he . . . left he went past the kitchen door and said: "Thanks for a nice lunch," and I said: "Before you go where is this place. This is what I want".

Reg Ward's personality and unconventional approach were influential factors in the decision of many who joined his team. They were an eclectic group whose "approach was pragmatic, owing little to 'new right' or any other ideology" (Hebbert 1992:124). This was important in terms of the approach they adopted, which was often far broader in its focus than their rather limited initial remit suggested. One of Reg Ward's team said that, when the LDDC was created, he "wasn't confident" as he had "been involved for so long".

> The turning point . . . was meeting Reg Ward. He was very very different from all the other people who had been involved in the Docklands process. They tended to be rather grey people, bureaucratic people who'd been in the same place for years and years. Reg was a bit of a maverick. He was very charismatic and he had all sorts of ideas. I can best express it when I spoke to him about where we would have our office. Previously the DJC [Docklands Joint Committee] had their offices . . . in County Hall and then a hundred metres away from County Hall because it was very

convenient for people . . . I don't think any of them had actually been to
Docklands. It was planning by remote control and Reg I remember said
"Well we'll be in Docklands, why do you ask?" I said "Where in Docklands,
have you ever been there?" He said "Oh yes I have" and I said "Well there's
hardly anything down there." He said "Well I've found a building" and I
said "Which one Reg?" He said: "Fred Olsen's building, the old shipping
line building. It's been disused for two years." I knew that building well
and we went down there and it was all covered in pigeon droppings and
papers left by the PLA [Port of London Authority], there was broken
windows and I said "It's going to cost us a bomb to repair this." "No", he
said "We'll buy some paint and we'll get ourselves a bit of carpet" and I
thought is this chap on the level? I mean is he for real? And he was. . . . I
remember certain people saying "I'm not going down there to join Reg
Ward, he's mad, off his rocker", and that appealed to me because I thought
my God if someone's going to change all that previous thinking, just push
it out and set it alight, it's this chap . . . so I gave him my full confidence
and enthusiasm . . . Out of the original team [I worked with] I only managed to get six of my colleagues to join me in this adventure.

For those accustomed to working in the West End with facilities and amenities
on their doorstep, the move to a building in the almost derelict docks was something of a culture shock, but one which added to a spirit of adventure:

We were in West India House which is gone now . . . it was a lovely little
building . . . [designed by] Norman Foster. . . . architecturally . . . very attractive . . . and quite an important building . . . in the history of the docks . . .
[but] locationally it was literally miles from anywhere . . . It was very, very
isolated. We managed to . . . add to our own isolation because . . . there
was a wood yard operating and they bought timber in from Russia and
we had to open the old swing bridge once a week to . . . to let the ship in
and on one occasion we opened [the bridge] and the machine just broke
down. We couldn't close it. That bridge was the only road to our office
from the West and as most (of) our visitors [came] . . . from London that
was where they came in. . . . The engineers looked at it. They said "This is
not capable of being repaired, we'll just have to build a new bridge" which
obviously took some time . . . [So] we managed to negotiate a deal with
the wood yard at the other end of the dock where people coming to visit
us were allowed to drive into the front of the wood yard, through the wood
yard, out the wood yard's backdoor and then down to our office, so we
had a sort of procession for some months of Ministers of the Crown [and]
chairmen of major companies interested in Docklands had to drive through
the wood yard to find our office, so it was quite hairy.

Another of the initial team said:

I remember standing outside the office and you could hear a pin drop
and you used to drive round the docks in the fog wondering whether you
were going to drive over the edge! I mean, it was silent, derelict. A great

beauty about it in many ways. One remembers standing there with colleagues and saying bloody hell how are we going to do this, impossible . . . But we did it. There was an energy about the place, a will.

The enormity of the task, and the excitement of the enterprise, contributed to "a great sort of . . . wild west frontier culture among the original team" — a relatively small, dedicated group. Reg Ward said:

> The extent to which I was able to build a team of individuals around me in a very unconventional way and also to keep the team extremely small meant that all our responses were direct and immediate. If we determined that we were going to try something and it didn't work, we could always change overnight. It was a very stimulating, a very exciting process to be going through, but it was one that was very difficult to justify in a conventional way.

"We had to change people's perceptions completely", one of the original LDDC team explained, and tackle scepticism. "Basically a lot of people in the UK property industry, investment industry were saying, 'Well there they go again'. Peter Walker had these consultants in 1971, Geoffrey Rippon set up the Docklands Joint Committee, Peter Shore announced urban partnerships, and here's Michael Heseltine with his new bright idea." What was required was action, and results, which had to be achieved over a relatively short time span. A report drawn up by consultants Coopers and Lybrand (1980) highlighted the need for the LDDC to begin a "very high level of programme activity" in order "to establish public credibility, gain the support of the local authorities and convince the private investment sector of London Docklands' potential" (quoted in LDDC 1998:8). There was also a "political imperative" though, as this LDDC executive explained:

> Our coming together had been delayed, and there was an election peering around the corner in 1983. So in effect we had 24 months to demonstrate to all our critics and our supporters that . . . this mechanism set up on an experimental basis by Heseltine in Liverpool and London actually can produce dividends. And it really was slaves in the galley . . . Reg used to whip us all into shape, you know, heads down, yes the local papers sniping at you or the local politician on the TV programme last night said . . . ignore it. We'll handle that. Just get on with buying that land, preparing those sites, getting that first development in. And it was tremendously satisfying to know there was this momentum that you know people really meant it and we had real money, real powers, and a real job. Whereas my experience of local authority is that people are frightened of that. I remember . . . if you actually said to your committee "Yeah we want to build that road, and we want to build it by next spring", hang on, hang on, wait a minute . . . I used to spend years and years as a young sort of surveyor/planner working on schemes and they used to go back in the drawing drawer . . . And that was the most satisfying thing about London Docklands, that in those years we actually did a plan to build three industrial units, or whatever, or to build an estate of a hundred homes or get a

design guide out for a particular site and get markets to make it desirable, it actually happened.

Transforming perceptions and "pioneering" change: 1980–1985

The first thing . . . was to change the perception of the place and to start painting pictures of what the docks could be like if the water was actually used. There was no question of filling any more of the docks. The water was the one quality that lifted us out from everywhere else and so one consciously went round . . . painting an entirely different picture from what the area actually was and projecting what it could be like, an area that, at that time, had no credibility in conventional market terms, no acceptability in local political terms and hopes which had nothing to do with local people's aspirations and needs. (Reg Ward)

The redevelopment of London's Docklands did not begin with the high-rise office blocks that now dominate its skyline. Its beginnings were rather more modest, and at the time few imagined, or believed, that commercial development would ever have been possible. "The initial level [of] funding projected for the Corporation was little different to the existing Urban Programme funding for the Docklands area [which was] £25–30 million." Reg Ward said "the aspiration for Docklands" involved "some low-cost housing, new industrial type of development, environmental improvements related to the new housing and industry" with "no concept of a major infrastructure improvement". "This remit", he continued, "only justified, in Michael Heseltine's thinking initially, an organization of about 35 staff. With some difficulty I got him to accept an organization of around 75 staff" (Reg Ward, personal communication, January 1998).

The potential of the 5,100 acres was not immediately obvious to many, including private developers, because East London was associated with a plethora of negative images typified, as a property agent said, by its industrial history and its associations with poverty and crime, which the experience of this established resident aptly illustrated:

When I first worked up the City they used to be saying awful things [about the Island] . . . dreadful things . . . I was always defending it. . . . When I was in the Civil Service . . . I was about 51 . . . there was this chap up there and he always said: "Everybody knows everyone East of Aldgate is crooks." So I said "That's funny I've worked here about eighteen months and I aint gone down yer pockets yet while you've been out of the office." They . . . had that attitude.

A businessman said:

You don't spend money on the East side of London, you go to Kensington or Mayfair or Knightsbridge . . . You don't spend money in Tower Hamlets.

You don't spend money in Newham . . . Poplar, Stepney and Bow. The names, you've got to admit it they all sound wrong . . . they're all shady Dickens areas or immigrant areas, which is true they are, from the Jews to the West Indians to the Asiatics they're all there and still [there's] that connotation.

Such images, as a commercial property agent realized, presented "a difficult psychological barrier to overcome". Yet in the space of just five years, the Isle of Dogs, the centrepiece of Docklands development was transformed from an area where the commercial potential was perceived to be "small" (Eyles 1976:26) to one where the landscape changed almost weekly as developers built larger and more spectacular commercial projects.

Nigel Broackes said that from the outset he, Bob Mellish and Reg Ward were convinced they could develop the area successfully "if it was housing-led on the basis of owner occupation". "The history of housing development since the war within our boundaries was abysmal", a senior member of the Corporation said. "It was . . . an area which didn't appeal to anyone", said another, leaving "a heavily dependent population" and "less than ten per cent owner occupation" when "the rest of London was averaging forty per cent owner occupation. There must have been a huge latent demand which was not being met." Broackes explained their approach:

> We sought . . . critical mass . . . to prove that past experience had been no indication of what could be done . . . Our plans [were] for roughly 10,000 dwellings, you could call that a population of 28,000 perhaps, on land that we could provide to the builders, and roughly another 5,000 on land belonging to other owners who would be stimulated to develop by our success. That gave us a programme of 15,000 houses, 40,000 people perhaps, more than doubling the resident population. . . . That was the fundamental plank of our plan and it was quite different from the new towns. The minds of senior people at environment were tuned into us wanting advanced factories, municipal accommodation for key workers and all the old formulae that hadn't been too brilliant. We thought if we could create positive land values that development would ensue and we thought if we could get home ownership in it would have a vast impact on the place. (Sir Nigel Broackes)

The emphasis on private housing development was such that only those sympathetic to this approach were considered for Board membership: "Broadly speaking if people didn't agree, they wouldn't be suitable for joining the Board", Broackes explained, "so we didn't kick off with anybody who was basically unsound in the context of this doctrine. That's a terribly important point in the genesis of the development, of successful enterprise, and we had to do it in a very short space of time."

While those within the Corporation might have been committed to private housing, the developers were not convinced, as this LDDC executive described:

> There was a great myth in Docklands that this wasn't the place for owner occupation . . . When we took over . . . something like ninety six per cent

of the housing was rented . . . the vast majority from local authority . . . and the . . . house builders . . . wouldn't build housing for sale because their argument was nobody would buy a house in Docklands and it was a self-fulfilling prophecy, obviously.

Attracting developers therefore was not unproblematic. He continued:

To start with the house builders were not interested but we managed to persuade . . . people like Wimpey and Barratts and Comben (major house builders) to try a pilot scheme. We arranged very attractive financial packages for those first schemes . . . we actually said the very first people that come in and take all the risks, if this works, you can take all the profit and the public sector won't claw a bit of that back . . . Even with those packages they were very very reluctant and we leant on them very heavily. They decided they'd have a pilot project done in Beckton and that was so hugely successful that everything was sold sort of before it was completed and they were back demanding land for their next sites and other builders wanted land for their sites and so on, the housing actually happened very very quickly.

"Out of that came the largest private sector building programme ever launched in the UK", another member of the team said, which "happened . . . almost over-night." "Docklands . . . instead of being a no go area [became] *the* new place to live, with all the strains, advantages and disadvantages that had." The pace was so rapid in fact that, as another member of the team said:

Having been set up in July 1981 Heseltine was able to come down in February/March 1982, some six/nine months afterwards to a total of 300 homes . . . where you could buy a house or a studio for £18,000 or a three-bedroomed house for less than £30,000. The land had been acquired. We'd got designs, the drains had been put in; if we'd gone into consulta-tion — shall we have housing, well what kind of housing — we'd probably be coming out with the first sketched drawings in that year.

A director of one of the leading house building firms involved in early hous-ing schemes confirmed the scepticism of the building industry. When I asked whether the area's development potential had been seen by the industry prior to the LDDC he said: "[I] don't think we did to be honest as an industry. I think we were just frightened to death by the scale of it to be frank with you. It was so vast and I don't think anybody thought there would be political will to make it happen. I think the LDDC under Reg Ward's leadership was a very, very clever vehicle to start things moving." I asked what had deterred private developers:

I think it was a question of was there a market there for housing for sale; I think the area was seen as a perfectly suitable place for social hous-ing, housing for rent . . . ninety five per cent [was] social housing . . . and I think that's what frightened away developers going into that market where quite clearly it was dominated by local authority housing. Also the infrastructure . . . was not there, there were no roads, there was nothing to get you in, there was no light railway at the time, so it was a major risk. . . . Certainly when we started in 'eighty-one . . . we paid something like twenty five thousand pounds per acre of land, and when you think land was going in Docklands in its heyday for well over a million pounds an acre, it gives you some indicator of the major concern that the housing industry had about being able to sell houses in that area. So what one had to do was to produce houses that were so affordable that people flock in and buy, and of course it was a major success story for a number of years as a result of starting with very cheap housing . . . The expensive housing didn't come until some time later. (Housing developer)

An independent builder who had also worked for some of the big national companies met Reg Ward in 1980 and was enthused:

I thought what a challenge . . . at last here's an opportunity to redevelop or to rebuild part of our wonderful city . . . I hadn't fully realised that there were 5,000 acres of dereliction. I went out there and just drove around and thought oh god what have we done as a society to just let all this lay dormant for so long . . . I thought what a marvellous opportunity for me as a builder to do a little bit of rebuilding, you know just a small bit, I'm not in the big league and so I felt . . . there was a challenge to see what we could do and so it was in early 'eighty-two we started out first (project).

Although this builder shared the reticence of his contemporaries in larger firms, he also felt that the location was advantageous in itself: "I thought constantly, Janet, just standing in that dereliction and looking towards St Paul's and Tower bridge, this is London, it's the Thames. There isn't a lot of land left around the Thames and whilst, from a purely commercial point of view I saw all the problems, how do you attract people . . . there was no DLR [Docklands Light Railway], there was no infrastructure . . . you attract people there initially by pricing your product accordingly. We were producing . . . good housing at very (competitive prices)." Persuading the members of his Board was another matter though: "When I showed them the site they said 'You'll never get people there'. I said 'Yes I will'. I just had a firm belief it was London. Okay there was dereliction, but [there was] water". (See Figure 2.3.)

Criticisms of the Corporation's emphasis on private housing in a climate of considerable social housing need were numerous and persisted throughout the 1980s and 1990s. But the LDDC did not see social housing as a priority and in fact regarded the density of existing social housing to be part of the "problem" that regeneration should address, as one of the LDDC Board members explained:

[When] we came in 1981, you were not coming to what I would have described as a balanced community such as you would meet in any other part of London. A feature of most parts of London is that you get all sections of the population living more or less integrally cheek by jowl . . . wherever you go in London you get a balanced population. Now I

Figure 2.3 "For Sale this View". Marketing the view of Greenwich from Island Gardens on the Isle of Dogs. © Mike Seaborne

don't think by any stretch of the imagination could you say that the Isle of Dogs, or Docklands, was a balanced population . . . there was only 4 per cent of private housing in the whole of Docklands, all the rest was local authority housing; that alone was a measure of the problem you have in having a balanced population, which is obviously a much healthier thing because it gives people a better chance, I think, to achieve their aspirations if they see success as well as failure around. (Lewis Moss)

Reg Ward reflected that "there was a fundamental underlying view, which could never be crystallized and could never be made in public debate, that what this area required was not more public housing, it was actually making far better use of the existing housing, refurbishing it, looking after it, getting a source of pride into public housing." Certainly local authorities had shown themselves to be poor managers of the properties they already possessed, as this private developer commented:

If the old way was so bloody successful, why have we got all these problems? You know the local authorities don't manage the units properly. The people that live in them respond to that mismanagement in a very natural way in my view and it's not until you get a proper rigid management constraint . . . that people respond to it and start liking where they live and start liking themselves . . . I think it's just stupidity to suggest that one marched on with the old way of doing it . . . I don't think local authorities know how to produce housing and why should they? I certainly think people like us know how to produce housing because we do it for a living and if we get it wrong we don't sell, so we go broke.

"I haven't really had a satisfactory dialogue with those who are sort of big protagonists on social housing", a later senior figure in the Corporation said,

> about how . . . [they] think cities work, I mean, we all know and you can see it from the experience in Paris who now have enormous problems with large out-of-town estates with a kind of mono-class culture, usually Moroccan or Algerian and . . . appalling problems of latent violence, racial feeling and all the rest of it. . . . That was, if you like, the product of building social housing on a mass scale and it does seem to cause, if it's not managed very carefully and allocated very carefully, some very difficult problems, and we have our own examples . . . It's a very difficult problem to manage and very few societies have managed that sort of mass subsidized rented housing successfully and so by all means make a plea for housing for those who cannot afford owner-occupation but there was a deliberate strategy by the government that just because there was so much subsidized rented housing from the local authority in the area in the first place when they arrived [not to build anymore].

Whatever the rationale, it did not change the fact that social housing was in desperate need of attention at a time when local authority budgets were under severe financial pressure, and in an area in which "all three boroughs . . . had more than enough trouble tackling the problems of some of the most deprived wards in the country without concentrating their efforts in the Urban Development Area specifically" (LDDC 1998:14). In fact neglecting those areas might serve to put pressure on, or embarrass, the LDDC and provide political capital.

Officially, even if they had wanted to invest in social housing, "The LDDC could only assume housing powers by a special Parliamentary order, which was never made", so the Corporation "was not a housing authority and could not build, improve, sell or manage new or existing housing" (LDDC 1998:6). However, the LDDC did get involved in shared ownership, affordable housing schemes and housing association initiatives. Reg Ward said: "Outside our strict operating regime rules, we found the means to 'use' the private sector to provide social housing units, through the back door as it were, of around 2,000 units over a period of five years" (personal communication, January 1998). "We were at great pains to try, by a series of quite ingenious schemes, to ensure that there was a reasonable percentage of affordable housing, either for the local population or for their children", a Board member explained, and one of the volume builders described the types of schemes in which his company were involved:

> We mixed housing for sale with rented and with shared ownership . . . which we introduced, they weren't placed upon us. . . . We did a very nice scheme in Wapping . . . I think there was about seventy, eighty units there of which the first thirty or so were shared ownership. Several of the Beckton schemes . . . had a social housing element either rent or shared ownership . . . Isle of Dogs we did them an affordable scheme there very early on

[19]84/85-ish one-bed houses at £35,000, two-bed houses at £39,000, . . .
[and] we sold I don't know seventy four of those to a housing association
for shared ownership. We tended to roll along sort of mixing and matching
to suit.

However, this "mix and match" also influenced the quality of housing, as a
Docklands estate agent explained:

Affordable housing schemes . . . were set prices . . . so obviously they (the
house builders) build to that price because the builder still wants to make
a profit out of it so . . . you're gonna find the whole thing's been primed
down and that is what happened with quite a lot of schemes on the Isle of
Dogs.

At the outset, land prices were low and the type of housing that developers
built allowed some local people to take advantage of the schemes. But many others
found it more difficult. As this Island couple explained:

We went for a house at . . . [on the Island]; £30,000 they said they were
and we went down there and they were £36,000 there were no £30,000
houses. It had an advert "Ask about our special scheme for low income
earners" . . . I phoned up . . . [name of company] and said "What about
this scheme for low income earners?" "Oh that doesn't apply to our
[development on the Island], only the one at Beckton", which was typical
you see cos they were building the cheaper houses at Beckton so people
were moving off there . . . That really made it so obvious that if you want
to buy an house as an Islander you had to move off.

A Docklands Forum report suggested that even the original affordable housing
schemes were beyond the reach of most local people, a situation acknowledged in
the mid-1970s which explained the emphasis in the London Docklands Strategic
Plan on social rather than owner-occupied housing (Docklands Forum 1987:22). A
Greater London Council survey conducted in 1985 revealed that in Tower Hamlets
three-quarters of households in the borough earned less than £8,500 per annum
(GLC 1986) and to purchase a £40,000 home required an income of £12,000 or
more (DCC 1988:13). "The incomes of too many of the original East Enders, even
those who had jobs, were simply too low" (LDDC 1998:4).

The LDDC's social housing schemes were also abused, as the same estate agent
working in the area at the time described:

We had a terrible racket because [local people] got preferential treatment and anybody who had a rent book was first in the queue . . . All these
people were getting rent books who weren't council tenants and they were
buying these properties at cheap prices and then flogging them on for
another twenty grand or something like that. There was a desperate abuse
of the situation, but that's why we've ended up with some of the housing
we have here. It was trying to do something for local people which went

drastically wrong. They just didn't have enough controls . . . it was an absolute free for all . . . a lot of people made a lot of money doing that.

A member of the LDDC confirmed that "some of the houses . . . reserved for council tenants" led to "an unhealthy sale of rent books". "These included people from outside the area giving local addresses in order to qualify, and tenants selling their rent books to speculators" (DCC 1990:50). "So our problems with the housing programme", an executive continued, "were all about sort of keeping some sense of order and realism in it cos it just exploded." A new resident on a development with an affordable housing element said:

I think the system went wrong . . . because on this development there's 150 properties and 60 of them (were) low cost housing that if you are a council [tenant] you could get it for £38,000. I know for a fact there are a number of people who have their own homes and weren't even council and still managed to get properties and have sold them on or moved on, and I think that caused a lot of ill feeling . . . You hear stories of little old-aged pensioners being given fifty pounds to have their rent book taken off for a couple of hours and then the person sort of gets a house that way . . . People that didn't need a low-cost house actually got them and I think it was a big fiddle . . . and people who needed them never got them and I think that caused a lot of ill feeling.

It was not only outsiders who sought to exploit the system. One group of residents on a neighbouring development took their revenge on a greedy tenant abusing the system.

In [name of street] . . . one day . . . the corner house had all this paint over it and things had been written . . . The people in [the road] . . . had done it because the person in the . . . house, owned their council flat, was doing a self-build on the Island and had [a] housing association . . . house. So the people in the close all got together one night . . . [and] sorted it out for themselves. But they should never have been allowed to have got one.

Reg Ward firmly believed that providing low-cost private housing would generate social mobility for local people, and in theory it could. But it was not long before even those who were able to purchase affordable housing at the outset were priced out of the market as land values rose. Christopher Benson, the second chairman of the LDDC, explained that this was an inevitable result of development that was geared to elevating land values:

The one thing that needed to be done in the residential sense was to keep land values low in order to keep affordability or rents that were affordable. So the two things were absolutely in conflict. Now people will tell me, and maybe I've got some sympathy with it, that had there been the grand plan, you know the great town planning exercise, that all that could have been accommodated; yes it could if we'd known today's lessons yesterday.

Nobody had the slightest idea there was going to be the level of interest in that area to build the office blocks and the places of work that have actually gone up there.

Attracting investment: business moves in

[Prior to the development] the perspective was an area of London which property people normally hardly think about at all. It was, if you like, in property terms a no-go area with minimal values on the land and an E postal address was property death in value terms. And it was an area which . . . Central London property people rarely went to, they wouldn't think about, despite the fact that there were these vast acreages. . . . My vision was that the objective of the exercise property-wise was to make Docklands a part of the normal London property scene. (Lewis Moss, LDDC Board member)

While progress with housing moved rapidly, the same enthusiasm was not initially reflected in the commercial sector, where development according to one LDDC executive "was much more difficult". The reluctance of established London firms to consider moving to the area initially led the LDDC to court companies in other parts of the UK. "We managed to find one or two developers from the north of England," one of the team recalled, and then canvassed heavily abroad, "because they didn't have the same prejudice about the East End of London. To them it was, what it was to us, . . . just two miles from Tower Bridge where for the City it was beyond the pale. . . . Tower Bridge being the pale."

"The major developers would never in a month of Sundays even come down to look", Reg Ward said. "So you actually had to try and persuade individuals, small companies and so on to share in the dream." His entreaties were certainly taken to heart by this businessman:

Reg Ward has a fantastic place in the history of Docklands . . . he did this fantastic PR effort . . . [He argued] you must forget the British idea that this is the East End of London. It's not. It's a city, it's an area which is a mile and a half from the very centre, Bank, and all you've got to do is to get some people to start investing money there, encourage the improvement of the infrastructure and suddenly you'll have an incredibly accessible [area] very close to the City with buildings that can meet the needs of the financial world. You could not get the institutions or the estate agents or any of those closely involved in property selling and property investment to come and look. . . . They said: "Look, it's no good saying to people come to the East End of London. They won't come." They're not interested and the people who came in were the Dutch, the Americans . . . , the Japanese . . . the Swedes . . . but in the early days persuading anybody to come down and commit themselves was incredibly difficult.

A former banker did not believe it was simply a rejection of the East End as a location that influenced the profile of companies who moved into Docklands:

> If you've worked here all your life and rents have grown from £10 a foot to £50 a foot ... that's just what it costs for an office. Whereas if you come from New York and offices are much nicer [and] cost half as much you're more aware of the fact that you're not getting the same thing when you come to the UK, whereas the UK residents says well there's nothing wrong with this it's what we've always had ... So ... the behaviour that we have here is first and foremost human behaviour, it's nothing to do with traditional UK class structure or anything else ... A friend of mine ... who runs a company ... I approached him and said would he like to come [to Canary Wharf] and he said "Well frankly no". He said "I'm running a successful company ... we're doing very well and I can afford to pay the rent and be wherever I want and I want to be in traditional areas of London" and so he's paying £55 a foot and doing that. He happens to be in a business where he can write off ... most of the cost of that office space so it really doesn't make a heck of a lot of difference to him where he is ... that decision was not being made because of any image problem, it was just that's what he wanted to do ... The East End of London ... it's been true it hasn't been the best place in town, it's not the place where your spirit soars when you drive down the street the way it does in ... other parts of town ... I think it's been easier for us because we didn't know enough to know what was good and what was bad.

The Corporation, according to one of the original team, sought "to persuade people by all sorts of methods" to encourage commercial development:

> I remember on one occasion Wimpey [one of the major house building companies] were very anxious for some more housing land and we [said] to them "We'll give you some more housing land if you do a little industrial estate as well," so it was much slower, much harder and you can see it in stages, the early stuff that was done, quite simple, quite a low cost (See Figure 2.4) and then on the base of that, there's a little bit more confidence and the next wave being slightly better development and so on. When we started building the Docklands Light Railway then that unleashed overnight much more interest in building commercial space on the Isle of Dogs. (LDDC Executive)

Reg Ward suggested that "the potential higher focus on the Isle of Dogs"

> was sponsored somewhat arbitrarily ... by two "personal" decisions (a) to position the office, and my residence, right in the middle of the problem area, as a statement of intent and identity; (b) to project an interim strategy that, by starting in the very middle of the area two miles in the wrong direction east of Tower Bridge, it opened up the ability to come back towards the higher profile area of St Katherine's Docks and to jump towards the Royal Docks. (personal communication January 1998)

Figure 2.4   Open land, new build. © Mike Seaborne/Island History Trust

The Docklands Light Railway (see Figure 2.5), which eventually spanned the entire Docklands area, was a very important lever for development that might never have materialized if the planned approach had been adopted. As Reg Ward explained: "The view expressed by the established authorities who looked at Docklands" was that "all it needs is a new express bus route. So you can well imagine what would have happened with a conventional planning scheme." Ward's dream was to use the railway to promote development:

> (In) 1980–81 there was no transport infrastructure on the Isle of Dogs. The bus services were totally inadequate. There were no rail systems. There was no way that local people really could get about with ease and that applied to the whole of Docklands. It was, as it seemed, totally detached from the rest of London. From late 1980 in the shadow days, I had been pursuing the concept of introducing a light railway, re-using the old railway tracks that actually existed and dramatically going down over the dock system. No way could it be justified as a piece of public transport. The only way you could hope to justify it was as a piece of development infra-structure. The existing land values were negative or at least nil. As you know, one of the perverse realities of development is that the higher the land value the easier it is to get development to take place! You can't

Figure 2.5 Light Railway under construction. © Mike Seaborne/Island History Trust

get development to happen and work when there isn't any land value. So I ... worked terribly hard ... on selling this "irrational" concept of a Docklands Light Railway in order to create residual value and to attract development. I was terribly lucky despite the lack of support from London Transport, the local councils, or the GLC to get outline approval in July 82 ... This was the next major factor ... which began to change the tone and the possibilities on the Isle of Dogs. (Reg Ward)

A former Dockland Light Railway employee described the scepticism about the need for a light rail system in the area:

Everybody said fancy building a railway in Docklands ... and then when the bridges went up over the docks ... each time ... the price of land went up. The developers got more and more convinced cos they could see something happening and that's why the Development Corporation wanted a railway that made an obvious visible link with the centre of London ... I can remember ... when the Government said you can spend £77 million on a railway and I can remember people in the LDDC and GLC and London Transport all saying to me: "We're worried whether we're gonna get enough people to go on the trains — that sticks in mind. There was this

Figure 2.6   The Docklands Light Railway, Janet Foster

fear that having spent £77 million it was gonna take years for the railway to take off. How wrong have they been.

The importance of the railway was summed up by this property developer and LDDC board member:

> By building that (the DLR) and building it fast we established a reality which couldn't be reversed and we established a position from which inevitably we could put up a good strong argument and a successful argument to government to go further, and the proof of that is we've been able to upgrade it; secondly to extend it . . . Now if we had said at the time: "Oh no, this isn't going to be good enough, we must wait until we get the right thing," we'd be waiting today. Because with all the claims on government money and the pressures and the negotiating between government departments, it would never have happened . . . Instead of which we've achieved all that with the DLR . . . because we'd seized an opportunity and made that a reality. (See Figure 2.6)

Another key factor in the commercial development on the Isle of the Dogs was the creation of the Enterprise Zone in July 1982, which offered attractive tax and other incentives for businesses to move into the area, minimal planning and other controls (see Figure 2.7). The brainchild of Geoffrey Howe, Conservative chancellor of the exchequer, the zones were perceived as "test market areas or

Figure 2.7 London's Enterprise Zone. © Mike Seaborne/Island History Trust

laboratories in which to enable fresh policies to prime the pump of prosperity" (from Open University 1982:42, quoted in Brownill 1990:32). This could not have been a more apposite description for what occurred on the Isle of Dogs. Originally conceived as an attempt to attract light industry, this kind of Enterprise Zone as, Reg Ward explained,

> had nothing to do with the dream that one had actually fashioned for Docklands . . . So philosophically one was actually against simply using the Enterprise Zone to re-establish the past with all the lack of horizon that that had meant . . . You needed to generate an entirely new economic base with a whole range of activities that would permit it to grow over the next twenty years, not ones which were possibly stagnant or declining. . . . So against the . . . intentions of government and civil servants I directed the Enterprise Zone really at [the] commercial sector and we were very fortunate in pulling that off.

Nigel Broackes was not in favour of Enterprise Zones theoretically: "I was quite critical of a lot of it", he said, but "we had to go for it . . . because if we didn't get one Wandsworth was going to get one. Wandsworth was the alternative sort of blue-eyed sparkling hope for the Conservative doctrine, so we had to play very

Figure 2.8   "The crinkly sheds", Janet Foster

hard for it and not quibble . . . Anyway we got it and that intensified all the pressures on the Isle of Dogs." Reg Ward viewed the Enterprise Zone as

> a very powerful element in encouraging quite a tentative activity to take place initially, quite small buildings, the crinkly sheds that people tend to criticize from time to time; Heron Quays, Millharbour schemes and a high point, Northern and Shell offices . . . But that was in reality all import-ant in projecting the new economy of the Isle of Dogs because it actu-ally stopped the Isle of Dogs from being pushed back into its past, which most people would have been comfortable with. It provided a heavy prag-matic base from which to build onwards and upwards. Despite the fact that the designs were all a bit twee . . . they stood out a mile in the market place at that time. They were very attractive to a whole range of new types of activity. They sold and leased extremely quickly. That in turn had the effect of actually increasing land values and made it progressively less difficult rather than very easy to attract other people to come in. (See Figures 2.8 and 2.9)

A businessman whose company moved to the Isle of Dogs explained the advantages:

> The tax breaks were fantastic . . . I mean in round figures if you spend £100,000 then the days when tax was 60 per cent you could immediately wipe out your entire tax bill in the first year if you happened to pay that

Figure 2.9 Northern and Shell, Janet Foster

much tax and your building had cost you £40,000. You could then let it and in three or four years you can cover the whole cost. Then you've got 850,000 square feet of property in Docklands which would actually cost you absolutely nothing by three or four years time. So it was well worth doing.

Small businesses thrived in the Enterprise Zone. One businessman attracted to renting space there had "to build a tailor-made operation" for a prestigious catering contract in the city. "I had eight weeks to find and build a factory ... Everybody was saying 'it's all gonna happen in Docklands' and I thought close proximity to the City ... so I grabbed the last available unit and started up my business here." A businesswoman inspired by Reg Ward also found premises in the Enterprise Zone. She said: "It was a very good jumping off point for small business and the ... tax advantages and development advantages ... were a tremendous help to get going."

"The overall development strategy was totally open ended", Ward said, "We would attempt to secure development *wherever* we could make it happen". This flexibility allowed the Corporation to exploit opportunities as they arose:

The next major breakthrough (was) ... at the end of 1982 [when I discovered that] Docklands was the last remaining sector in London which was actually totally free of electromagnetic interference ... It suddenly occurred to me, why couldn't we promote Docklands as the new telecommunications centre in the UK? ... This happened to be at the time when

British Telecom were actually looking to build satellite earth stations . . . I went to them and said "You know the normal planning problem?" I said "Come into Docklands and you can build as many antennas as you like and as big as you like." And so in a very short period of time we persuaded British Telecom to build the satellite earth station in North Woolwich. Mercury was being promoted as a competitor so it wasn't too difficult to play Mercury off against British Telecom. We got the Mercury satellite built at Wood Wharf in the middle of the West India Docks . . . By the end of [19]83 we had actually persuaded British Telecom to build a complete fibre optic ring main around Docklands when in a sense you can say there wasn't a single end user in sight [as the development had hardly begun]. (Reg Ward)

From relatively modest beginnings the LDDC continued to capitalize on opportunities as they emerged. Although the deal with Rupert Murdoch, the publishing magnate, to move his News International empire to Wapping had been struck with Tower Hamlets before the LDDC began, and the *Daily Telegraph* and Associated Newspapers "identified sites in Docklands in 1979" (see DCC 1985:20), they moved in during the LDDC's ownership of the area, when the introduction of new working practices and technology were rocking Fleet Street. "We were here at the right time, at the right place with the right acreages of land", one of the team explained, "and that really created the first impetus." "*The Telegraph, The Guardian* and all the other newspapers followed and we were able to build up this publishing, printing media village." "It was very difficult to rationalize and to explain in advance", Reg Ward said, but "probably eighty per cent of what we did in those first four or five years was the result of total accident, creativity and belief, accompanied by the level of the effort needed to actually make it come true."

The enthusiasm of the LDDC for their regeneration task was reflected in the experiences of the business community too. "There's a spirit about this place", a female executive told me, "It gets under your skin." This was something which an established business man who had been on the Island for forty years or more found a little difficult to comprehend: "The extraordinary thing was that the new people who came down here seemed to get enthused about the whole Docklands idea . . . I'm not sure why. I think the word Docklands has got something to do with it . . . Docklands or docks is a fairly emotive word isn't it. You think of ships arriving and there's a spirit of adventure . . . I think to stick with that description was an inspired choice really." "Docklands is addictive" was the way a former LDDC consultant described it.

A businessman who went to work on the Island very early in the development process recalled how he felt at the time:

When I went down very end of 1983 nothing had really started . . . It was very quiet . . . terribly desolate and deserted and depressing I suppose . . . It was an area of tremendous unemployment and need. Basically that remained the situation for quite some time . . . There was a lot of activity in the LDDC trying to get things started holding regular seminars . . . so at least I felt part of the development community. There was an incredible sense of isolation splitting off from the big firm. Personally I found it much tougher than I could possibly have imagined and then suddenly I discovered that by crossing the road . . . to the West India docks they had water skiing going on there . . . about 11 o'clock instead of having coffee I went for a ski and came back and nobody knew I'd gone and life suddenly looked up. Quite genuinely that made a tremendous difference. I suddenly realized there was a life in Docklands which didn't exist in [the West End] . . . From that initial feeling of dereliction . . . as it developed you had this tremendous feeling of pioneering and this wonderful large village atmosphere. That was the thing I liked most about it . . . everybody initially knew each other and you were all part of a community, and it was a developing community, and you were all doing something very exciting.

An LDDC executive who joined the team in the mid 1980s said: "Whereas I'd slightly dreaded the idea of working there, I actually liked it. I liked the water and . . . I've certainly found through experience that I prefer working there than I do in the West End." This trend was one that others commented upon. Another businessman, for example, said: "You talk to people who came in [to Docklands] even at a senior level from the LDDC; you see after about six months they've got that spirit they begin to live and die the place and to think about . . . promoting Docklands as a whole." "It was extraordinarily exciting for those five years I was down there", a businesswoman told me:

It really was . . . The whole atmosphere of the business getting going and the sort of enterprise thing . . . and you watched it being built . . . and it had this beauty with the water you see all the time and the historic feeling about it, and it was transforming something into something else which when you were there was extremely exciting, and the fact that everybody else was in a small business like you were made for a great community amongst the business people and the Docklands Business Club of course was the centre of that.

An executive in one company that moved to the area surveyed staff experiences and discovered that "staff liked it. They felt as though they were pioneering. They felt they had something to give to the place rather than being a very small wheel in an enormous cog that was already winding. They actually felt here they were getting on the bottom of the escalator and helping it generate." But their initial impressions when the move to Docklands was raised were quite different: "It was like going to the moon to some of them; they felt that if they got back alive they'd be doing very well, but when they got there they really enjoyed it."

Many of these feelings were generated by the dynamic style of the development, which, as one of the team recalled, required a complete rethink of what he learnt about the importance of planning years before:

When I went to planning college, the essence of a town planner was to . . . produce a plan, once you had your plan you waited for something to happen. That was fine when the economy was expanding. After the war things were growing and you had to control things. Development control was the essence of the planner, but actually initiating change, attracting investment isn't part of the curriculum, so when the crash came planners really had no expertise in how to bring about development . . . That's why we shocked a lot of people under Reg Ward when we said: "We're not going to have a plan" because a plan to us under Reg . . . was a constraint. Why constrain people, let's see what the market will bring back and try to control it, but don't try to set the parameters before you even know how the market would respond.

The speed of development for the business community was advantageous:

> If you wanted a railway, you built a railway . . . there was no question of saying can I, can't I or asking everybody. It would have been five years before anything got done. You see that again was a most exciting thing, to be part of that scene where things happened was lovely . . . There was a nice feeling of optimism [which] was extremely exhilarating at that time. . . . special to Docklands.

Those who believed in carefully developed plans were horrified. But Reg Ward was unrepentant: "All the best places have actually happened organically over time. We've got trapped into this notion that there are a group of people, whether they're planners or whoever, that have a crystal ball and can actually forecast the future and predict what will happen. I do not believe that that is reality" — open-ended, flexibility was the key. He continued:

> On a personal level, I always had a number of alternative routes that I could pursue to get to a particular end that I wanted . . . If suddenly my preferred route was blocked, then I could change tack immediately and find another way of doing it. Now you can do that very quickly on a personal basis. You can share it with two or three people, but getting an organization to understand . . . is actually very difficult . . . Suddenly you go off at ninety degrees . . . and the organization used to get quite exasperated with me . . . Progressively the organization enjoyed the fairly fundamental and subtle shifts.

This course was justified as it was "a way of getting where we wanted to go and rather more quickly". When I asked one of Reg Ward's team whether this approach made him a difficult person to work for in some respects, he replied:

> Oh the most difficult person . . . on the face of this earth, yeah, I mean Reg is a . . . really nice guy and we all liked him very much as a human being but he is . . . appallingly difficult to work for . . . Reg would have ten ideas an hour, some would be brilliant and some would be loony . . . they all had equal ranking to Reg. If you went back to him and said "This is a rotten idea", I mean he's very easy to talk to I'm not suggesting he was some sort of office dictator, it would be very easy to have that conversation, a very friendly conversation . . . In that sense he wasn't a difficult man to work with at all but very difficult to . . . persuade him to change his mind.

Another member of the original team said: "He was an absolute terror. He would drive people to distraction the things he used to get up to, but then you knew where his heart was. He was committed. It wasn't a job, it was a crusade and somehow or other everybody would walk on water for him . . . There was an energy about the place, a will."

Although Reg Ward was a central figure in the Docklands story and, as one of his team said, "Docklands is the way that it is . . . because Reg is the way he is. . . . it wouldn't have happened at all [otherwise]", the dedicated team of people working with him were important too, as it was they who strove to turn his dreams into reality:

> Reg is not an implementer, he's a dreamer . . . If you put all the papers and ideas together you could fill a whole room with ideas of Reg's which have never got off the drawing board because he couldn't persuade anybody else to pick them up and run with them. So I suppose in that sense we all had quite an impact . . . a whole host of people picked up Reg's idea and put flesh on the bones. And without those people, . . . it just wouldn't be there, and Reg would not interfere at all in the way you put the nuts and bolts and things together provided it . . . respected the original vision. If you tried to go back and change the vision, then he wouldn't have it . . . nothing significant happened that Reg didn't want. (LDDC Executive)

The Corporation's strategy was deeply controversial. There was no consultation and in some cases even the Board and the Department of the Environment were not fully informed. "In the early days we used to do things and not ask", one of the team said: "A lot of things we did were risky, and you know lots of things we did they thought we could get more money for this or that bit of land, or we shouldn't have done this and we shouldn't have done that. The fact is that we did it and then got rapped over the knuckles afterwards. But without that kind of risk-taking initiative nothing would have happened." "Quite often it was an issue of actually doing it, following our instinct, and justifying it in retrospect", Ward said, "and as a chief executive that's something which you've got to take the responsibility for doing, but you can never actually defend it if it doesn't come off."

A former consultant for the LDDC said:

> His [Reg Ward's] door was always open to anyone that worked for him or anyone from the community . . . Wherever he was you could always get to him, you could always put ideas to him and he would always give you a decision when a decision was needed and that was what enabled Docklands to get going I think. A lesser person perhaps wouldn't have succeeded. But of course he upset a few people. The government for a start, he upset local councillors [but] he was very well liked among the people who worked for him. He led from the front, he knew where he was going. His breadth of vision changed constantly . . . he was doing it by the seat of his pants . . . There's a lot of stuff here that should not have been put here and it shows doesn't it . . . [but] in those early days the LDDC was good. It was very progressive, you had a lot of talent around.

Reg Ward's personality, style and approach to developing the area did not make his relationships with the Board unproblematic, as he explained:

> An effective organization is always one in which there is a group of individuals who are given freedom to think, to do and to take risks within the parameters which are provided by the Board ... There was a reference point to the Board. The only issue was the timing of that reference. . . . The Department of the Environment used to get terribly cross with us, because their view of life is that they need to be in a position to direct and decide even though they say "It's not for us to do". It's the nature of government, it's the nature of the Civil Service. So, we had to be extremely diffident about surfacing something too early which, if they then applied the Civil Service [approach], meant that [it] could never survive, or it would take so long that the opportunity had gone.

"Reg was always wild", Nigel Broackes said, "and he needed somebody with judgement and experience and authority all the time ... He would have ten ideas of which ... some were excellent, some were just silly. . . . It was very important to sift those things because even a single mistake would please the adversaries." Reg Ward felt he was "very well managed by Nigel Broackes":

> I would come up with another of the mad Ward ideas and he'd say "Reg, Reg, Reg, Reg, Reg, Reg, no that's not what we really should be doing at all." But what effectively he was saying was I don't believe you can pull it off. I'm not going to say you're not to do it but if you proceed and it does fail it's egg on your face. If you happen to pull it off the Board will step in and take credit ... That also means too, and it's not an easy one to defend, that I would actually not ask a question of the Board to which I couldn't guarantee that I would get the answers I wanted to hear. Now you can only do that sparingly, but again, I think it's the role of the chief executive in a regeneration situation to be able to do so. I suppose from time to time this caused great irritation within the Board when they perceived what I was about and that strictly at that time it had not been debated and approved. We were always taking it back post-event with all the comfort of actually having succeeded. One was allowed to run in a fairly tense way on that basis but I think that's the whole issue of a creative organization. There has to be that tension if you like. There has to be . . . the ability ... [to] pursue ideas.

However, Ward also acknowledged that he had never found lengthy consultation easy:

> I'm conscious that throughout my career I have had difficulty in asking for consent to think and test out ideas. I do find it difficult. If I'm passionately committed to something I don't really want to be totally dependent on the extended consultation process in persuading people. I'd rather get on and do it and attempt to persuade and carry people with me. But it was actually inevitable, in the Docklands situation in which you needed to

pioneer new ways of doing things in which there would be perhaps very little support around even for trying, where the market place itself actually didn't believe it either. So you had to actually find people in that market place who were prepared to respond. And it was when you actually found somebody then you used it as an anchor. But often there was no way you could ever go to the Board or the DOE before that.

Conclusion

Without doubt the task of redeveloping Docklands required a directed approach with vision and imagination beyond that which the Docklands Joint Committee had demonstrated. The passion and enthusiasm for creating change in Docklands of those in the Corporation at the outset (no matter how differently that was perceived) was not mirrored in the bureaucratic and conflictual nature of the Docklands Joint Committee. The London Docklands Strategic Plan, for all the careful consultation, was not a visionary plan, and its policy of reclaiming land by dock filling deprived Dockland areas of the very foundation of their history, the docks themselves. Years of planning and relative inaction led to resentment and demoralization of the local population and the plan itself was not rooted in the economic realities of the day as it was based, as Fainstein (1994:232) argues, "on an obsolete vision of a manufacturing based economy". Change had to come. It was how change was managed that should have been the issue.

Reg Ward certainly had vision and a dream for Docklands. He and his original team put in motion a breathtaking pace of development. The difficulty, as Ted Johns a well known community campaigner pointed out, was that one man's dream is another's nightmare. The road on which the LDDC embarked was not one that many in the community, who had been promised more active involvement under the London Docklands Strategic Plan, could comprehend or sympathize with, as it did not seem to include them. Its approach was alien too because it appeared to deny everything that had shaped their attitudes and culture. Given that some community groups did not even accept that the docks needed to close or that manufacturing industry was in terminal decline, neither the LDDC with their emphasis on attracting private investment and service-sector employment, or the community who wanted more public housing and industry reflecting the skills of the local population, was in for an easy ride.

Nigel Broackes left the Corporation in 1984 so that his company could take advantage of the growing property boom in Docklands, from which they had previously been exempted because of his chairship. Knighted in the same year, he had fond memories:

I left it with a mixture of elation and wistfulness. It was one of the most important and exciting things that I've done in my life and I still regard it so ... I think of [the Island] as bathed in sunshine. It was always a nice day ... even when I was going there four times a week ... It was better than central London. When I left, the industrial era was drawing to a

Figure 2.10 "Cranes beside dock". © Mike Seaborne/Island History Trust

close . . . where plot ratios through my regime were 0.5:1 they were entering the 3:1 and even the 5:1 . . . and the whole mass of office development
apart from Canary Wharf was about to start. It was at its pregnant peak of
attractiveness in 'eighty-four [see Figure 2.10]. Since then some of the
development really has been too intensive.

Christopher Benson, who headed the Metropolitan Estate Property Corporation
(MEPC), a leading UK property company, succeeded Nigel Broakes as Chairman of
the LDDC, and after his appointment the development momentum that had been
difficult to establish in the initial years but had rapidly mushroomed eventually
reached fever pitch, with land values multiplying almost weekly and developers
flocking to take advantage.

Without the fetters of a conventional plan, lengthy consultation, and other
controls, the Corporation exploited market opportunities leaving many, including
its harshest critics, stunned by the breathtaking nature and speed of the development and the image of Docklands for a brief period became the very antithesis of
its former poverty-stricken past.

Chapter Three

"We didn't have time to be nice to people"

The LDDC when they were set up . . . the prevailing sort of argument was that Docklands was 5,000 derelict acres and people believed that it was just some sort of industrial waste land and nobody lived 'ere. . . . [But] there were communities in the Docklands not just on the Isle of Dogs, in Wapping, in Southwark . . . [The LDDC were] being directed by the Department of the Environment . . . to redevelop the area . . . and in that scheme of things the community didn't exist only in as much . . . that if the community could be bought round on their side then good, you know, it'd be good for them . . . I don't think the Corporation deliberately as a matter of policy said: "Well you know let's grind these bastards into the dust." I don't think they did . . . Reg Ward and the LDDC weren't necessarily anti-community they just wasn't for people. It wasn't about people it's about property. (Ted Johns, community activist, quoted in Foster, 1992:170, 172)

I think it was the cruellest thing that could have ever happened in a community that the work should just have been taken away from them. . . . Despite all the Dock Labour Acts and . . . and the history of strikes, there's no shortage of hard working people in Docklands but they'd been utterly demoralized and I think what we failed to discover was how deep was that desolation and I think we just regarded them — I say "we" I'm talking collectively as "we" my predecessors and myself — just regarded them as people who didn't understand what we were about. I think it was deeper than that and I think we should have taken a lot more time. It's easy to say all this in hindsight, we were not experts in sociology, we were chosen to do particular jobs, to go and build things and perhaps we should have started with the community feeling right on the Board. (Sir Christopher Benson, former Chairman of the London Docklands Development Corporation)

The competing visions of regeneration that dominated the political debate about the redevelopment of Docklands in the decade before the LDDC's arrival left their mark on local people's perceptions as the Corporation embarked upon its "vast task" (House of Lords 1981:13, para. 8.1). Labour's plans to retain the docks and attract manufacturing industry not only appealed to the dockland communities but also raised expectations, or reinforced existing beliefs, that this was possible, sentiments ably demonstrated in the comments of this former docker:

> I went in [to the docks] on . . . February 29th 1960 . . . and Friday 13th [June 1980] was the day I finished . . . and I only finished then because of a conspiracy of getting the docks here moved . . . You've probably been told a lot of lies about regeneration of a derelict area. It's total nonsense . . . They say it quite openly: "containerization, palletization, moderniza-tion", that is why they moved. Total nonsense. We were doing that since sort of middle sixties because the place got flattened, especially the docks, so everything was brand new and every modern equipment that was com-ing onto the market we was picking up . . . If it wasn't for [the LDDC] we'd still be working in there.

"I was asked what I felt" about the LDDC, he continued, and "my answer is what do you think of a murderer . . . they killed my livelihood, me family the whole bit . . . so I didn't view them in any other light other than anger at that time". "It was a totally horrendous time really", he recalled, and with every building that was put in the docks "it was like being raped all over again." At its simplest the LDDC arrived and ruined everything.

No matter how important the LDDC and its policies were as a catalyst for changing the face of the Isle of Dogs, they were not responsible for the closure of the docks. In fact, Sir Nigel Broackes said that, although they knew that the West India and Royal docks were going to close, the actual closure of both "came . . . as a shock". "We were completely unprepared for either of them", he said, "and should have been warned" by government and other public sector agencies "in the interests of co-ordinating the greater London effort." But the history of militancy, and a firm belief about the inevitability of dock closure generated little sympathy for those, who seemed in the face of all the evidence to want to recreate the past. As a businessman working in the area at the time said: "Part of the resentment was that the docker had been king down here — easier to get into Eton and the Guards than it was to get into the docks. I think there was resentment that they were no longer kings. If you know the Isle of Dogs you just accept the moaning . . . [but] a lot of it was grossly unfair because anybody would think that the LDDC had made the place derelict" the way the locals behaved.

"Local people were deeply hurt and deeply angry", a senior LDDC figure said:

> but it seemed to me that . . . the bitterness and the anger about the LDDC was . . . largely because of history. It was because the docks closed and East London lost its *raison d'être* and that anger was then directed at the LDDC. The LDDC didn't close the docks. The LDDC was actually trying to do something about it . . . The place became a complete no-man's land, enormous loss of jobs and no replacement. . . . The whole thing seemed to be handled very badly . . . there was no training, [and] . . . no real effort to bring new economic life in until the LDDC got into gear.

The LDDC's arrival, like the war and slum clearance before them, became another important symbolic marker that invited idealized contrasts between the area pre and post development and oversimplified very complex issues. The years of decline were conveniently forgotten or simply never perceived to be as serious as they were. "I don't think anybody even realized [it] was that bad to be quite frank", one local recalled. Another said: "When you have an LDDC tour they start off telling you it was derelict, but there is no way it was. It was very very busy. Everyone had jobs. You'd go out of the factory walk along and go in another one that same day. You were never out of work. It was very such a busy place." "Right up to the LDDC's arrival?" I asked. "Virtually", she replied. Yet, as I described in Chapters 1 and 2, there had been a rapid loss of jobs and, in 1981, at the outset of the development, Millwall ward on the Isle of Dogs had a 17 per cent unemployment rate, and the neighbouring Blackwall ward a rate of 21 per cent (Roger Tym and Partners 1987:7).

This chapter describes the confusion and contradictions embedded in local people's experiences of the early stages of the LDDC's development of the Isle of Dogs, the initial scepticism about the latest in a long line of "plans" for London's dock areas and whether this agency would be any different from its predecessors, the diverse opinions about to what extent the development was "good" or "bad", and whether a pragmatic or confrontational approach was the best means of fighting it. How much — if anything — could, or should, local people expect from it and did they have the skills to fight it once it became evident that the development was going to have an impact? Feeling angry and excluded, local people looked towards their elected representatives for support. But they had embarked on a road of non-co-operation refusing to be party to a process in which, as this Docklands Forum representative said, "their decisions . . . were marginalized" and "the electors were disenfranchised in terms of the amount of development that was going to occur, how it was going to occur" and "what kind of decisions were made."

The docks are *not* coming back

Given the uncertainty of the future and that the docks and its related industries were all that existing Island residents knew, it is not surprising that some fought for their retention even after the LDDC was established. Although community activists recognized that they were perceived "as a relic of the past" and that the Corporation's plans did not include a role for working docks, they still lobbied and "argued that they should be encouraged to stay." They were disappointed. The LDDC "wasn't interested in ships coming in and out of the Docks", a prominent campaigner said, because those at the helm did not have "any sense of place".

When the Enterprise Zone was announced, which offered companies tax breaks and other incentives to attract them into the area, some locals interpreted this as a prime opportunity, as an Islander explained, for "low level industrial units to cure the unemployment that was caused by the docks shutting down, to bring employment back" and utilize "the sort of labour that was available". But the end-stage of the development on the Isle of Dogs, and the LDDC's plans for the area, as I described in Chapter 2, could not have been farther away from this type of activity. However, as Reg Ward argued: "It would have been hopeless to assume that one could have persuaded local people that their future didn't lie in the past but lay in an entirely new range of activities which they saw themselves as not suited for and had nothing to do with their own aspirations."

"One realized that one had enormous problems with the community here" one of his colleagues said:

> This time it was for real and we were quite arrogant, we were saying quite openly I remember, saying: "We're not gonna kid you that we're going to bring the port back and those jobs back the . . . world is changing and we've got to try and bring the kind of jobs that they can provide here." But if you're thinking we're going to bring back the port somehow bring back shipping, . . . forget it. I remember feeling pretty bad saying that on the one hand and on the other saying "Well there's got to be some honesty for the first time" because people hadn't had much of that in the past because there were constant promises made. They were being used.

Even after the development was well under way, and it was evident that the area had changed dramatically, disbelief still permeated many people's experiences perhaps because, for some, change was simply too painful to contemplate. "People were very very cynical," the same LDDC executive said, "and I think people here were terribly depressed that there was going to be change and it was going to be change which they had very little conception or knowledge about." The past was something with which locals felt comfortable and in which they had a place. However as Reg Ward argued: "Whilst it was socio-economically attractive to espouse bringing back the docks [and] . . . manufacturing industry, the reality was that they had both gone and there was no reason why [they] should ever conceive of returning." Furthermore, "Reproducing the past really did nothing for local people." So the Corporation

> embarked on this very difficult, totally uncertain and, locally very contentious road in which the activity that one was trying to attract and generate had as little to do with the political aspirations of the local authorities and organizations that existed as it had to do with the previous experience [of] . . . the people. They were all quite reasonably apprehensive and not persuaded. (Reg Ward)

And the stage was set for the initial conflicts between the Corporation and its local population with their fundamentally different visions for Docklands. "It comes back to . . . a view of what the country's objectives for the Docklands" were an LDDC executive said:

> Certainly the Corporation's starting point . . . [was] that Docklands is a great at least regional and arguably a national resource and that what needs to be built . . . is a thriving part of the metropolis in both business and personal terms . . . and what you want there is an economy which is thriving and people of all sorts of . . . backgrounds and social classes and all the rest of it . . . and that within that the needs of local people are . . . an essential element but simply one element . . . If your starting point is that Docklands is primarily a place full of people that have lived there . . . for

decades shall we say and the first objective is to improve the quality of their life and . . . and then you could add to that anything for other people that you could do, but provided it didn't change the prime objective of improving the quality of life of existing residents, then I think you have a whole raft of valid criticisms to level against the way the programmes were mounted.

Just as the House of Lords Select Committee had characterized the Docklands communities as "parochial", the Corporation with its path set on a wider vision, in which locality was a physical asset to be exploited rather than a place populated by people with needs and attachments, had little time (and perhaps inclination) to address "local" preoccupations directly, even though by any stretch of imagination there were very pressing problems of unemployment and demoralization. In fact, a senior LDDC figure said that they acknowledged from the outset that "for part of one whole generation . . . there was nothing we could do to alter the future" and that "there would be a significant sector of people who would gain nothing" from the development. But this course was justified because the losers would

be greatly outnumbered by the people who would gain and one of (our) responsibilities was not only to the people that were there at the outset but to the people who would be there in the future. And to the people who were there at the outset it included their children and their grandchildren. For the people who weren't there and didn't at that time know that they were ever going to be there we were only the trustee.

"The only way to function in the modern market place where you've got a major problem area is actually to harness change", Reg Ward said. However, change can be a painful process, and as he noted, "To some extent it is easier for someone who's actually controlling it, generating change, rather than the imposing of it. It's a far different story if you're at the bottom end impacted on it." Indeed it is.

This chapter describes the experiences of those who were not controlling the development, who were "at the bottom end" whose experience was overwhelmingly one of powerlessness. As this veteran campaigner aptly articulated, contrasting his own experience with that of his father and grandfather before him: "The struggle for working class people . . . has never changed." "The struggle is exactly the same. They had no control over anything, no real control . . . no control over their immediate environment and neither do I." (in Foster 1992:171). A businessman who had worked in the area for some years sympathized with these feelings: "The people have this terrible sense of frustration", he explained, "of not being in charge of their own destiny, of not being able to influence . . . [the LDDC] in any way . . . They just want a genuine crack of the whip." But they were not going to get it.

Will anything change?

After the LDDC's creation, there was no shortage of groups, on a Docklandswide basis, for example the Docklands Forum, Joint Docklands Action Group, the

Docklands Community Poster Project, and the Docklands Consultative Committee (what Brownill (1990:111) referred to as the "DJC in exile"), protesting against the development and many of them had links with inner London Labour councils. There were also more local groups, from "those specific to one area or . . . tenants' association" to "umbrella organizations in the different localities such as the Association of Island Communities." This "plethora of types of groups at different levels . . . took up the issues raised by the LDDC in a variety of ways" (Brownill 1990:110) and some of them received generous financing in the early 1980s from Labour-controlled local authorities keen "to support them politically and financially" (Brownill 1990:110). "Quite a lot of them were salaried from the GLC", an LDDC figure said, "to infiltrate Docklands. They weren't Docklands people."

Despite the ongoing formalized opposition to the development from the local authorities and activists, the years of demoralization, indecision and debate during the 1970s left many outside this arena deeply sceptical about whether this new organization would have any more impact than its predecessors and, although the pace of change in the early stages of the development was rapid, it took time for these changes to manifest themselves on the ground. "For the first couple of years most people still hadn't even heard of the LDDC and what it was going to do", a resident explained, or were simply "not interested, let them get on with it . . . it's nothing to do with me. It's not going to have any impact on me." She continued: "The LDDC was hardly mentioned by anybody . . . we heard virtually nothing from them at all and during the second year there was nothing obvious that you could get a grip on and see what they were doing. To some extent there was this new organization set up by the government and . . . most people on the Isle of Dogs couldn't see the relationship between a government organisation and themselves." "It was only . . . when things started to happen that people realized that this agency that was there was going to do things that affected people's lives" and, as another local pointed out, by then "it was too late".

There was also "an awful lot of apathy" that activists felt the LDDC exploited: "Their strategy was," as one said, "to ignore the few people who're making all the noises because they're troublemakers and the others won't do anything because they're apathetic. That was their [the LDDC's] view." The apathy of course may itself have been a reflection of powerlessness and, certainly for the poorest and most marginalized people, issues relating to the development were peripheral alongside more immediate and tangible difficulties. So while Reg Ward marketed the Docklands dream, local people were often disinterested or incredulous (Foster 1992:172). "We were surrounded by scepticism", an LDDC executive said:

> I don't think people realize that . . . in those early days . . . the actual overriding attitude we . . . faced from local people [was not] . . . opposition . . . It was disinterest and scepticism. The overriding feeling of everyone you spoke to was . . . people have been talking about doing something for twenty years, we've seen six different plans, nothing's ever happened. This won't be any different, not interested in what you're talking about, nothing will happen just don't bother us. (Quoted in Foster 1992:172)

Gans' (1962:288–90) study of an area of Boston (Mass.) due for redevelopment suggested that the time lag between the publication of plans and their impact being felt influenced residents attitudes there too. They "did not feel themselves sufficiently threatened to be alarmed", he wrote, "nor could [they] really conceive of the possibility that the area would be torn down . . . the idea that the City would . . . turn the land over to a private builder for luxury apartments seemed unbelievable." Many Island residents thought this too, who after all had been told on too many occasions in the past that things would change, only to find that they stayed very much the same.

There was also "an enormous groundswell initially that we would go away", an LDDC executive explained, a feeling fuelled by the activities of local politicians who were "saying 'hang on, don't co-operate. . . . In [19]83 we've got in our local manifesto that we're gonna abolish this nasty corporation'". So "a lot of people waited for the [19]83 election". But the Conservatives won, the LDDC stayed firmly in charge, and Docklands, as the emerging "flagship" of private Tory enterprise, became a medium for further expressions of anti-government hostilities.

We don't want you here

Despite the powers vested in the Corporation to pursue development in any way they saw fit, without recourse to local residents, opposition was not deterred, not least because experience on the Isle of Dogs in the 1970s had shown that grassroots pressure could influence policy makers and encourage them to change their minds. Some of the key activists like Ted Johns (who spearheaded the declaration of independence in 1970, gave evidence at the House of Lords Select Committee, and had served both as a Labour councillor in the 1970s, and chair of the Association of Island Communities) had been involved in the battles for a very long time and were neither cowed by the might of a development corporation, nor ready to give up their fight for a better deal for local people. He and some of his fellow activists were accustomed to struggle. Island people have "always had to fight for their existence", I was told, "and they fight for what is theirs . . . It's their Island and they don't intend to let anybody take their Island away from them" or certainly not without a fight (Foster 1996:149). Passionate statements of this kind were, as I described in Chapter 1, fraught with contradictions and the legitimacy of such claims was openly challenged by the LDDC.

One of the activists described the gulf that existed between their views and those of the Corporation when he "made a statement about *our* land" at a public meeting. The LDDC representative replied:

> "That's not your land, you didn't pay for it." That's the major difference. They don't understand that *our land* is a gut feeling . . . On another occasion I made a comment [about] indigenous Islanders. This LDDC bloke got up and he said: "How long do you have to be here to be indigenous?" . . . I said "*Indigenous* means to belong or to come from . . . It's a feeling and that's what you're missing because I suspect that you didn't have strong roots, that perhaps your other family . . . are dispersed . . . and I can understand why you can't comprehend but please do understand what it

means [to us]. But he laughs and he says "Oh that". That is what you're up against. This is just old hat, that is dreaming, that is reminiscing, that is historical . . . you missed the boat, go away. (Peter Wade, part quoted in Foster 1992:174)

"All the Docklands villages are pretty insular", the LDDC executive involved in the exchange above explained,

but the Island's a special case . . . we had real battles over defining a local resident. To us a local resident was a resident of Docklands, but if the house was on the Island and the local resident was from Wapping they were an outsider. There's a certain charm about that I grant you, but it has some very much less pleasant undertones as well, particularly on the Island . . . where racism is very very strong, another element of this insularity and a far less charming one.

There were others who challenged the "rights" to which some established Island residents felt they were entitled including, as I describe later, their own elected representatives, who would have been in complete accord with the LDDC about the less positive sides of Island life. Although Ted Johns emphasized the importance of unified opposition at the time the LDDC was established, it was almost impossible to achieve because of fundamental differences between the Association of Island Communities (AIC), a key organization in the battle against the development on the Isle of Dogs, and those of more militant local Labour councillors and groups like the Joint Docklands Action Group (JDAG). This produced an uneasy, and sometimes openly hostile, relationship throughout the 1980s and 1990s so that, while others fighting the LDDC and the development were actively funded by the local authorities, the Association of Island Communities was excluded. "In this area, . . . sadly we've got the kinda head banging sort of Labour councillors", one activist said, who "are as careless as the LDDC . . . They're not about people they're about Party and they're about Politics — with a capital P" and stood accused of making political capital from people's fears and grief over the development. The high-profile differences between the AIC and local councillors (which at one point led two AIC members to stand as independents at the local elections, and the councillors to set up another umbrella organization in direct competition with them) led many activists to feel as neglected by their own elected representatives as they were by the Corporation itself. As one said, it felt as if they had "been neglected by everybody that could help in sort of like a statutory [way]" and this was a view shared more broadly in other parts of the Island community too.

The Association of Island Communities, among "the oldest, broadest-based and most effective community action groups in London" (Cole 1984:314), had in different forms been a vehicle for fighting issues that affected the Island for a very long time. Their independence from councillors in the 1980s also had historical parallels. Unlike other areas of the East End, the Isle of Dogs had never been particularly politically active despite the strength of trade union support. Cole (1984:194) suggests that geographical isolation and the "politically deadening" effect of having local councillors who served for very long periods (Alice Shepherd, for over thirty years (1928–62) and Nellie Cresnall for 46 years (1919–65)) "inhibited the introduction of new people and new ideas into local politics" and "helped to perpetuate apathy and dependence within the local electorate" as one or two councillors, in the words of one resident "sorted out problems" (Cole 1984:196). On the Island it seemed that "local action groups . . . replaced the Labour Party as the chief expression and instrument of class solidarity and collective action" (Cole 1984:340). The AIC was the most prominent of these groups at the time of the LDDC's creation.

Although generally viewed as the bastion of established Island power, the Association of Island Communities' membership also included some very committed middle-class members who settled on the Isle of Dogs in the 1970s, before it became a trendy location for the more affluent. Ted Johns, former chair of the AIC, openly acknowledged the "crucial and vital role" that these individuals played in their different ways, who, from the outset, favoured a pragmatic approach to Docklands redevelopment. "I took the view that the best thing to do with the LDDC was to get all we could out of any developers who came", one of the middle-class newcomers explained. "My own feeling was it [Docklands] was such a big thing, it required national decisions and there was no way the local community was going to prevent it. I mean the thing politically was far too big. My argument was let's see what these people can offer us, what can we as a local community get out of it, rather than be hostile to them and try to prevent them because it would never happen." "That wasn't the prevailing view", he continued, "The prevailing view . . . among our activists . . . was that we had to fight them."

This approach was well understood by another middle-class resident involved with the AIC at the time who said that "working class people in the East End" had been

> denied . . . any opportunity to use all their abundant skills, their initiative, talents . . . for so long . . . (that they were left) only to protest. . . . all they can do is say "no we don't want that". Whereas I suppose my reflex is how can we despite everything make it happen, find a way round it . . . now that reflex has died in working class communities because for so long it's been hopeless . . . they've actually lost hope.

Pursuing a more conciliatory path, one of these newcomers formed an action group and Nigel Broackes, Chair of the LDDC at the time, was invited to meet them:

> He sat on the flea-ridden sofa . . . this tycoon who was in charge of the LDDC, and I said to him: "Don't fill in the dock". He said "That's the last thing we're going to do" . . . He said "I agree entirely with that. We must keep the docks" So I thought well at least now we've got somebody whose got some idea, you know. I don't know whether it was imagination or whether it was a developer's natural gift for seeing what the opportunities were . . . In many respects he had the right idea.

This was not an opinion shared by key activists in the AIC who felt the LDDC lacked any "sense of regeneration" and were viewed as no different from their former dock masters. Even the LDDC's offices alongside Millwall dock had powerful and symbolic associations. "Most of the time when we met Reg Ward was . . . where we . . . used to get our pay", one of them said, so "as well as going to meetings in there and talking to 'em I had the added frustration of sitting in the place where I'd worked for 20 odd years . . . when there was no dock-related work going on." Ted Johns, in an account of a meeting with Reg Ward, demonstrated the extent to which past experience influenced their interactions with the LDDC:

> Reg Ward had a little office . . . he saw us, he sort of waved and he said "Let's have a cuppa", picks up the phone and then started reading through some papers . . . and then put those down [and] spoke to his secretary . . . I said to [a middle-class newcomer from the same organization] "I'm not hanging about . . . he's treating us . . . like nothing, just kept waiting." So after about ten minutes I just got up and walked out. [She] came out with me . . . and she said "Ted, very often you're late at meetings". I said "Yes but I don't do it to insult the people." I said "It was the sort of thing I learnt when I was 14 you know in the Unions. The Guv'nor would sort of say . . . "I want to see so and so" . . . you'd go in and you'd wait, you'd be standing there waitin' for 'im "Oh yeah just a minute", pick up the phone "oh so and so yes alright yes", put down the phone "Be with you in minute". He'd pick some papers up "Right lads sit down now I won't be long." Now he's actually got you on his ground cos he's got you waiting and I said . . . "That's the oldest trick in the book" and I said "I'm not falling for that one" [so they left].

Those within the LDDC who worked closely with key members of the AIC recognized their history and their tactics. They "have been fighting bureaucracies for a quarter of a century or more", one of the LDDC team said,

> and they are very very good at it and they have a particular style which is in public terms very confrontational. Ted will never say a good word about a local authority, or government or the development corporation in public. His strategy is to hit them hard, make them hurt and get them to pay you off, not pay Ted Johns off but pay the community off and I understood that very well.

"It was actually very difficult for them not to take a combative style", said another, "because that's the only way anything ever happened." Reg Ward said that, like him, Ted Johns was "a dreamer" but "could only see one way of arriving at it and it wasn't the way of an imposed non-democratically-elected body . . . espousing . . . a set of objectives . . . which did not grow out of what was seen as the needs of local people. It was bound to be incredibly difficult, bound to be frustrating."

Public and private differences

The public opposition to the LDDC was characterized as a battle, but behind the public face some relationships in the early phases were not as acrimonious or difficult as they sometimes appeared. "It was a battle but it was . . . in a sense a formalized battle", one of the LDDC's executive's said, "To say it was a game would be in a sense to cheapen it because they weren't playing games." Their feelings and emotions "were very real". "But" he continued,

> in one sense it was a game, not in what they were fighting for but the way they were fighting for it was quite a ritualized game . . . a lot of politicians have chosen to represent it . . . as being a real war between the corporation and local people because that helps them to make a political point. I don't blame them for doing that but they're wrong. It wasn't like that at all.

"There's tremendous toughness in the East End", Reg Ward said, "but underneath there's also a very helpful tolerance, a warmth, a kindness." "I could be hammered all evening at a public meeting but it wouldn't stop them taking me back into their homes for a cup of coffee or stop for a drink and so on . . . People would say 'well we don't agree with you Reg but thank God somebody is at last trying to do something'." Ted Johns admitted that he "quite liked" Reg Ward and both agreed they were "romantics", but Ted Johns characterized Reg Ward as a "dreamer with both feet planted in the air!".

These more affable emotions never entered the public arena, as one of the LDDC team described:

> We'd be in a pub one evening sorting out the details of a new project and that would be fine and we'd shake hands and so on and the next day you know Ted would be holding court with the reporters telling them . . . what an invidious organization the corporation was. . . . That didn't worry me, I mean that was Ted . . . He never made any secret about it, he would say "I'm going to see the press tomorrow and tell them what rats you are." . . . When I left Docklands . . . the local community groups came along to my sort of farewell beano and they bought me some nice presents and some nice cards and said some nice things . . . It amuses me that people who should know better don't understand that. . . . That's the way they work and that would have been the way they worked . . . with local authorities before we came along when the corporation closes down and whatever takes over . . . Ted and Peter will still work that way and they'll be very good at it and they're doing it cos they have a genuine desire to improve the lot of their friends and neighbours.

A consultant who also worked for the Corporation recalled his frustration when he first went to the LDDC: "Reg . . . was always wrapped up in negotiation with Ted Johns and Peter Wade [but] . . . I'd watch him [Ted] on television . . . saying 'Trouble with the Corporation is they don't talk to you. You never see them. They're like ships in the night.' I thought, you bloody bastard!" Their style, he said, "was constant, direct, confrontational".

As in all of the dock communities, it was a handful of key activists on the Isle of Dogs who "provided inspiration and kept organisations together" (Brownill 1990:113). "I used to sit at the AIC", one activist said, "and say 'Oh, let's have a demonstration' and everyone'd go 'Yeah' . . . They used to say 'let's go out there and have a go at 'em'." Demonstrations not only visibly displayed opposition to the development but made people feel that they were *doing something*. "They were excited about demonstrations." Yet as Brownill (1990:113) noted, it was difficult to achieve "sustained involvement" over long periods and while "large numbers of people" might willingly form for a one-off protest march or meeting, keeping the interest was "a different issue".

Furthermore, while opposition was presented as though it was universal, many divisions existed between, and within, different groups and individuals on the Island that were never very far from the surface although they seemed less apparent when criticism was focused on the development and the LDDC. The complexities were summed up by the comments of one newcomer who said: "Nothing adds up" on the Island "and that's because the subtext is a totally different agenda." Consequently, despite the vocal and at times visible public opposition to the development, it is questionable to what extent the small group of activists who led it were actually representative of the views of Island residents as a whole. In fact, a survey conducted in 1986 (see Wallman et al. 1987:31) highlighted a far more ambivalent attitude to the development. In a question about the effect of the "developments" over 40 per cent of respondents said there had been no effect, almost equal numbers said that there had been a good effect (16 per cent), a bad effect (16 per cent), and a good and bad effect (15 per cent), with 10 per cent saying they were unsure (see Urban Change Group, undated: 4; Wallman et al. 1987:31). Some indication of the minority involvement in any group, let alone an action group, was provided by the same survey in which 73 per cent of respondents *did not* "belong to any clubs or association".

The few committed activists, however, felt that they were drawing on a great deal of silent support. So although they were fully aware that some locals were not opposed to the development, "the community" was perceived to be unified in its opposition. "They [the community] did feel very strongly about things", one activist said, "they were unified about the LDDC. There were some people who thought the LDDC were the best things since sliced bread but that was only because the LDDC was manipulating these groups by givin' 'em what they wanted . . . But the community was united." Yet, despite an AIC voice that was, as a community development worker pointed out, "powerful . . . to the section they serve", it was also partial.

The comments of the established Island residents below demonstrate the more positive perspectives about the development and their disapproval for the activists, whose actions they believed would gain nothing. "I would say 70 per cent were against" the development, one local woman said, "There was an awful lot of antipathy against it . . . they didn't really want it. . . . There was lots of anti-groups . . . the Association of Island Communities were against it, the Church was against it . . . at that time . . . very much against it." "My view was that if it didn't happen what was going to happen to the place?" Another woman drew parallels between the Island's economic importance in the past and this latest wave of development:

> Back in history the Island has always played quite a large part in world wide things . . . the ships used to be built here . . . after that we had the docks and of course the London Docks were some of the most important

in the world. . . . So it has always played a large part . . . worldwide and I feel . . . if it comes up to be a financial city we're still going to play a part in world affairs, so I like it from that point of view . . . I know there was a lot of resentment . . . a lot of people don't agree with it but I do.

"I think in many ways it's an improvement on derelict docks" a third agreed:

> I mean, when the docks were there, there used to be these high walls all around and only the people that worked in the docks went into the docks, but now you see anybody can go through can't they. There's access all the way through. We often walk through and alongside the water Sunday morning if it's nice and watch the buildings going up. A lot of the buildings are atrocious but some of them are nice. It's caused a lot of inconvenience but you can't have any sort of change without some inconvenience can you? I don't know, I'm rather glad I'm here and [can] watch it all happening.

When I asked one Islander to describe how she felt when the LDDC was established, she said:

> Well it's progress innit really. It has made a lot of difference I mean I've worked in the City and it was a hell of a job to get to the City. It was a hell of a job to get off the Island because [of the] bridges . . . and there was always boats and that going through and barges going through . . . you never knew when you were gonna get to work. But now you see they've got the Dockland Light Railway and . . . the bridges are open and they've re-routed the buses so you can go straight to town . . . from the light railway . . . it's a lot better now. Yeah I'm all for it really. It's a pity the jobs had to go in the docks but then that was containers that did most of that because . . . well once again it's progress and there's advantages and disadvantages in all of it isn't there.

As I describe in Chapters 4 and 8, the uncertain, ambivalent and positive feelings about the development rested alongside more negative and critical ones throughout the Docklands regeneration. The Isle of Dogs, however, continued to register the highest levels of dissatisfaction of all the dock areas throughout the LDDC's reign (Wallman et al. 1987, LDDC 1991 and 1996).

Property versus people

Despite their insensitive and uncaring public image, described by one Islander as "running rough shod" over local people, the LDDC were conscious that their approach was problematic and painful for the community and some were initially inclined towards consultation as this LDDC executive explained:

Some of us felt that perhaps we ought to open up a dialogue with the communities here but in the end we took the view that once we got involved . . . that pretty well every proposal, policy formulation had to be debated, countless meetings, we would never get anywhere. We would fall into exactly the same traps our predecessors [had] that there would be a lot of debate, a lot of talking, a lot of plans and very little action and government was determined that things should happen here . . . [So] we kept our heads down. We didn't have time to be nice to people, to consult with people because we actually had to demonstrate that this experiment set up by Michael Heseltine was the way forward.

"I accepted that" he continued,

because having been involved in two years of LDSP (London Docklands Strategic Plan) . . . each chapter we spent three months on consultation before we could go on to the next chapter and the world was changing but there we were, we had this fixation in our little world that we were doing the right thing, we were consulting . . . so when I joined the Corporation . . . I recognized that you had to sort of keep your eye on the wicket . . . otherwise you'd go all over the place and you just wouldn't achieve anything. So we took the view let's keep our eye on the ball in terms of physical regeneration and once we've actually got the momentum on the way . . . then we can afford ourselves the luxury of time to bring some more people with different kinds of skills, open up a dialogue.

Consultation also, as Reg Ward admitted, conflicted with "dynamic" development and he "didn't really want to be slowed down or to go through this painful exercise of trying to get agreement in areas where quite clearly agreement couldn't have been reached:

If things hadn't happened as quickly perhaps that would have been easier to do. But once you've got a development momentum running the last thing you do is stop it. You have to keep fuelling it. And therefore . . . even if on some things we were doing, you might have won the debate you could not risk doing so at the pace at which we were actually going. (Reg Ward, part quoted in Foster, 1992:173, quote amended by Ward 1998)

This further disempowered the Docklands Forum, which the House of Lords had said must "know what it [the Corporation] is thinking of doing, so that they can express their views on any proposal" and whose "views, even if not accepted, are always seriously considered" (House of Lords 1981:para 8.7). A year after the LDDC was established they produced a report *Consultation What Consultation?* highlighting the problems of the LDDC's approach, which was likened to "a public relations exercise" rather than "a genuine canvassing of views" (Brownill 1990:51).

The composition of the LDDC itself exemplified its priorities, as this employee said, and they were not focused on the community:

The whole climate and the whole emphasis given by Heseltine . . . was concentrating on attracting investment . . . so it was very much what I would

describe as a welly boot brigade type, physical regeneration, and I think
there was a perception too that we had to get the first developments
underway before we actually tackled what I would call software type activities, so . . . the recruitment concentrated very much on planners, architects, civil engineers, surveyors those kind of people who could actually
take stock of what needed doing.

Local people, then, were an obstacle best avoided and little was done to assist
or include them in the regeneration "vision". One senior figure characterized the
Corporation's adversaries across all the Dockland communities after the creation
of the Corporation in 1981 as "very disorganized, pretty woolly-headed and rather
demoralized because we came in with firm authority and ideas and no obligation
to take them terribly seriously". The fact that some were "paid activists" by the GLC
or local authorities did nothing to enhance their credibility; they were seen as
ideologically opposed rather than as having legitimate complaints. Nigel Broackes,
Chairman of the Corporation, characterized the LDDC's relationships with the community during his era as "harmonious coexistence and fairly courteous exchange
of information and of views".
   "The Corporation . . . I think in those days, this is a personal view", one of the
team said, "used to ignore its opposition and it worked. It didn't actually join the
battle too much and I think it must have been very frustrating for those seeking to
oppose but you only really sort of got into those debates when it was necessary
because it stood in the way of something you wanted to do." This approach left
many locals feeling excluded. "The argument always was", one woman explained,
"if you're talking about regeneration we want to be regenerated too." "Absolutely
nothing" was done for the indigenous community who had "vital needs" a prominent middle-class figure lamented: "The worst housing statistics in the country,
inadequate leisure facilities and play space." There were "huge" but wasted "opportunities". But these potential immediate and tangible benefits for local people fell
outside the Conservative regeneration ideology. The Corporation "were not to be
a social organization" an executive explained; "We were not to have responsibility
for housing, . . . it wasn't our task to support community groups". Their task was
physical and as Island residents were to find out this meant maximum inconvenience with very little in return.

It is going to affect us

Something had to be done obviously, but not on the vast scale that it has.
When they first come they were full of promises. It was gonna be just light
industry and small buildings and things. Then it . . . got bigger and bigger
. . . as it went along. (Island resident)

Although from the outset the Corporation's strategy relied on generating positive
land values and attracting private sector investment (though neither they nor
anyone else anticipated quite what impact this would eventually have), the visible

representation of their approach did not become apparent for some time. "In 1984", a local clergyman said, "the LDDC were still very much getting into their stride. There was very little up . . . and it was lovely walking in the Enterprise Zone because you could see vast open areas of dock and water."

As the commercial development began in earnest, therefore, it was "a big shock". "I don't think it hit yer at first", a couple whose house was adjacent to the Enterprise Zone (where the most intense development occurred) said, "You didn't realise this would happen. I don't think any of us did".

> You'd have a letter to say that somethink was gonna be built and do yer have any objections and I used to think what do they mean? . . . It seemed to creep up on yer . . . Sometimes they'd started (work) before we got the letter . . . we wrote several letters . . . and very often we didn't hear another word . . . and I think you carry on with your daily life and before you know where you are it's gone up and it's too late. (Part quoted in Foster 1992:173)

"When the LDDC started to do things that really did have an impact on people's lives", a local councillor recalled, when "they were interfering with people getting on and doing the ordinary everyday things like sitting in their homes watching television and suddenly there would be a pile driver shaking your house; that's when it started to sink in," "People don't seem interested", an exasperated resident involved in opposing the Corporation from the outset said, "until it really hits them and [then] they start shouting when it's too late." This was exemplified by the comments of an elderly Island man who said:

> All of a sudden plans was given permission for unlimited office developments and that's when the snowball started . . . Then you start saying "Well we don't like that" and then you realize that you've got no say in it. That's the first time . . . it comes home to you that this is no longer a democratic idea you've got no democratic rights at all. . . . I can't vote them out if I don't like it. . . . They're an autonomous authority . . . you can tell 'em until you're blue in the face what you want but if it doesn't happen to coincide with what they want then there's no chance.

"Dynamic" development, which for the Corporation and business community was the source of excitement, led to all kinds of problems for the residential population. The LDDC "really deserve their bad image . . . for what they did a few years ago" a man involved with the light railway explained:

> They just used to have road works in the middle of the night and laying pipelines at 2 am. There was no consultation and no liaison, nothing. They just used to carry on . . . they put everybody's backs up from day one and that was a problem for the railway because in trying to build up the railway's identity and acceptance with the locals, who are a pretty jaundiced bunch, we were always tagged with being linked with the LDDC and it got so bad that we would never go to a public meeting if the LDDC were present . . . We kept well away from them cos they were awful . . . They were the lords and masters you see especially on the Island. It doesn't take

Figure 3.1   "It's right on top of us", Janet Foster

much to upset people there, and they got very stroppy, and things got from bad to worse.

Is it surprising some residents felt the LDDC was uncaring and ran rough shod over them? "To be honest with yer the LDDC couldn't care a damn", one local said. "We've got that big development going on just over there — that's 120 foot high and, I mean, it's right on top of us. We objected, we opposed it, we didn't get anywhere . . . so what can you do? They're their own planners — you can't win . . . you just can't win. All the objections you put up they just veil over." "They are bandits all of them" another said "their whole aura is one of mistrust" (see Figure 3.1).

The bitterness and antipathy felt towards the LDDC were, as this local estate agent said, entirely understandable:

> It's their home and it was a neglected area and nobody gave a monkey's about how those people got around and how they lived their lives. And then as soon as the money comes, you know, "OK we'll provide the buses and we'll provide this . . . we'll provide all the facilities for the people with money at prices that people with money can afford" . . . the local community aren't considered at all.

As Ted Johns wryly commented: "A working-class community can be a working-class community for 100 years, 200 or 300 even; suddenly it's socially unbalanced

...when the communities happen to be living on prime land...not because it's...fallen to pieces but [because] the land has become extremely desirable."

A more benign view, however, was offered by one of the original LDDC team:

> At the time I joined (late eighty-one) there must have been about 35 people. There'd been a great deal of public debate about this quango. Lots of opposition from the boroughs who saw all their authority being taken away from them, opposition from local people and my own view, and this is a purely personal view, is that we were actually as scared of them as they were of us, which I think is one of the reasons why in those early days we were accused of secrecy. I don't think it was so much that we were seeking to be secret as a kind of tactic; it was more sort of emotional...[but] the perception was that we were secret.

In spite of their treatment and understandable anger, only a very small minority protested, and even among these groups and individuals confusion existed about what exactly local people's rights were, as a woman involved in one of the Island action groups explained: "The conflict seemed to be about whether we had any right to make a demand and should we be asking nicely, or should we be demanding...were we asking for our rights or were we looking for favours?" Those without experience in union or community politics often felt they could change nothing, what one affluent resident referred to as a "curiously cowed mentality". "Tenants would take up arms against the council over the repairs...and rents", he said, "but anything outside that it seemed...very difficult to get tenants to take a wider interest."

This was partly because they felt they lacked the tools with which to fight. "I do think that once people like that make their mind up it doesn't matter how you object or what you do; they'll still go ahead and do it", an established Island woman said. "I don't think ordinary people matter a great deal to people in power. I really don't." She continued:

> I just think ordinary people perhaps don't know enough about what you can do. You know you get people like with this link from the Channel Tunnel to London where it's going through villages, where houses are privately owned and people have got plenty of money, those sort of people seem to, or probably do, know more about what they can do, as a community...we're just sort of ordinary people here, I don't know...it don't seem to sink in until it's [too late]...well even then I wouldn't really know what to do. We've never had to deal with anything like that.

The comparison with the Channel Tunnel link was mentioned by others too. "Most of these villages [involved in the Channel Tunnel dispute] consist of very important people who commute up to London and that and they know all the tricks of the trade and...they can afford to fight it", said another.

When I asked a woman who had lived on the Island for over twenty years what she would say about the development to someone who knew nothing about it, her words spoke volumes about the neglect of local people: "I think I'd like to say there's ordinary people. I think if you're wealthy you seem to have more say...you could move if you didn't like it, you could go and live somewhere else. But that just to remember that there are poor . . . people" who have no choice.

By the mid-1980s, as the impact of the LDDC's policies were beginning to be felt, and it seemed local people were not going to receive a slice of the regeneration cake, opposition to the development began in earnest. The "immediate period of anger in the early eighties" was rather disparate, a community activist explained, with "a lot of people running off in different tangents". Later opposition became more focused and drew on "a much stronger base" with a determination that "*we will* and *can* influence the situation . . . And it was very, very strong, very strong."

There was an obvious "enemy" around which different sections of the community could organise, and the more threatened some felt, the more convinced they became that the LDDC were trying to drive them out of the area, the more determined they were to oppose the development despite the odds being severely weighted against them. "I've had it said to me when I've moaned [to the LDDC]", an Islander explained ". . . 'Well would yer like to move?' I said 'I wouldn't like to move thank you. I'm sticking right where I am!' . . . The fact that they want to get rid of us would make me all the more determined to stay . . . You think to yourself you're putting up with all this just to spite them but you do!" There was "a bit of siege mentality in the people who live here", another Islander explained: "over my dead body am I gonna move sort of business." People were angered and exasperated by the LDDC's policies but determined to stand their ground even though the might of big business and government made such opposition seem futile. It was as if the LDDC condensed all the community's discontent in one image. Yet as Suttles (1972:62–3) argues, small community organizations are greatly disadvantaged as they do not usually command broad-based support, cannot be regarded as representative and lack the status of big business and government. The result is that "with a narrow support base and limited access to higher administrators, many traditional local groups seem to be frozen out of regularized and orderly negotiations with big government and big business" and are "ignored".

Maintaining the moral high ground

While the battle against the corporation was being fought by a diverse range of local action groups of different kinds and with different agendas across Docklands, local politicians maintained their isolation refusing to co-operate with the LDDC, or to provide local authority representation on the Board. As I described in Chapter 2, their opposition was integrally linked, not only with events in Docklands itself but also to the broader political agenda being pursued by the Conservatives, which in many ways the LDDC and its approach personified. The stance of Labour politicians in the area was "very much a matter of principle" a representative from the Docklands Forum explained:

> In retrospect it's easier to say, oh perhaps they should have got involved and try and change it from the inside. But at that time . . . we could understand that they thought "Right, we're going to make a stand here. It's

utterly wrong what's happening here. They [the LDDC] snatched control of the process from us and they've given our local electorate absolutely no involvement at all, no input." And, you know, faced with that I think it's a kind of a logical course of action that you say we're going to have absolutely nothing to do with you . . . We're very disappointed and we're ashamed of . . . the setting up of this and hoping that at the end of the day the elections will bring home an alternative government.

Councillors were "caught between the devil and the deep blue sea", a local authority officer said,

because if they had gone on to the LDDC Board . . . would local people have perceived them to be selling out? One of the reasons that local people have elected them is because of their standing up to the LDDC . . . so for them to go hand in glove to be seen working alongside them might not go down too well with the electorate . . . In some ways what they're doing is maintaining the moral high ground but not getting anything from doing that apart from being on the moral high ground. Can you see any further and clearer by being up there? I don't know.

Notwithstanding the principles underlying their approach, it is arguable that the LDDC's lack of accountability made it even more important to capitalize on the few mechanisms that existed to express a different view. As a senior member of the LDDC Board said, the local authorities "probably knew they had achieved very little and I suspect that that increased their dislike of this new body coming in". But he also said:

I was disappointed it lasted for so long . . . and I think the councils would have been well advised to have said "We don't like your existence here at all, we'd rather you went away but on the other hand [now you're here] let's be pragmatic" . . . I think we would have been more successful in meeting the needs of local people had we had more effective links with the local authorities, and I think at the end of the day the people that have lost have been the local people . . . I don't think with hindsight that councils did their constituents any good . . . the situation is much better [now] but . . . we did go through some bad patches when relations were difficult.

"The local politicians have a lot to answer for", a senior LDDC executive said.

It does take leadership . . . it does take vision on the part of local politicians and I know from their point of view the LDDC was a very, you know difficult, untrustworthy and damaging partner to work with but they should have done better . . . they should have been on there [on the Board] a long time ago. Things wouldn't have happened as they had if those politicians had had the courage to get their people on the Board and not take this holier than thou attitude, they're gonna have nothing to do with it . . . They could have done the right thing by their people and . . . it wouldn't have been run so much as a secret society.

Many Isle of Dogs residents, including committed Labour Party members and some activists, agreed with this analysis. "The Labour regime [local councillors] ... are a *militant*, militant", an activist said. "I mean I've been called a left-winger in me time but they are out of sight." "They took a vote of non-co-operation with the LDDC because it was undemocratically sound", said another Labour Party member. "Well yes we know that. We recognize that but they're here ... you aint got much option ... you've gotta try and get what you can out of it and you don't get that by not talking." "As far as they [the councillors] were concerned the LDDC didn't exist" a third party member said "They wasn't gonna talk to 'em, undemocratic body ... so we said 'Well ... we know what you're saying but they're there and they're doing the damage' but we couldn't get through to 'em." "Their policy was don't talk to the Liberal administration, don't talk to the LDDC. So what are yer gonna talk to, the brick wall? How are yer gonna get anything done if you don't argue yer point?"

There were some very practical consequences that resulted from Labour's chosen path. A member of the Association of Island Communities explained that they couldn't even take planning applications to the Borough, "to their professional departments for advice, because they wouldn't even talk to us about the LDDC". Others, forced to liaise with the local authorities during these early years, also found them obstructive. One of the Docklands Light Railway team said: "They were very hostile as individuals and councillors, some of them still are, but to the very last man they were hostile then. The railway was completely wrong because it was something they could latch on to, it was a very tangible thing they could have a go about and Newham and Tower Hamlets were awful to us." A house builder described how one local authority held up work on a scheme in which he was involved:

It was terrible ... I got planning permission for a ... site ... and it took me over a year to get building regulations [because those] are still vested with the local authorities. [They] said we have no moan with you ... but we will not give you building regulation because we do not agree with the Conservative government's policy to release land and we took over a year [to get it] whereas normally building regulation is a technical process, you get an approval in two or three weeks. It's not a political issue of whether you can build or not, it's to make sure that the design is structurally sound that's all ... It was a shame their whole attitude was so negative. When I tell you on this particular site we were building from one- to four-bedroomed houses and we were selling four-bedroomed houses for £40,000. That was [nineteen] eighty-three. It was for the locals, for people who were resident and lived in the area, that was the condition.

When I asked one LDDC executive how difficult it was dealing with authorities that would not communicate, he replied "That depends what you mean by difficult":

I mean on a day to day basis it was actually very easy, arguably too easy, because in the Corporation's terms we'd want to do something, we'd write to [the authority] and say these are our ideas and we'd like to discuss them with you and they wouldn't reply to our letter. So you could argue it

> was very easy, wasn't difficult at all. We just went on and did it. But in terms of actually making sure that you heard what people had to say and could consult and could respond and could discuss, I mean, it just didn't happen through the council at all.

Council officers tended towards a more pragmatic view about working with the LDDC, but their political masters would not entertain it, as this officer described: "On officer level . . . we'd be quite willing to work up good projects to go the DOE [Department of the Environment] and get approval. But we are constrained politically by the Labour members". Nevertheless, the non-negotiation policy served a useful purpose as this local authority employee explained: "We tend to throw our hands up in the air and say 'Oh well, it's the LDDC, we can't do anything about that' and in terms of planning I must say that's true because we don't have any planning authority for the neighbourhood at all, but in other areas I think we could have done better."

The refusal of local councillors to consult resulted in the LDDC "bypassing the local authorities and setting up direct consultation procedures with tenants' groups, residents' groups and so on". "And the local authorities really really hated that", an LDDC executive said: "I don't think they'd ever admit that publicly but that was what it was really, you're being bypassed, using somebody else as the spokesman of the local people rather than the democratically elected Members, really stuck in their craw." "And that caused tremendous resentment", which had its impact on those who were seen to be consorting with the "enemy". The fact that they talked to the LDDC made an already uneasy relationship between the Association of Island Communities and local councillors even chillier. "It made us almost like Conservatives," one of the executive committee explained, "which is crazy." One activist described their situation as "piggies in the middle" wedged between "a local authority what were not supportive to its community and the LDDC that we felt was not supportive to its community", leaving them nowhere to turn . . . for help. So the Association of Island Communities and other organizations were left to liaise directly with the LDDC in the hope that they could persuade them to fund projects and introduce policies to benefit the local community, a process one local described as "battling for the crumbs". Yet, as one of the Islanders commented, "if the Borough had come with us we'd have got more, if the LDDC had been an elected body we'd have got more, if they'd have bloody listened we'd have got a lot more!". "Whatever this community has gained it's gained for itself and that's precious little", another established resident said "So much more could have been done."

Although activists believed that councillors had "done the greatest disservice to the Island people possible" and that "without a doubt the first five years" was the time when "if you cultivated . . . the seed . . . then you would get something out of it", it is difficult to gauge what impact the local authority might have had if they had consistently been represented on the LDDC board. It would not have shifted the emphasis or shape of development. However, it might have meant local interests were more regularly aired and taken into consideration. "If they'd have come to us and said 'look let's work together and let's start that by setting out a set of objectives and principles', that would not in my view have been capable of achievement", an LDDC executive said, because "the differences of philosophies between the Corporation and the local authorities were so huge they couldn't have been breached." Nevertheless this would not have prevented their working together "on a day-to-day basis in terms of do we need a community centre here, or what sort of employment do we need there, should this road go in here or in here, at that sort of level they could have had a huge impact; most certainly". Another LDDC employee agreed: "There would certainly have been more pressure on the Corporation to look at things in rather rounder ways than it might have done in those areas."

Although they were not communicating with the LDDC, the local councils and the GLC had been busy producing different "dreams and schemes" for Docklands after its inception. The difficulty, as a report from the Docklands Consultative Committee (1985:23) said, was that "the LDDC is winning hands down because it has far more legal, political, and financial power: not to say the advantage of being the principal landholder". Furthermore, as Brownill (1990:112) notes, "The lack of ability to translate the alternative vision onto the ground was a major drawback and made opposition seem only theoretical in response to the LDDC's ability to grant planning permission, sell land ... and fund infrastructure works".

It was not until 1986, when political control of Tower Hamlets council transferred to the Liberals, that the political climate changed centrally and negotiations were resumed with the LDDC. The Liberals saw the behaviour of their opponents as "a very cynical ploy" that "scored a lot of points simply by being anti (LDDC) even if being anti got them nowhere". A councillor said:

> It's a lot easier saying things are going to be disastrous and they're not gonna work because if they turn out not to be working they claim credit for it (saying told you so) ... What we tried to do was try and negotiate a reasonable package to try and make the best of what's there. I think that's the only responsible line to take given the circumstances, but one which carries far greater risk because there was a really serious risk of failure.

"Serious harm" was done by Labour, he continued: "There's no question in my mind about that ... Just a mountain of lost opportunities. There was quite a lot of council-owned land there which ought to have been developed seriously. The potential from the extraordinary market there was completely lost because of the conceit of refusing to deal with the LDDC ... and not being in general political terms [willing] to be innovative in any sense really."

Local Labour councillors were unrepentant. One of them told me:

> Docklands presented an opportunity to solve London's housing crisis and it's just been squandered ... We don't trust the LDDC [and] I don't think they looked at the potential of the area for everyone. They were just interested in the commercial highest bidder and that was their goal ... They had no interest in what the local community wanted.

As I describe in Chapter 7, disillusion with elected representatives and their neglect of the feelings of "locals" did not end with their approach to the LDDC, and after years of antipathy towards the policies pursued by the local Labour councillors they found themselves ousted from power temporarily, not by the Liberals, but the British National Party.

Conclusion

It is easy to see why Island residents felt powerless, angry and neglected both before and after the redevelopment. "People could have accepted that you had to build for the city", a local woman explained, "if they had at least made a nod towards recognizing the local people who had lived here", but the LDDC kept their heads down and stuck to their task and for a while this strategy worked. Although pain was certainly inevitable, there were other ways of handling dynamic development without the insensitivity that characterized the LDDC's approach. There was no talk at this stage about compensation or community gain, which would at least have helped to ameliorate some of the physical disruption, and in their quest to attract developers at any cost the LDDC made it very plain that the feelings of local people did not matter. Their view of the opposition that did occur? According to a senior LDDC figure it "wasn't an impediment, it was just part of the landscape". The LDDC were "in no way harassed and not even distracted by it".

Meanwhile, outside of the LDDC, "People were feeling extremely angry", a former councillor said, and the Island Tenants Action Group (ITAG) which was established in the mid-1980s, "was an outcome of that anger". "I suppose what happened to ITAG was a good reflection of the whole of the Isle of Dogs", she continued, "that people were angry, they tried to do something about this anger but it was deflected so the anger almost turned back on itself because the LDDC were quite impervious to it."

Reg Ward said that the task they had been given required a particular kind of response:

> By 1981 Docklands was well down the spiral of terminal decline, physically, socially and, above all, economically. The reality was that this could not have been halted, let alone reversed, by community regeneration programmes no matter how needed or justified. Neither was there anything in the economic environment of Docklands that provided a springboard for recovery at that time. Dereliction was widely visible. In property market terms, it had become a non-location. It needed total change to break the mould and point the area upwards again.

The LDDC certainly turned the area "upwards", but not in a direction that sought to encompass or include the established population, and they, not unnaturally, did not like it. After years of waiting for something to happen they wanted to see tangible benefits. Although the outcome of the battle might have been predictable as the development began in earnest, local people's frustration grew and the LDDC had yet to see the full extent of community opposition. As a senior LDDC board member said, relationships between the Corporation and local residents "got a bit uglier later" because people's expectations were raised, initially by Reg Ward and later Michael Honey, both under the chairmanship of Christopher Benson. "It need never have happened", he said, "and the reasons for it happening didn't just relate to the quality of the adversaries it related to the weakness of the Corporation." In fact, it was an inevitable consequence of embarking on a development programme that ignored the locality and where the frustrations of Island residents could only be contained for a limited period.

The LDDC discovered the hard way that a delicate balancing act was required to keep government, developers and residents on side and that without all three problems were inevitable. In their enthusiasm to exploit market opportunities at breathtaking pace, the LDDC left not just the powerless local residents critical of them and their approach but, as I describe in the next chapter, powerful developers and various parts of government too — interest groups who could not so readily be ignored.

Chapter Four

"Grab and greed"

The tragedy is immense: Docklands was the world's choicest building site. Between the Royals and St Katherine's Dock, an area was freed for new use larger than the whole of the city of Venice. At the centre of a great capital was the chance to plan and build with vision. It didn't need to become a jumble lot of giant offices at the whim of land prices driven dizzy by speculation and recession. The possibilities were there. . . . Instead of what the architect Richard Rogers describes as "a hymn to greed". (Widgery 1991:225)

What in the beginning didn't happen was that there wasn't enough emphasis on the social side from the LDDC. They were purely interested in getting people in. Now you can argue for and against that mentality. There's one argument which says that to get the [thing] rolling you have to be mercenary cos there's nothing here, and so maybe it wasn't such a bad thing, although they weren't thinking about the local community at all. They just went straight in to build these mega buildings and sod everybody else. And that was the mentality of the LDDC at that stage. And it worked. You know at the end of the day, however much we may not like the situation now . . . [it] is vastly different . . . It's been quite an achievement to carry through. (Docklands estate agent)

Introduction

Whether people like or loathe the development in Docklands, they cannot fail to notice it. The dreams and schemes that received such a sceptical reception at the outset suddenly became a visible reality as a "Wall Street" on water rapidly emerged (see Figure 4.1). Even opponents of the London Docklands Development Corporation admired the pace and enormity of change that occurred: "Let's be honest", one critic said, "what they've done has been spectacular" (in Foster 1992:170). "The landscape changes by weeks not by months or years", a property agent said, "It has been such a radical change over such a short period of time . . . you just can't fail to get excited by it." "Suddenly everybody wanted to be down there. It became totally attractive."

Figure 4.1   "Wall Street on Water", Janet Foster

The pace and rapidity of development was on the one hand the LDDC's great-est success, responding to and capitalizing on market forces, themselves boosted by a buoyant economy in the mid-1980s, to attract private sector investment. But it was also its greatest failing as the consequences of market-led development without an adequate infrastructure to support it, insufficient planning and control, and lack of consideration for the local community became all too apparent. "They get all the credit for kinda starting the process", an American banker explained, but "they also probably get most of the blame for not having perceived how successful it would be."

"Nobody anticipated the market would take off quite like that", a builder recalled, but the result was that an area that had been perceived just a few years earlier as having little private residential or commercial potential became a prime site for speculation. As a businessman explained: "They started putting up huge buildings and everything changed out of all recognition. [19]83/4 land values for residential . . . on the Island were £100,000 or less. You got the commercial develop-ment going up to four, five million pounds an acre. It just shows how things changed."

This chapter describes the events surrounding this dramatic transformation, first in the housing market and then in the commercial sector, and the way in which the LDDC, previously geared to attracting investment and fuelling develop-ment, found themselves overwhelmed with its intensity — "All of a sudden it was tumbling round our ears" an LDDC executive said. It was also "tumbling round" Island residents too, but in a very different way. As business boomed and enterprise flourished, they remained firmly on the sidelines; and as the pace of development and the associated disruption increased, so did their anger, until opposition to the development machine eventually spilled visibly on to the streets, capturing media attention and generating concerns among the development and business community itself. The contrasts between conspicuous greed and relative neglect were simply too great and left even some developers pressing for a more enlightened approach. "Docklands has made me a fortune", a property developer explained, but

> you've got to have a social conscience as well. I just do not believe you can build things just for building's sake. I'd rather not be in the game than just do that, . . . just to go in there and make a few million which a lot of people have done and walk away from the situation and really just leave the legacy of a building that bore no relationship between that and the rest of the community facilities required.

"A social problem of a different kind"

> It was as though . . . we'd . . . unleashed some sort of tiger [and] . . . for a while we were coping with the problems of the success and overheating of that market and that happened very quickly . . . prices were soaring [and] it was hard to keep a quantity of the houses at prices that local people could afford. (LDDC Executive)

Despite the initial reticence of the major house builders, the residential housing market very quickly became established in Docklands. As demand grew, land values rose and the relatively low-cost housing originally built (see Figures 2.2 and 2.3) was rapidly usurped by more expensive and sometimes luxury housing catering for a highly selective market (see Figure 4.4). In 1982–3, 99 per cent of housing in Docklands was sold for less than £40,000. By the first half of 1985 just 43 per cent of property sold was in this price range (Docklands Forum 1987:17), an indication of just how quickly the housing market was transformed. Nigel Broackes' estimates to the House of Lords Select Committee, before the LDDC was established, indicated a housing balance with approximately 50 per cent owner occupation, and 25 per cent shared ownership and house association respectively. In actuality, by 1985, over three-quarters of house building was for private ownership (LDDC 1998:16).

Although the Corporation recognized some of the problems associated with rising land values, they were in a contradictory position since the higher the land values the more revenue they received. Consequently, as a builder explained, "There was a natural inclination for them to try and increase the values of land and push them up so their residual was greater, which obviously went back in their coffers to pay for . . . major infrastructure." By early 1987, the LDDC had recouped £57.7 million from the sale of land (DCC 1988:13). However, land values were rising so rapidly that between the purchase of land and the eventual sale of houses built on it, developers were making massive profits, which provoked criticisms that the LDDC had not done enough to claw these back. "All the people who bought land

Figure 4.2   Clippers Quay development, Janet Foster

Figure 4.3   Clippers Quay from the dockside, Janet Foster

Figure 4.4 The Cascades development, Janet Foster

in Docklands for housing paid the open market price for it at that time", an LDDC executive said. "Some of them made huge profits, I don't think there's any question about that", but

> the way those profits impacted was that because the market had got very excited and prices were shooting up . . . if they bought some land from the corporation at day one and paid the full price for it at that point, by the time they had finished their housing and were ready to sell it . . . it was worth a lot more to them than it had been on the day they'd paid for it.

"The residential market at that time was . . . an investor-dealer market, not [an] owner-occupier market", a property agent explained, "so you actually had a very distorted market." It was unique, he continued: "In all my years I've never experienced anything quite like it and you also had the sharp wide-boy coming in to make a fast buck." The differences between speculators and other types of buyers were outlined by a Docklands estate agent:

> We classify three different types of buyers: the speculator, investor and end-user. A speculator . . . never actually wants to own the place, he just wants to put down his percentage to reserve a property and, before he can complete it, sell it [for] a lot more money. The investor does want to own . . . but . . . doesn't want to live there, he wants to rent it out. And the end-user just wants to buy and live in it. . . . Well, predominantly in the

mid-eighties it was the speculator who was buying off plans. . . . They were able to make twenty, thirty per cent in six months . . . Not bad and for that you'd need an investment of 15 to 20 thousand High risk . . . cos it was a very short-lived market . . . A lot have made their money and gone. But because they actually put down money before even a brick was built, it was good for the developers' cash flow and enabled them to build a far higher quality scheme here than in the past.

Speculation was so intense, said another agent, that "on some sites there were six or seven contracts out on the one property" before it was even completed, which "bumped the prices up". Inevitably, where large sums of money were concerned there were abuses of the system: "There were these dead unscrupulous, I can't call them estate agents, property dealers who were in operation", she recalled. "They had these rings of dealers organized . . . How these people got away with it I just don't know but they did . . . and they cleared off juicy great profits." "There was all sorts of unsavoury practices", one of the LDDC team admitted: "People were buying houses and then selling them on the next day for huge profits."

People who bought homes on the early developments were all too aware of the speculation taking place around them, as an architect who moved to the area in 1985 described:

It certainly became in vogue to live in Docklands and suddenly the prices started rocketing. People were just asking silly money for land and developers were putting up luxury developments and it was attracting, OK, the sort of yuppy market I suppose which hadn't occurred in the first instance. When I moved in we were all either such people as myself who just wanted a bit of extra space or young couples who just wanted their first home to start having families and things, but then it sort of developed into . . . really quite wealthy people who just wanted to make a killing out of it. I suppose that was when . . . there was some justification of the local people sort of saying that . . . it was perhaps changing the neighbourhood in a way they didn't like.

"You had people coming down here buying two or three", a newcomer on another development said. "Certainly on this estate people bought two and just sold 'em. In about [19]87 two-bed flats like this [went up to] £130–135,000" — for a flat which cost £80,500 in 1985. On a further development, a woman who had benefited from her husband's purchase of two properties, one of which they subsequently sold, reflected on the injustice of a situation that while others struggled to afford one home they had been able to profit from the purchase of two: "We had the advantage of buying (in 1985) . . . we made almost three times what we'd paid for it . . . it's sick isn't it that people with a bit of money always get more." A trebling of prices between 1984 and 1987 was not uncommon and in some circumstances increases were even greater. On the Clippers Quay development on the Isle of Dogs, for example, record rises occurred. A two-bedroomed flat which cost £39,495 in 1984 rose to a staggering £199,950 by 1987 (DCC 1988:12; see Figure 4.5).

CLIPPERS QUAY   E14

Resale prices
2 Bedroom flat

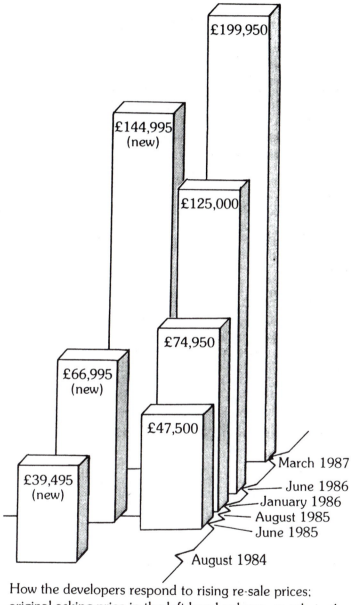

£199,950

£144,995
(new)

£125,000

£66,995
(new)

£74,950

£39,495
(new)

£47,500

March 1987
June 1986
January 1986
August 1985
June 1985

August 1984

How the developers respond to rising re-sale prices:
original asking price in the left-hand column, re-sale in the
right.

Figure 4.5   Resale prices on a two bedroomed flat on the Clippers Quay development.
Source: Docklands Consultative Committee, Docklands Forum 1988:12

Figure 4.6 James Town Harbour, Janet Foster

    Although the market was dominated by speculators, there were also investors like this newcomer, who, with the proceeds of the sale of his business, invested in Docklands housing:

> When I moved here [1984] I didn't just buy this house, I bought a flat on the river. I was buying [another house] . . . I had a deposit on that. Subsequently I bought a small studio [flat] . . . then I bought one on [another development] . . . We've got five properties . . . I didn't ever speculate . . . I was looking for long-term investment.

The rise in property prices was not only happening in Docklands. Many of those who benefited from their investment in the area were keen to highlight that other parts of Britain in the mid to late 1980s were also experiencing rising prices because, as an investor commented, "Our economy just went through the roof too quick [and] for a period . . . the greed of the money market" dominated, which "spoilt Docklands and it also spoilt the UK economy." However, the situation was perhaps more intense in Docklands because, as he continued, "There was a higher amount of speculation in London [because it] is the money market of the world."

    What had originally been conceived as a plan to generate social mobility and a more "balanced" population started to have all the hallmarks of intense social divisions. While private housing was way beyond the purse of most Island residents still in council accommodation (see Docklands Forum 1987:22), developers proceeded to build for an increasingly select "niche" market (see Figures 4.6 and

Figure 4.7 A luxury development overlooking the low rise council homes, Janet Foster

4.7). Not surprisingly, the Corporation came under increasing criticism from a variety of quarters for not controlling the profiteering and concerns grew about the consequences that inflated land values had on the type of housing. A property agent working in Docklands at the time said:

> I think the property industry itself was wrong in that they hyped the residential market down here. It was that pioneering instinct syndrome, but the net result was that the value of residential down here at one time was actually going up to, and I think outpaced, square foot for square foot rates in Chelsea and Fulham. Now there is no way you can actually command that sort of price down here compared with up in that part of the world which has historically decades of [high prices].

The LDDC should have prevented "the dramatic increase in land values", a builder said, to prevent the "greed" of housing developers and speculators. "The sheer pressure on land values" meant "you're pushing up and producing buildings just to make a commercial equation work rather than looking at the market." As the director of a large house building company explained:

> On the Isle of Dogs there is an awful lot of water and for developers to make the most money out of the water . . . you had to build high-density schemes. High-density really means . . . flats . . . This silly yuppie thing wanted a particular niche market which was a no-maintenance . . . very

Figure 4.8 Social housing, Janet Foster

sort of sexy units, with very sexy sort of entrance halls but that was really where it stopped . . . They weren't "real" buyers in the long-term home ownership chain.

Whether they were "real" buyers or not the impact on the housing profile of the Isle of Dogs of "sexy flats" did nothing to address the paucity of family housing, either rented or private, in the area. "Tower Hamlets has the highest proportion of flats to houses in the country", (see Figure 4.8) Michael Barraclough, the initiator of the self-build housing scheme (see Figure 4.9) said, "so if you were a planner and you suddenly got all this free space — you would see what any community would want in this area is to have more family housing . . . instead of which nearly all the development is yuppie *pied à terre* flats, so in an over-flatted environment you're stuffing still more flats. The effect of that in the long term is of course disastrous." On a plot of land opposite the house he built himself, developers proposed yet another block of flats. Notified under the planning legislation, Barraclough went to look at the plans and produced an alternative design

in which I got 26 houses with gardens where there were the 30 flats and I wrote to the LDDC and said "Can't you get the developer to build houses because his responsibility to the community of the future is enormous, because the difference in that community of 26 people with families and

Figure 4.9 The self-build homes. © Mike Seabourne

children . . . putting down roots and 30 transient *pied à terres* who merely look at it as a convenient lock-up shop is enormous.

The LDDC were unmoved. But others, including the construction industry themselves, were becoming concerned. "If you want to create a balanced community, where do children go if you're in a two-bedroomed flat?" a property developer asked.

Where do they play? Most developments tend to be like Stalag 17 with great high walls around them . . . in terms of social mix I think that's my biggest single criticism, that's where you come back to planning controls. [The] Corporation had residential [controls] outside the zone and in my view didn't use it sufficiently to say "Hang on, no, you shouldn't be building something like Cascades", which I think is an absolute abortion . . . It's just not suitable housing. It's great for business people who want a weekly flat — they've got no part in the area — they'll come from Monday to Friday, they go away to their country places at the weekend. So there's no heart.

An estate agent also recognized the problems created by the lack of family housing on the Isle of Dogs:

There is going to be polarization. There's gonna be two distinct markets and there's nothing to bridge it and what we need . . . are really town houses, big family houses so that somebody can move through the social market and economic market within a given area. For example, if you live in Chelsea there's no reason why somebody can't live in a council flat, perhaps buy his council flat, sell it, buy a little house, and move up within Chelsea. You've got the entire sort of micro population mix there . . . You've got everybody. You don't have that down here. You won't have families here, which is very sad really cos they're not building enough houses.

Although the developers played a pro-active role in creating this situation, many of them argued that it was the Corporation's responsibility to restrain them and the market sufficiently:

They were not restraining . . . the price of property . . . to stop the speculation . . . [It] has been done in . . . new towns where on corporation owned
land they can make it available at a sensible price, so the builder can
build a sensible house at a sensible price . . . The cost of lime and brick in
London costs the same as anywhere else but the land element in London
[is the real cost] . . . There was an essential point at which the Corporation
as the landowner [could have said] that whoever bought that house couldn't
sell it at vast profit . . . Over a five-year period if [a householder] sold they
would pay the Corporation, as the landowner, an element of the profit
. . . This is what . . . many authorities [have done] . . . trying to balance out
what does happen where suddenly people are attracted to an area. [In
Docklands] where simple balance of supply and demand was available
. . . prices went through the roof. (House developer)

Ironically, then, although developers believed in the market being given a free hand they also felt there was a role for planning and controls to prevent the excesses of the market from producing the kind of skewed development that occurred on the Isle of Dogs because of the long-term problems it might cause. "The situation could have been controlled a little bit better by the . . . Ward regime; it's about my only criticism", a volume builder said:

Land has a value only on the basis of what you can place upon it. So you
can stick a particular planning constraint upon it by saying you can only
build this, that, or the other. Now there was far too much encouragement
from the Docklands' planners to have this double-sided wall right along
the Thames on both sides, very high rise flats to maximize on the views
of the Thames. There was very little pressure placed upon developers or
landowners to ensure that every so often you'd have a site with just houses
on it. Now if you were trying to sell me a piece of land, and it had a planning consent on it that could not be fought through the courts as being
incorrect to put low-rise housing on it, all you could expect is me to value
that land and pay you the best market price for that particular product
mix. But at the time we [developers] were all getting higher and higher

Figure 4.10 Riverside developments, Janet Foster

and higher and higher and higher to beef up the density to pay for the land value, because Docklands [LDDC] were not vicious enough in their control of the mixture between houses and flats. (See Figure 4.10)

When I replied that such constraints might have been viewed by the Corporation as a deterrent to developers, the builder said:

> Well, I mean, it's a question of timing isn't it. I think when the boom was really going you could've done all sorts of things to control it . . . Docklands were able to control things in that way on their own land and they did to some extent try to do it by putting housing at forty thousand pounds capped value. Later on, you know, a certain social housing element . . . things like that it's quite possible. But there was an awful lot of ego in it as it really got going with the planners wanting to build these edifices that they'd seen elsewhere in the world and I think there was an understandable rush of blood to the head with these guys thinking they were going to produce a *La Defense* style situation, a Fisherman's Wharf . . . [or] Vancouver Island, where I think actually there was room for all of that and some standard housing as well.

Sir Nigel Broackes, former chair of the Corporation, said the speculation was very damaging: "That wouldn't have happened under our regime because our policy from the beginning was to saturate the market so that would not happen."

There was a hiatus after me when the production rate slowed down and the price inflation got completely out of control . . . things just sort of stopped happening and then the . . . million pound penthouse and so on started hitting the headlines. . . . Some of it was developers' work, I mean quite a lot of it was, and we had no control over that, but in my time there we nourished our developers and lent on them, you know, all at the same time to make sure that they were being sensible to be successful but not to create long-term problems and it was getting a bit out of control about [19]87/88.

Although many developers undoubtedly pursued short-terms gains rather than long-term goals, there were differences between developers. One, for example, contrasted the approach of volume builders and those who had their own companies. Talking about one of the developments on the Isle of Dogs, he said:

That's overdense. My site, which is next door, when I put . . . the last . . . plans in they said "You're not using the density up as much as you could." They have a ratio of 2:1 and I've got 1.5:1. I can put a lot more square footage down and I said "No that's enough" . . . The highest I'm going is seven storeys and that's on the river. The rest is three and four storeys. . . . if you talk to the builder who it is *his* company, compared with the executives of the big nationals, there is a different approach. But if you talk to many of my colleagues who've got their own private companies . . . it's their company, it's their family cos I always used to say a long time ago that whatever I built I could take my grandchildren [to]. I haven't got any yet [but] one day perhaps [I'll] take them round and be proud of [what] I've built.

This particular builder had won a number of awards for his housing schemes, and in fact one of his daughters eventually moved on to a development that he had designed and built on the Isle of Dogs. His sentiments were exactly what this executive had in mind when I asked whether he had one wish that might be granted in relation to the development in Docklands:

I would say make the riverside really wonderful and alive . . . Improve the residential quality and the environment, that would be my greatest wish . . . Make it a nice place, make it a place people want to be not just another 50 or 100 flats where the ceilings are too low and the quality is too cheap and it says this is all you can afford so get used to it kid . . . The cost of this stuff boggles the mind. You can buy an apartment in 5th Avenue in New York for the average flat in Docklands. So the stuff is junk when objectively measured against anything. I don't understand why that has to be.

Even the volume builders became concerned about the restricted nature of residential development on the Island, presumably because it would in the long term be bad for future business. This influenced some of the schemes in which larger companies chose to invest. For example, one housing executive told me that his firm took the view that there was a need for some balance in housing development which included an element of social housing:

If we were trying to respond to the market we wouldn't have done any of it because we'd have made a lot more money by selling them all on the open market. We took the view here that what you needed to do was to have a balanced business and the only way to have a balanced business was to do some social housing and some housing for sale, and . . . that's what we've done throughout . . . and it was also about frankly putting the right type of tenure in to certain areas and it struck us that what you were getting was a sort of gentrification. So you'd got these sort of little ghettos not of social housing . . . you'd got ghettos of up-market yuppie boxes and it seemed to us that that was all very well so far but if you went too far with it you started potentially to have a sort of a social problem being created, so we went in to the mix and match, which worked very well for us.

Although the LDDC's public image seemed to suggest a lack of willingness to tackle the problem of an overheated housing market, the Corporation had, in a review of housing policy, expressed concern that by 1985 their "affordable housing programme was proving less than effective in providing original residents with a step up the ladder" (LDDC 1998:16). In fact, Reg Ward said that while "there was no change over the period" of "the very clear, and very limited, remit given to us in 1981", behind the scenes "The Corporation did find ways that were well disguised, [and] deliberately not commented upon, of using its "success" with the private sector to generate the cross-funding of nearly 2,000 social housing units across Docklands."

Furthermore, in an internal LDDC document in 1986 it was recognized that many poorer Docklands residents were not only denied access to local private housing because of the cost, but also had "very little opportunity . . . to move out of older and deteriorating council housing". The report also highlighted the "marked contrast between bustling house building on Corporation land and languishing inactivity in existing housing areas on land earmarked for public sector housing" (quoted in LDDC 1998:16). In the light of this, the recommendation was for an "adjustment" in the LDDC's housing strategy to incorporate "new council house building, refurbishment of existing council housing and for housing association provision for rent and shared ownership", acknowledging that all such schemes were "constrained by a lack of public sector funds but . . . also from a lack of concerted and co-ordinated programmes by the authorities and agencies concerned" (quoted in LDDC 1998:16). This was political dynamite. It was also, given the guiding philosophy of the Corporation's political masters, far from easy to achieve.

Big Business arrives

Although the changes in housing made their mark on the Isle of Dogs, it was the commercial development that had the most dramatic impact on changing the landscape of the area and was the issue that fuelled greatest opposition. The scale of the office development surpassed even the LDDC's own expectations and "the thought in [19]84", as a businessman said, that "someone would come and build Canary Wharf would just never have occurred". Yet by 1987, just three years later, a contract to build the largest office development Europe had ever seen had been signed.

As in the residential market, rising land values, and this time the attractions of the Enterprise Zone's minimal planning regulations, and tax-free and rate-free periods, bought successive waves of business interest beginning, as I described in Chapter 2, with very small developments but rapidly being over taken by larger- and larger-scale projects (see Figures 4.11–4.13). It did not take long for them to realize what they could do with the Enterprise Zone, a property developer said, "and it was actually the tax-driven side which in my view was creating the density".

"The first developers in the main", a Board member explained, "were pioneer builder developers", "and then from that one got the first owner occupiers" and "following that there was the first very large speculative office building . . . put up without a tenant in mind but to be there to be occupied when a tenant came along." From there "the thing took off and you started getting the public property companies". The progression sounded smooth but it was very rapid. One developer said it was "like rolling down hill, suddenly it gets faster, faster, faster. That is what happened". The process was fuelled by external economic forces. "Looking back . . . there were a number of disconnected effects which all came together", an LDDC Board member said:

> The use of the Enterprise Zone, which helped the Isle of Dogs get off the ground . . . the cheapness of the land here because it wasn't being used for anything, . . . the rapid rise in rents in the City of London brought about by "Big Bang" [which changed the working practices of the London Stock exchange, and created the need for offices with large computerised dealing rooms] . . . all the foreign banks moving into London . . . and . . . the [financial] deregulations.

This was the view with hindsight. Reg Ward said that in 1984 when "we made our breakthrough in quality office-type speculative development" with "South Quay and Hertsmere House" (the latter sold for "just under £300,000 for the site!")

> There was nothing upon which one could predicate any relationship with the pressure on space within the City with the beginning of Big Bang. Any such interests were confined to the Square Mile and gradually and reluct- antly towards Westminster. It had not crossed anyone's mind, at that time, of considering going East. None of the main developers, institutes and banks had the slightest interest in Docklands as a major office location, and in the Isle of Dogs in particular!

Those who did venture East, however, were excited by the development and the rapidly changing landscape. A small businessmen who acquired premises in the Enterprise Zone in 1985 said:

> The first two years here my business was a gold mine, literally. I mean, the business we were doing, gosh . . . it was exciting. I mean, everybody

Figure 4.11 The different phases of development, from low rise to high rise.
Janet Foster

Figure 4.12   Low level developments now dominated by Canary Wharf, Janet Foster

Figure 4.13   Low level developments now dominated by Canary Wharf, Janet Foster

referred to it as pioneers. I think that's a very good definition. . . . it did very much have this very very exciting atmosphere about the place. It was electric, no doubt about it. People moving in wanting a service, phoning you up, you go and see them, the place was certainly injected with some-thing . . . The buzz was definitely there.

Another businessman called it "the Klondiker spirit", while a colleague said it was "a bit like the Wild West of America". "It gets under your skin somehow", a prop-erty agent explained:

When we started we didn't know what the businesses were going to be, who was going to come, and the early buildings were built around any sort of user . . . Now it's changed. Now we're into the top quality office type of accommodation being provided all the time and we're creating the third business district of London and that's a fascinating process and one which is a remarkable achievement over such a short period of time.

Despite the "success", many firms still proved difficult to persuade. "The 'UK Establishment' find this area much harder to grasp than the foreign companies or the companies that are over here with foreign parents", a businessman said:

. . . The solicitors, accountants and the English banks, which intrinsically like the security blanket of the City area, they're the hardest to persuade. Whereas the foreign companies come over here, see the qualities of the buildings that are here, look at the cost and think well what's the prob-lem — it seems like a good idea to me. In cost parameters you have to go to Swindon in the West and up towards Peterborough, even there the rents are still in excess of what you can get in Docklands three miles from the City.

A senior LDDC figure pointed out that many important financiers never invested in Docklands: "There are thankfully entrepreneurs in Britain and they're the ones that have come here. You don't see the great names in property here — there are no Laing Securities, very little London and Edinburgh Trusts. There's no Greycoats, there's only a mere whiff of a Broadgate in the shape of Rosehall Stanhope. It's been entrepreneurs that have done things, entrepreneurs who give finance."

The City's lack of enthusiasm for Docklands was in part due to the competi-tion that office space on the scale and quality developed in Docklands created. "The threat that Docklands poses for the City is enormous", a Board member said. Yet Christopher Benson, former chairman of the LDDC, argued that both he and Reg Ward had many meetings with key City figures to emphasize that they were "not trying to shift the City eastwards" and that they wanted "to break this myth [of] . . . competition with the City of London". "We're supposed to be building things", he continued, "not breaking down the fabric of this whole area." The City, however, remained unconvinced and "in order to 'combat' Docklands . . . took a lot of its controls off and started . . . allowing international type buildings to go up." "That was a bit of a knee jerk that wasn't necessary", Benson continued;

"[they] should have been more consider[ed]." In Paris this kind of competition was avoided in the development of *La Defense,* a new business district, by controlling and then preventing all new office development in the central areas of the city until *La Defense* was established (Fainstein 1994:210–11).

For a while the commercial story like its residential counterpart was a "success" and the sheer pace of change was astounding. In 1985 approximately 28,000 people worked in Docklands (only 1,000 more than in 1981, Census of Employment). Just two years later, in 1987, this figure had increased by 67 per cent to over 42,000 (RBL/LDDC Survey 1985/1987, in LDDC 1995:3). The difficulty was that promoting massive office development without an adequate infrastructure to support it, alongside the celebrated rejection of conventional planning and the speed at which development was occurring, was stacking up trouble. The Docklands Light Railway, for example, which Reg Ward and his colleagues had fought so hard to get in the first place, was almost moribund before it opened in 1987, the system unable to cope with the volume of passengers. Road improvements could not keep pace with the development either and complaints became commonplace. The LDDC even found themselves blamed for things they could never have envisaged from the outset: "One of the things that baffles me is how could such a big development have so little infrastructure", a businessman said:

> You can understand mistakes like this if it was an underdeveloped country ... but here in Docklands bearing in mind right from the embryonic stage you had a situation where you knew and the plans were well under way that there was going to be a Canary Wharf one day which required massive changes to the services ... I mean it's only just approaching a situation where you can see a network of roads actually coming together ... My biggest frustration of being here has been the traffic, [it's] absolutely terrible.

"There wasn't enough forward planning done by the Corporation", a property developer said. "In the Isle of Dogs there was none ... They had started on a design philosophy which was low-rise, low-density development [but] the scale and density suddenly just went out of all proportion ... we started to get 1.2, 1.3:1" and "then ... came ... this grandiose scheme for Canary Wharf" which pushed the densities up higher still.

"You've had a laboratory of capitalism with all its good and bad qualities displayed clearly, openly", a businessman explained. "You say 'gee there must be a better way' and then you think about a socialist period where nothing happened at all, it just deteriorated." "The problem with capitalism", he continued, "is that you frequently get ugliness, blight, one thing doesn't relate to the next ... [and there is no] overall sense of plan." As land values soared and investors poured into the area, many, including developers themselves, felt the process was not sufficiently controlled and that the LDDC were "culpable for an anything-goes policy".

Even if the LDDC had wanted to "plan", it was practically very difficult to do in a market-responsive model, especially when the market itself was continually evolving and developing. Nevertheless, although there was no formalized "plan", attempts were made to impose controls, according to one of Reg Ward's original team:

The idea that . . . what you have is what you get for better or for worse, if you let the market forces [dictate] simply isn't true, . . . It's true we didn't use our planning powers in the Isle of Dogs, but that's in the Enterprise Zone; outside the Enterprise Zone we had the same planning powers as a local authority and used them very strongly. In the Enterprise Zone it's true we had no planning powers, but we owned all the land so if anybody wanted to buy land from us then one of the conditions of disposal was that we approved the design and Reg interfered with design . . . all the time both in detail terms and in big things . . . , so the idea that that just happened that hotch potch, if that's what you regard it as, simply isn't true. It happened like that [because] that's an expression of Reg's views on design. Now whether you think that's good or bad is a separate debate. But it didn't happen because that's the way the market dictated it. It happened because . . . the architects working in Docklands were the architects that Reg wanted for the main part, the sorts of buildings they designed are the sorts of buildings that Reg likes.

"The planning in the early days was relatively relaxed in terms of the detail", a builder said, but confirmed the LDDC's "influence [and] control . . . on the element of design".

Planning really was only identifying the use of the land, whether you put offices, houses, etc. . . . [The Corporation] . . . relied on the skill of the builder [and] . . . it was a design-led competition . . . They went to builders who in their opinion were capable of producing good design rather than just let any builder [submit a] design.

An architect within the Corporation had this to say about planning:

There are four components . . . a city should have . . . the area has to be *accessible* and we're building roads and railways to make it accessible; it needs to be *diverse*, you need the corner shop as well as (the large developments); it needs to be *intense* [vibrancy] . . . you get that in London you don't get that in Milton Keynes . . . ; the last one is *flexibility*. There is a danger of starting off with a grand plan of foreseeing exactly what the needs and aspiration of society will be in 10 years . . . and you go through this physical planning process with this single vision, this is the Christopher Wren plan. What you find is that by the time you get half way through you don't need big dealing rooms anymore — the stock market has gone electric and you need a lot of little offices with plenty of plugs in which to plug PCs. You've gotta make a decision somewhere along the line but you must make decisions at the point when you're actually gonna do something.

"If society doesn't change, then you can have plans", he concluded. However as this local authority officer pointed out, there were some really practical issues that did require planning:

The postal facilities do not match in any shape or form the size of development as it is now and as it's going to be when Canary Wharf gets going. It's absurd. People cannot get their mail away. The larger companies have their own arrangements, but the medium and smaller companies it is absolutely hopeless . . . the level of postal facility here is still the same as it was ten years ago. What it really needs is a big main post office . . . What firms have been doing is either secretaries or the directors taking the post in their cars and posting them outside Docklands in their own home area, which is absolute nonsense.

The dawning of a new age

It was only the second-generation or third-generation developers that really understood what the potential was . . . The signing of the Canary Wharf development . . . was the turning point for Docklands. They'd been piddling round until then. . . . This new vision . . . was so much bigger and grander and more international than some of the high-tech sheds they'd been putting up . . . Canary Wharf . . . really got it right in terms of the scale. A lot of the locals were very anti Canary Wharf but . . . perhaps they didn't have the vision of what it needed. It needed critical mass. (Business journalist)

The Canary Wharf development, probably the best known in Docklands and in many corners of the world, more than any other changed the face of the Isle of Dogs. It began entirely by chance, as Reg Ward explained:

The Chairman of Credit Suisse funded the Roux brothers [chefs and] restaurateurs. The Roux brothers needed a 4,000 square foot "shed" to process their meal package for Harrods, Simpsons, etc. They had spent 18 months looking in West London unsuccessfully. Someone suddenly suggested why not look in Docklands? Michael Von Klemm was despatched to Docklands. We entertained him for lunch on a Thames Barge moored alongside Canary Wharf, shed 31. Over lunch he lost interest in the 4,000 square feet shed. He was attracted by a story about. . . . interest in shed 31 . . . He was from Boston and remembered the Warehouse conversions around Boston harbour. Three months later the study into the possible location of two American banks on Canary Wharf was launched. In September 1985, a potential scheme was submitted.

The proposal, constructed by a consortium of three American banks, outlined the biggest development Europe had at that time seen, offering a mixture of office, retail and hotel accommodation over a 71-acre site on the West India dock (see Brownhill 1990:54–6). Three towers reaching 850 feet into the sky were to form the focal point of the development and, as the majority of the planned development area fell within the Enterprise Zone, it "needed no planning permission, no public inquiry or public discussion" (Brownill 1990:56). It is not surprising that a development of this scale excited the business community and horrified the local population, who would not only have to live through the years it was under construction but could barely grasp the full impact that a development on this scale would have on the area.

An LDDC Board member called the events surrounding the Canary Wharf development an "extraordinary saga". Indeed they were. "Spearheaded in the first case by Ware Travelstead (an American banker) and Reg Ward with their dream of this enormous . . . development", the scheme "didn't take off because the people who were handling it . . . had to get it pre-let in order to finance it." But, the idea was so appealing "that Olympia and York came along with their enormous financial strength and . . . said 'We will build it cold' . . . a great act of faith and of tremendous importance to the Isle of Dogs and to Docklands as a whole."

In commercial terms the Canary Wharf development represented "a quantum leap in quality of office design, construction and environment", as Sir Nigel Broackes explained: "It's not a slight improvement on the last one, it is vastly better than anything else you'll find anywhere else in the world. And [it] is a terribly important thing that it happened at all. It's a welcome thing that it happened in Docklands." The 800-foot tower that Olympia and York placed in the centre of their development put the area firmly on the map, punctuating the landscape of East London with a landmark for miles around (see Figures 4.14 and 4.15). But the development itself, both in its initial scheme and that later built by Olympia and York, was very controversial indeed and, alongside the other frenzied activity in the Enterprise Zone, placed tremendous strain on an inadequate infrastructure and finally provoked local people into vocal and visible opposition.

On the sidelines

If many local people had not realized the impact that the LDDC and its policies were to have on the Island at the outset, by the latter part of the 1980s they were in little doubt that the LDDC was to be taken seriously. As the "glitsy postmodern image" was imprinted on the landscape, poorer Island residents stood on the sidelines, waiting for some indication that the energy and activity that had benefited the residential and commercial sector might turn in their direction.

Although the LDDC had appointed a single community worker in 1982/3 for the entire Docklands area, funded some social and community projects from the urban programme they inherited (including groups like the Association of Island Communities that opposed them), and later increased their community relations team, the local population remained a peripheral rather than integral part of the development process at the very time when arguably most community gain could have been exacted from developers cashing in on the boom. Consequently, as the development gained momentum and it became clear that the Corporation had no directed strategy for regeneration outside of the commercial and private residential sector, as this local woman said: "The disenchantment set in . . . when the people living on the Isle of Dogs started to realize that all of this glossy new

Figure 4.14   The tower on the Canary Wharf development, Janet Foster

Figure 4.15 Canary Wharf and the Docklands development from Greenwich, Janet Foster

utopia was not for them. It was for somebody else . . . not for anybody who'd lived here before, not for any of the people who'd put up with the noise, the dirt, the squalor . . . not for those people who had already lost their jobs as companies moved."

Not only were local people excluded from the development but, as it gathered momentum, they felt that they would eventually be driven away by it as developers got ever greedier and as it became clear that the poorer sections of the community, especially those in run-down council accommodation, did not fit the emerging Docklands image. "In effect", one of the Island's clergy said, "the dominant tone is the Island's getting a bit too 'nice' to have people like that [the poor] around." A couple who lived adjacent to the Enterprise Zone and had a large office development rising into the sky at the bottom of their back garden summed up these concerns: "I often think I'll wake up one day and we'll get a letter to say they're pulling these houses down . . . you begin to wonder, just six houses [in our terrace] and it would give them another big site wouldn't it." These fears were not unusual, as a consultant who worked on the Docklands Light Railway project explained:

> They (the community) didn't trust us. . . . and they didn't trust anything to do with the Development Corporation. They had real fears about losing homes and local jobs and about seeing the price of property rise . . . Once Canary Wharf came along . . . there's nothing we could do that anybody

believed anymore . . . there was so much resistance and resentment about it — very difficult.

Suspicion became a prominent emotion in many local people's dealings with the LDDC, and much of this was of the Corporation's own making. Attracting development "with no strings attached", especially in the Enterprise Zone, afforded local communities no protection. Reg Ward said: "We were not ready to dictate to the private sector." All the LDDC wanted to do was to "encourage them to come into Docklands". But this permissive approach carried a severe cost for the residential population, not only for the poorer sections of the Island community, but increasingly for more affluent owner occupiers too, as I describe in Chapter 5, who bore the brunt of the problems posed by the development and whose needs were never adequately considered. It may be, and this was certainly argued by Reg Ward and many of his colleagues, that any constraints placed upon developers would have made them look elsewhere. But given the advantages of the Enterprise Zone it is at least feasible that planning gain agreements could have been reached.

Given their differing agendas, the sheer scale and pace of change, and the historical context, relationships between the Corporation and local people were bound to be fraught even if it had been better managed. They "felt there was a steam roller . . . it was a steam roller . . . [and] . . . it was perfectly natural the whole time for that tension . . . or frustration of the non-listening bank, or the non-listening corporation if you like, exploding on key issues" (Reg Ward). The LDDC became synonymous with images of distrust, secrecy, broken promises and an organization motivated purely by profit (exemplified by the images generated by the Docklands Poster Project, see Figure 4.16). "It's all for them, it's all for theirself, for making money", an exasperated resident said. Another agreed: "When all this started . . . everybody was grabbing everything; this is how we felt and we were very bitter." A Gallup survey in 1988 (LDDC 1988a) strongly reinforced these perceptions with respondents expressing the belief that the LDDC was "colluding" with developers. "1 in 5 (21 per cent) of respondents spontaneously gave 'making money and profit' as the main role of the LDDC" (LDDC 1991:122).

But that is only part of the story. Some of those closely involved in the development process genuinely felt that their endeavours would benefit the poorer residential population in the long term and resented their uncaring image. When I asked an LDDC executive if he had one message he wanted people to know about Docklands, he replied: "We actually cared about what we were doing."

> We didn't come here to earn a lot of money, you know, individually. We didn't come here to make a profit. We didn't come here with any particular political motive as individuals, that we were as concerned and still are about the worries people have. Some of those we can deal with, some of them we can't. Some of them are the proper responsibility of other authorities, some of them are . . . part of society, [and] what's happening to society.

But there was an understandable frustration that development with a profit motive could happen quickly and rapidly and yet sometimes very basic needs, for example improving public housing, could not be addressed, as this local woman explained:

Figure 4.16   Big Money is Moving In. © Mike Seaborne/Island History Trust

A lot of the development round here has been done on land that was vested in the local population. It's been done on land where people used to work ... that land belonged to those people and it was taken away and given to Wimpeys, Barratts, John Laing ... The paint factory ... it cost (the LDDC) £2 million to clear the site and they sold it for £400,000 to a property developer who built houses on it ... it's just the unfairness of it all. ... People would say "Look you're regenerating this area, our council houses are falling down."

Despite the close physical proximity between the residential properties on the Isle of Dogs and the Enterprise Zone where most of the commercial development took place, and the fact that one could hardly fail to notice the changing skyline as the development got underway, for some the development remained strangely distant — aptly expressed in the comments of this resident, who said

I really feel that ... it's nothing to do with me. I mean I go in there [Enterprise Zone] to meetings ... but it's not for the likes of you and me missus, you know. It's ... a bit like that ... I don't resent it, but I'm only speaking personally, I think there are a lot of people who do [but for me] ... it doesn't have any bearing on my life ... I like a lot of the architecture there, I love walking round it, but at the same time I feel totally divorced from it and sometimes envious that people have got nice new homes, that have got fewer problems than we have but it's envy rather than a bitterness or a real jealousy . . . it's there but it doesn't really impinge on my life.

The notion that the development was "not for the likes of you and me" was both symbolic and spatial in its representation, as this businesswoman explained. "One of the things that makes it difficult for local people [is that this is] . . . the first time since 1800 when the dock wall went up that that area has been open to the public . . . there is this very real feeling that [the Enterprise Zone] is not an area that is part of the Island." Furthermore, the development did nothing to generate feelings of inclusivity, as this established Islander explained:

> We keep hearing we're gonna build so many homes and so many flats and there's gonna be squash courts and swimming pools and all wonderful things you know and shops [laughs] . . . In the housing boom the houses got put there all right but what we was promised to go with 'em — you know it'll contain walks round the side and restaurants and all this sort of thing — all very well [but] its meaningless to the average working class person around here. I've had a look at some of the restaurants and I couldn't afford to eat in 'em unless you saved up on your pension for 10 years and then you might be able to have a night out. But to put that sort of thing under yer nose doesn't make 'em any more popular you know . . . The prices I see here are out of reach for us.

Rumours abounded about the lack of respect or understanding the LDDC displayed towards the poorer residential communities: "some of the condescending remarks we've had to put up with", an activist said:

> The LDDC tour, they don't do it anymore, [but] they always [had] a tour bus go round every Thursday, round Docklands, some are business people, some are tourists, some are school children, whatever, and this person on the tour bus, this tour person, LDDC employee said: "Now this area round here used to be all rats and flats." I mean, these are the people that are vested with the regeneration of this area.

Such accounts probably lose nothing in the telling but were interpreted as being indicative of the LDDC's disrespect and neglect of local people and actively fuelled resistance.

Very visible opposition

Irrespective of whether they represented a minority or majority view, the role of a few key players within the AIC, the more affluent "newcomers" behind the scenes (see Chapters 3 and 5) and the more "in your face" interface of the core activists, was very important indeed; and as the intensity of the development increased, activists, determined to make their presence felt, stepped up the nature of their front-line activity:

> When we did the first things about Canary Wharf and what shit bags the LDDC [were] we had a little demonstration which was to see how far we could actually push 'em into giving the Mudchute and AIC funding and it didn't work so I went and got a barrel of manure and tipped it right outside the door, red hot manure right outside the door of the LDDC and said . . . what one of yers gonna step over that then? . . . They're throwing shit at us now they can have some back and it went in the papers . . . like you know dung thrown at the community, community throws some back. (Community activist)

On another occasion the approach was even more extreme:

> [An LDDC official] did a paper . . . to . . . say that we [Association of Island Communities] shouldn't get [our] money . . . Well somehow [we] found out and we went with fighting irons on that night. We all walked in slammed the door and they really knew they was in for a row . . . they started a discussion and [an LDDC official] went "Well you don't deserve it anyway" so . . . [we] jumped up and grabbed him called him all sorts of things . . . [The officer] had actually said that the community had no clout and he's sick of hearing people can turn cars over and burn 'em, he's never seen any proof of it. So I went up to him and I tapped him on the shoulder and I said to him "You want to keep looking over your shoulder cos very soon somebody's gonna have you." He said "Are you threatening me?" and I went "Not me. I'm just telling you once word gets out just keep on looking over your shoulder and you'll see what the community can do." Well it was four days later there was a petrol bomb gone through the window in a meeting. Well it was absolutely scattered. I mean the place was alight. . . . then we got emergency funding.

The LDDC officer involved in this incident did not work for the Corporation at the time of the research and it was not possible to verify these events. It is plausible that this account was less dangerous and dramatic than it sounds. Language and imagery were certainly important weapons in the battle against the Corporation, especially in a situation where it was important to demonstrate the community's strength and that they were determined to get what they wanted. A description bordering on terrorist activity makes that point very effectively.

Brownill (1990:111) notes that "The lack of formal input into the LDDC meant that local and direct action was one of the few ways open to groups to get their voices heard and to bring about change. It was also a very effective way of getting 'bad press' for the LDDC, which given the Corporation's stress on image and hype was another effective weapon." Suttles (1972:63) argues that if community groups are ignored by those in positions of power they sometimes resort, as they did on the Island, to "more militant tactics" to draw attention to their cause even though their opposition might "seem absurd and forlorn". In so doing they can also invoke "widespread sympathy" and this is what happened when the developers announced their intention to build Canary Wharf.

The proposals for the Canary Wharf development, which led to the most vociferous public opposition and formed the pinnacle of protest on the Isle of Dogs, came at the end of a period of mounting anger and frustration and marked the point at which this finally spilled over into a broader base of opposition. People who had previously had little or no involvement in community protest were drawn into the battle. An established resident who said she was "not a kind of banner-waving person. I have my opinions but I'm quiet" decided to go to a public meeting about the Canary Wharf development that left her seething:

> Well, I went to this meeting and he [Reg Ward] was telling us about all the developments that's coming and how wonderful it is for local people. ... We'd read reports, little pieces in the papers, that the people on the Island didn't have their sunday roasts until the development started, things like this — people's backs was getting put up. ... somebody said: "There's gonna be lots of jobs for local people." [Another said] "How can yer say this?" because already *The Telegraph* had moved in and were bringing their own work force. There were hardly any jobs for the local people, so they said "What about educating the children so that they're ready for these jobs." He [Reg Ward] said "The East Enders ... didn't know the meaning of education until the Docklands came." I was raving you know, especially I was quite proud because my daughter she's got into Cambridge and she's come from a comprehensive school. I thought how dare he ... course we can't afford to send [our children] to private education but we still care ... I think that sort of attitude put a lot of people's backs up from the beginning; how dare they come and take over our space and treat us like we're nothing, that's how you feel.

There was something about the sheer scale of Canary Wharf that galvanized the opposition. Brownill (1990:120) says that this development "was a good example of the dilemmas posed to local organizations when faced with major commercial developments". "Initial reactions to the scheme were tempered by two factors. One was the negative experience of past opposition" in other areas where inquiries had been held (for example, over the proposed airport in the Royal docks) and been unsuccessful, as well as the self-evident differences in "power between local groups, the LDDC and a multimillion pound consortium ... As a result at initial presentations the AIC and the Docklands Forum did not express opposition" (Brownill 1990:120). As Peter Wade, former chair of the AIC described:

> Canary Wharf came on the scene via a consortium of Americans and Midland Bank ... They came into the area, we met with them ... Ware Travelstead ... was the guy leading ... We were asking what benefits one could expect if this was actually built and the cost to the community and it was abundantly clear even after the first meeting that ... There was no way that we were going to win.

Figure 4.17   Marchers with flags. © Mike Seaborne/Island History Trust

Even if it appeared that they could not prevent the development, a few certainly did not intend to let it happen without expressing their discontent, and once it became clear nothing was to be extracted from the developers the tack changed. "We didn't want Canary Wharf and we demonstrated that very, very visibly", Peter Wade said. This was something of which the LDDC were fully aware: "They were busy trying to extract from us the best price they could for Canary Wharf", one of the Corporation's executives explained. "What do we get out of it? What funds are you the Corporation and what funds will Canary Wharf pump in to community affairs and that was part of the pressure making it clear to us that if we didn't write them a big cheque they could cause us a lot of difficulty." And that is exactly what they did.

Although the protest still involved a minority of Island residents, the Association of Island Communities was able to draw on a ready supply of disgruntled and angry people as feelings were undoubtedly running high. A demonstration around the Isle of Dogs in which a coffin was carried to symbolize "the death of the community" (see Figure 4.17) if the development went ahead captured the imagination of many Island residents, as one of the participants described:

> There was hundreds and hundreds of people . . . Peter [Wade] was walking in front [in his undertaker's gear] and there was . . . all . . . the parish priests at the front and then the coffin being pulled behind. And then . . . as it went all the black banners and then the big Canary Wharf banners. We was up until twelve and one in the morning planning and planning and

planning . . . it was incredible. It was a feeling of total faith in the AIC by
the community because as we walked along people were following us. . . .
people were lining the streets . . . they were clapping because they thought
it was absolutely marvellous.

A report in the *East London Advertiser* (30 May 1986) with the headline "Islanders
go into mourning" said

The Last Post rang out across West India Dock yesterday as Islanders
mourned the "death" of their community — killed off, they claim, by a
"disease" of developers and property speculators. Dressed in black, a
solemn procession, heads bowed, circuited the Isle of Dogs before staging
a mock "funeral" at Canary Wharf, proposed site of "the final nail in the
coffin" — the £1.5 billion "Wall Street on water" banking development.
The demonstration which coincided with a whistle-stop tour of Docklands
by Prince Charles, was organised by the Association of Island Communities to highlight the twin crises of high unemployment and bad housing which it blames on the policies of London Docklands Development
Corporation.

Although opposition to Canary Wharf was considerable, some, including AIC members themselves, did not participate in the march or approve of it: "All of us [were]
really against it", an Islander said, "I mean they even went so far as to march down
the . . . road with an empty coffin which I didn't agree with. That I thought was a
dreadful thing to do."

Whatever people may have felt about them, it was the work of a few AIC
activists that produced the most embarrassing and in some respects successful
exhibition of opposition to the proposed Canary Wharf development. Peter Wade,
one of the protest's initiators, takes up the story:

We got wind . . . that the Governor of the Bank of England, Sir Robin
Leigh Pemberton, was coming down to Canary Wharf to turn the first sod
and he'd invited bankers from all over the world, there was about 200 of
them, . . . they'd broken a bit of concrete to expose a bit of earth and
they'd put a huge marquee up and champagne and all sorts. . . . they put
all the chairs out in front of the marquee and they've got this lovely big
podium (for the speakers) . . . and all these lovely flowers all the way round
the bottom of the podium . . . They didn't think of security until 7 o'clock
in the morning . . . but what they didn't know was that 11 o'clock the
night before . . . we drove in 40 sheep in the back of our farm lorry and we
let them out into (a warehouse on the wharf) . . . and three beehives full,
I mean, thousands and thousands of bees off the Mudchute and got them
into the shed there . . . The old warehouse . . . overlooked the marquee,
only yards away . . . At 6 o'clock in the morning we shipped in all our
demonstrators up into the warehouse . . . there was massive banners all
produced . . . I got Ted in a boat . . . with a ruddy great tannoy system and
. . . I walked down when everybody was seated and when Leigh Pemberton
started to speak I walked right up to the side of 'im and blew a whistle and

Figure 4.18   Sheep and Suits. © The Art of Change

then all these banners dropped over [proclaiming "Kill the Canary. Save the Island"]; course there was all the chants and there was a record had been made by a group of housewives in Southwark called "Give Us Back Our Land", t'riffic record it was . . . so that was playing and Ted was . . . behind 'em . . . with the speaker. The lorry . . . with all the sheep in and the hives . . . was driven down, went past the marquee; no-one took any notice of it, they was looking at all these people . . . It stopped the other end of the marquee and they got the big tailboard down and all these sheep come running down . . . and . . . charging up the side of the marquee . . . (see Figure 4.18) all of a sudden there's all these sheep going through the chairs and about half a dozen sheep flew over to the podium 'cos they saw all the flowers and start eating all the flowers . . . and Leigh Pemberton started laughing, he couldn't stop hisself . . . he couldn't do anything about it, and Ware Travelstead, I mean, if he'd had a gun he'd of shot me, no doubt about it, I mean the way he was looking at me was unbelievable and . . . Reg Ward, he was the chief executive in those days, I mean he was having kittens 'cos . . . he was the one in charge of setting it up for 'em . . . and while they was all watching the sheep the bee keeper come across . . . put one of these bloody hives right on the bit [where] they . . . got the sort of spade stuck in to turn it over and . . . as he left the hive he kicked it . . . over, and all these hundreds of thousands of bees just went and me included were stung. (Foster 1992:174–5)

"There was uproar", another of the participants explained: "We got these sort of high-up type people, you know, who think they're above it all, all scrambling around like the rest of us" (Foster 1992:175). A report in the *East London Advertiser* said

> One hundred and fifty thousand bees and a flock of sheep gatecrashed a VIP launch of the £1.5 billion Canary Wharf development scheme . . . But behind the smiles of the buzzing, bleating demonstration is a serious message: Islanders will no longer talk with docklands chiefs, but will fight them with a campaign of "civil disobedience" . . . Mr Wade [said] . . . "The demonstration was making a serious point. Our people are trying desperately to find a future in what's happening to the island. And it's abundantly clear that people aren't listening." (*East London Advertiser* 25 July 1986)

When I asked Reg Ward what he felt about these events, he said:

> I was both irritated and totally amused. Harmless point . . . very nicely made in a way. The only worry was actually the bees . . . that was . . . pushing it slightly too far . . . The Americans couldn't understand it at all, they were horrified, taken aback by it. The Governor of the Bank of England handled it beautifully with great aplomb . . . The Board were always . . . apprehensive about Canary Wharf. To those who weren't totally comfortable with it that was sort of another sign that perhaps we were pushing it too hard and too far. So there was a range of emotions. I think from those of us after the shock . . . very much admire Peter Wade and Ted for the idea and how it was actually handled and very amused about it but wished it hadn't happened. The Americans were all . . . very cross and not able really to understand it all or live with it.

When I asked whether the LDDC saw community groups like the Association of Island Communities as a force to be reckoned with, or whether their protests were simply an irritant that did not relate to the business of the Corporation, one of the team said:

> I don't think we were ever particularly impressed. Part of the difficulty was we understood the game and in terms of the impact of the protest on you as an individual when you know the guys that are marching towards you with the bees and . . . you've been in the pub with them the previous evening it's actually quite difficult to take it too seriously in that sense. But on the other hand we recognized them as a force as we knew the amount of support they had from local people and that a lot of what they said was very important and worth taking account of. You go back and trace our programme and you can actually see how, we built that there, we built that

there, we did that because the AIC told us it would be a good idea and when we thought about it we saw they were right.

For the participants the demonstration was a great success:

The week after the demonstration . . . the whole thing collapsed. We like to think, Ted and I . . . that maybe ours was the final cap, you know, with all the money problems and now this massive protest from a community that we've got to build it in, you know, it probably had some little thing to do with it. . . . That particular onslaught was the final cap . . . [There were] financial problems . . . but when [you're] trying to encourage the Bankers of the World, cos that's who was sitting round that marquee, they were from all over the world, it'll tell yer what sort of level it was with Leigh Pemberton as the main bloody speaker, you don't go no higher . . . [it was a] . . . pitch to the world for support. (Peter Wade)

Reg Ward denied that the Canary Wharf protest was significant. "It had no impact or no relevance other than as a token of local feeling", he said. Neither did it lead to any sustained debate about whether development of this kind was appropriate for the area or whether it should happen without a public inquiry. There were organizations like the Docklands Forum and the Docklands Consultative Committee that pressed on these issues, but at a local level they seemed less relevant, perhaps because of the absence of the local authority voice and a sense of powerlessness. After all, development was taking place all around them without their being able to influence it in any way. One former councillor reflecting on the protest said:

The demonstrations against Canary Wharf were sort of half a dozen people, I mean if you compare that to . . . the Channel Tunnel rail route and other big developments the people just weren't [involved]. You know, they felt powerless . . . If the debate had been started I'm sure people would have very quickly realized what we could have done, we could have built so much more, we could have had a mixture of office and light engineering and different sorts of jobs rather than this emphasis on banking and finance which is a very limited appeal . . . not everyone's gonna be a stockbroker, you know, there are people who want to do light engineering and want to work in service industries.

But this was not a debate in which the local politicians in the area seemed to be actively engaged either.

If the Corporation did not view the Canary Wharf protest as significant, Olympia and York, who eventually developed the area, certainly did, as one of their executives explained:

You have to [develop] . . . in a way that's sensitive to what's going on in the community . . . And the people that we had replaced on the project had

had some difficulty in the local community . . . There were some instances that had occurred where the local community felt that their interests were being ignored, so that one of the things we said to ourselves was that we were gonna have to keep the lines of communication into the local community open if we were not going to have difficulty. (Foster 1992:175)

Other businesses too started to express concerns about the neglect of local people, as this businessman explained:

The LDDC concentrated at least for the first five or six years of its life on the developer and the infrastructure . . . but all the people who lived round all they were getting offered was a building site and nothing to come in the future so there was a lot of aggro, a lot of ill feeling. Of course when the companies started to come down there was still no work because they were bringing their staff down with them . . . we were very wary of this . . . We felt that it wasn't doing our company or other companies or even the community a favour. This wall seemed to be coming up, so we started a programme of getting actively involved in things such as . . . training . . . Then we started to think what we can do to show Joe Public; there's all these people still live round the outside, they don't come in . . . to the middle here because it's the docks. So we started the Island Fun day . . . with the prime aim to get the community in.

A time for reflection

. . . we must consider the social consequences of always pushing the poor away from where the money is made and spent. If developers don't reflect on the implications of that, they may not realize their role in crystallizing the distinction between wealth and poverty which one sees in all modern cities. (Richard MacCormac, former president of RIBA 1993:23)

The shaky economic foundations that had produced the boom and led to the frenzy of development was inevitably followed by slump. This time the area went from boom to bust in just four short years owing to the type of investment and volatility of the economy. It was "not a natural evolutionary market", a local estate agent observed; "it rose up on the crest of a wave so it went . . . crashing down much faster than anywhere else." "In [19]87 this area went down like a stone long before the UK market started slowing down", another estate agent explained, because Docklands lacked the adequate infrastructure to sustain the type of development taking place there. "The City boys who were buying as investments . . . pulled out suddenly" and the prices that had so rapidly and inexorably risen fell dramatically.

"In the sobering aftermath of the boom, it has become clear that long-term profitability and growth were injured by the failure to rein in the speculators", Fainstein argued, and

Even though economic development was the justification for the pattern of investment that occurred, no one was responsible for calculating aggregate growth targets. Public officials frequently assert that such projections are outside their realm of responsibility and that it is up to private business to foresee demand and shape supply accordingly. Since public funds and deregulation underpinned redevelopment activity, however, the refusal of government to play such a role represented an irresponsible squandering of public resources. The unwillingness to plan comprehensively meant that too much space for the same kind of use was built on too large a scale, while there was insufficient production of needed housing, public services and infrastructure. (Fainstein 1994:237)

Although the consequences of the financial crash in 1987 were serious particularly in the short term, plunging the UK economy into crisis and recession, there were many who felt that it was a blessing in disguise, certainly in relation to the residential housing market in Docklands, that might help to create a better "balance" in the area. "I think that what's happened in the residential market has helped the area tremendously", a commercial property agent said:

I know that might sound perverse . . . but when we opened there were a lot of residential development under construction which was being bought by people not with a view to living in them but with a view to selling them off to somebody else and making a profit in the interim because prices were rising at such a level then that they could buy them off a plan, then sell them when they were ready and make ten to fifteen or twenty thousand pounds in the interim . . . That wasn't creating any occupancy in those dwellings it was just using them as a means to an end . . . [Following the crash, speculation] went overnight and it meant that a lot of that residential market became occupied, which is exactly what it was there to do [and that] helps build up the community.

A house builder shared this view, arguing that some of his colleagues "blame the Docklands markets" for their failure to sell. His response:

You look at what you designed, would you buy it? . . . They got carried along. So did I. I got carried along by the euphoria. We own nine acres of land on the Isle of Dogs on the River Thames, it's still virgin site, we've got the consent to build on it but at the present time we can't afford to sign up for it. I think, because of what's happened in the economy, I think that will balance itself out. The piece of land we bought . . . we paid too much money for it today . . . by the time we've finished that I think I shall get back what I paid for it but I'll never get back the carrying costs and that is quite a large sum of money. We'll make a profit but it'll be a long time.

Another developer said that the problems the speculation had caused and the developers' response to it would now be addressed:

There will be more and more houses built, in my view, on the Isle of Dogs because the market is now dictating that houses are what the public generally wants. They want houses with gardens, real people want houses with gardens . . . "Real" buyers are couples that get together, have children and raise families and then the children leave home and then they get old and . . . so it goes on, so I think there was enough property built to satisfy the market, the flats, and one is now looking more and more at houses.

Did this market suddenly emerge? Surely the need for family housing was there at the outset. The cynical view might be that developers were motivated more by profit than other considerations. When I asked whether builders would have felt the same pressure to provide family housing for "real buyers" if the market had not slumped, a developer said:

> Well it would have been far less acute for people to consider the build-
> ing of houses with gardens, and certainly Reg Ward and I can remember
> debating at length during the early eighties about the need to put into
> the better parts of Docklands higher quality houses with gardens and I
> recall very clearly . . . going to Reg . . . with a new house range which was
> immensely sophisticated, immensely sexy in all respects, and he was saying
> got to find you a site to build five or six of those on where we can get, you
> know, four-bed, two-bath executive homes which of course nobody built
> and it never came to fruition at all.

In the commercial sector, the experience of intense speculation left a bad taste for both critics and advocates of Docklands who had witnessed the worst excesses of unbridled capitalism. As Michael Barraclough, the initiator of the self-build housing programme on the Isle of Dogs said, "A unique opportunity for creating a fantastic city of the future" was "squandered in a sort of cheap grab" even though, as a developer said, "most of the greed spawned mediocrity".

The dream discredited

> Developing Docklands as an office center was not wrong. The availability
> of a huge tract of mainly vacant land in the heart of the London metro-
> politan area represented an enormous asset, and one that rightfully should
> have been developed for the benefit of all Londoners, not just the small
> number of nearby residents. It was the government's refusal to bolster the
> enterprise with major social and educational programmes, to halt compet-
> itive commercial development elsewhere, and to construct infrastructure
> in advance of development that created the ultimate debacle. (Fainstein
> 1994:238)

It was not simply economic fortunes that changed Docklands' image. The Corporation became discredited not only because of the policies they pursued but the *way* they pursued them. Although they were limited by their remit, and as Reg Ward argued, they did attempt to find ways round the narrow definitions imposed

Figure 4.19   Worlds apart, Janet Foster

upon them by central government, very little was done to compensate or to seek to protect the locality from the vagaries of market forces. Had they given the same effort to this that they did to attracting developers, the picture might have been different. The needs of those further down the social ladder were modest enough: they wanted housing for rent, or affordable private housing, and the refurbishment of existing council accommodation that was very run-down, a stark contrast to the new private residential developments (see Figure 4.19). Most of all they wanted jobs. All they got were promises of a better future (which in themselves might prove to be illusory). "Let's face it, the people on the Island have been sold down the drain all over these years", a businessman said:

> It was virtually wiped out during the war . . . then the tower blocks and the council estates were built. They were promised that they would be looked after, there would be a great tomorra . . . Docks started to run down they were promised yeah it's gonna be great and what happens? It was terrible. And what happened? It stagnated. Then the developers came in: "It's all right, don't mind living on a building site, it'll be great in a few years' time." It didn't happen. They naturally got a bit [upset].

Although it was self-evident that local people and the Corporation had different aspirations, in view of the enormity of the revenues the LDDC reaped from land sales, and the sums that developers pocketed, relatively small amounts, for example on refurbishing council accommodation and providing central heating, would at least have provided some improvement in people's quality of life and made them feel that in a small way they had benefited.

Those responsible for implementing change could rationalize their actions in the belief that they were operating in the interests of a wider vision and that their remit limited their ability to act in these areas. But meanwhile the poor, the powerless and the relatively undereducated were confronted with development on an unprecedented scale taking place at a pace unparalleled elsewhere without any detailed discussion of the consequences and no adequate means to fight it. The actions of a handful of activists drew attention to the contrasting worlds of rich and poor, of the plush office developments and apartment blocks alongside families with no employment living in cramped and sometimes crumbling council accommodation. The media brought the injustices of the development to a wider audience and by so doing influenced government and the business community to take remedial action.

Docklands' image became tarnished, and some went as far as to suggest that it "stank". "I have struggled . . . to get the City to fund my next project", a builder said:

> They don't want to know. As soon as they know it's Docklands — forget it. . . . They just don't think it's going to work. They say the infrastructure's not there . . . "There's always road works". I say, "Well if the gas board's got the road up outside Harrods there's a traffic jam, so what's the difference." Yet they still fight shy of it . . . they read the paper . . . I think the press have been very very unfair to Docklands; it's been a marvellous piece of enterprise with a lot of hard work by a few people to rebuild Docklands and they criticize.

While the media were undoubtedly used by the opposition to undermine the Corporation and by politicians to score points against the government, they had a lot of ammunition. "Docklands was bound to become a political football", an LDDC report said:

> The scale of redevelopment combined with its location so near the heart of the capital city and the media were bound to keep it in the spotlight. In addition, there was an in built conflict between Conservative intentions to inject private capital, attitudes and practice into the area and its historic continuing connections with the Labour party, the unions and its nostalgia for the past in a world which had moved on. With so many pressures, so much activity and the need for speed to retain the impetus of change, the LDDC had little hope of keeping everyone alongside. The concept of partnership had yet to seem obvious and activists consistently did their best to undermine the LDDC. Docklands was the largest redevelopment area in Europe and attracted worldwide admiration. Yet still the criticism mounted. To the government it also seemed extraordinary to be spending so much money and getting so little, if any, credit. (LDDC 1998:20)

Although many could not fail to be impressed by the breathtaking speed of development, others from a variety of different backgrounds began to feel that a more measured approach was required and highlighted that, in the reconstruction of a major part of London, haste, and for a short while developers' greed, should not have been the primary factor. "Let's get this one straight. . . . We're not talking about what we're going to have for lunch," a former councillor said:

> we're talking about something that's gonna have an impact on London in fifty years. We're still going to be living with the consequences of what's happened on the Isle of Dogs in a hundred years' time. Our great grandchildren are going to be looking at that Canary Wharf tower . . . How long did it take to build St Paul's Cathedral? It wasn't shot up over night and people are still walking round looking at it. Does it have to be instant? These people who put forward this premise it all has to be done now so somebody can get their profit today is what we're talking about . . . Is that the only thing that decides what's gonna happen to London?

In the unseemly "grab" little time or thought was given to asking what the role of regeneration in the area ought to be, and although the partnership of the private and public sector was the right one, the public aspect of the development, for a variety of practical and political reasons, had not been paid sufficient attention. "There is no doubt that the Islanders have suffered and are suffering", a businessman said "in eight short years their whole world has been changed before them." Outside the commercial development, part of that change was the influx of new affluent residents some of whom, as I describe in the next chapter, came to sympathize with the more established Island residents in a way that they could never have envisaged before they arrived.

The pressures for change from developers, the media, the "new" residents and their more established counterparts became overwhelming. A House of Commons Employment Committee, while praising the Corporation for its success in physical regeneration, said what was manifestly evident: "It is not good for the health of a community for the original inhabitants of an area to see others benefiting, as they see it, at their expense while they suffer from increased road traffic congestion, higher house prices and associated ills. Nor is it just." (House of Common Employment Committee, quoted in LDDC 1998:20)

It was time for a change. Reg Ward left the Corporation in November 1987, leaving behind him a Docklands that, whether one liked it or not, was certainly transformed. Major General Rougier replaced him, but only for a matter of weeks before resigning. Michael Honey, appointed in 1988 came into the Corporation with a different development philosophy and a commitment to social regeneration, reflecting a softening political mood at the time about the merits of development that was less socially divisive. However, as I describe in Chapter 6, he was not in for an easy ride either.

Chapter Five

Different worlds

An area with potential

Long before Docklands was seen as *the* place to be and a decade before the LDDC came into existence, a newspaper article reported that

> An East End construction company is building a block of 11 town houses on the Isle of Dogs — and the cheapest is £20,000 . . . Now it is the "in" thing to live near the river — and the agents handling the sale of the houses are confident they will have no difficulty selling them . . . The scheme has met with a cool reception from Island Councillor Ted Johns who said "The frightening thing is that this is just the start of what is bound to happen in the future . . . The riverside is gradually being taken away from the East End and this must be realised before it's too late." (*East London Advertiser* 29 January, 1971)

Ted Johns' comments were certainly prophetic, but it took 15 years before they were finally realized. "We sold our house" in the suburbs and "went to one of the first owner occupied houses to be built", a man still living on the Isle of Dogs in the 1990s said. "The word 'yuppie' hadn't been invented then." [Although] house agents were busy . . . persuading people to go the Isle of Dogs, nothing happened . . . it didn't take off at all". "It was too early for the area", an estate agent explained. "There was plenty the other side of London to be done . . . so you didn't need this large lump of land on the East . . . there was no infrastructure, no nothing." "People bought then . . . because they liked the area." A view exemplified by a newcomer who moved to the Island in the seventies attracted by "the nature of the people", "the community" and "sense of belonging". "The whole quality of the East End cockney is gutsy", he said.

In the 1980s the Isle of Dogs, now part of *Docklands*, and with affluent housing an integral part of the redevelopment plan, had an entirely different image that had little to do with embracing its East End roots or aspects of traditional working class life. "Docklands is a realm of fantasy that was created", a journalist who moved to the Island during the 1980s said. "This was *the* place where it was really gonna happen. This was *the* place where the streets are gonna be paved with gold and everybody drives a BMW or a Porsche and they work in the City and very very few people [do] of course." But "Docklands for a lot of people in that period I think crystallized some of those fantasies . . . Look at the sort of dwellings that have been built and what they symbolize . . . It was all fantasy architecture that was aiming to project this sort of lifestyle of gay young things who lived as if there was no tomorrow."

Those who moved to this new "fantasy" Docklands were called yuppies: a term "coined . . . in 1983 to refer to those young upwardly mobile professionals of the baby-boom generation" who "apart from age, upward mobility and an urban domicile" were "distinguished by a lifestyle devoted to personal careers and individualistic consumption" (Smith 1987:151). The beneficiaries of "the rise of non-manual and especially managerial and professional categories of employment" concentrated in London, these young professionals became "a powerful model" and "peg for advertising campaigns" (Short 1989:175). "To the popular press . . . which generally extols the virtues of gentrifying urban 'pioneers', the link between the two icons — yuppies and gentrification" was "irresistible" (Smith 1987:151).

Despite the imagery of plush and expensive developments occupied by rich, carefree and insensitive professionals living life in the fast lane, behind the superficialities and simplicity of the media moguls' images of the new Docklands there were more complex and conflicting processes in which the affluent newcomers were not always as "alien" to the East End as they might at first have appeared and, particularly at the outset, had more modest incomes and came from a diverse range of backgrounds. "You meet so many people from so many different backgrounds", a newcomer attracted by the development said. "People from all over the country are living here and it's fresh, it's nice . . . and of course you've got the indigenous population as well."

"It's not a stereotyped area", an estate agent explained. "There's never been a defined purchaser in Docklands. It's always been a complete mixed bag." Yet for a brief period during the 1980s Docklands and yuppies were synonymous: "We had some Germans . . . looking for 'zer yuppies', could I show them any yuppies!", a resident on a self-build housing development recalled. "I said 'There's no real yuppies round here' . . . and I just pointed them in the direction of . . . Road. . . . People talk about the 'yuppies' coming to the Isle of Dogs all the time and they came to find them!" On the estate to which the tourists were directed, a resident said: "All this yuppy nonsense. Yuppies didn't come to this estate . . . it wasn't considered a smart estate . . . [it] did not attract . . . *the yuppy brigade*. By that I mean . . . the City slicker, you know, living off his nerves."

In 1987, at the height of the property boom, an Island survey suggested that "29 per cent of all residents were born on the Isle of Dogs . . . and another 40 per cent came from elsewhere in East London" (Wallman et al. 1987:i); even the remaining third did not comprise affluent newcomers alone but included some of the poorer sections of the ethnic minority communities also living on the Isle of Dogs (see Table 5.1), which demonstrates just how distorted the focus on the "yuppies" as defining the Isle of Dogs really was. Nevertheless, the same survey revealed that in terms of length of residence, while 23 per cent had "always lived on the Island", and 30 per cent had lived there for more than ten years (i.e. before the development), the changing population profile was rapid, with 11 per cent having lived in the area for "less than a year" and 19 per cent who had lived there "for 1 to 5 years" (Wallman et al. 1987:i, 26, see Table 5.2).

Table 5.1 Where individuals were born

Place of birth	Number	Percentage of total
Isle of Dogs	1,731	29.1
Other Docklands	256	4.3
Other East London	2,174	36.5
Elsewhere in London	603	10.1
Elsewhere in England	473	7.9
Elsewhere in Britain	242	4.1
Continental Europe	56	0.9
Asia	195	3.3
Africa	79	1.3
Caribbean	57	1.0
North America	16	0.3
Other	75	1.3
Total	5,957	100.0

Source: Wallman et al. (1987).

Table 5.2 How long respondents have lived on the Isle of Dogs

Length of residence	Number	Percentage of total
Less than 1 year	244	10.8
more than 1 less than 5 years	439	19.4
5 to 10 years	384	17.0
More than 10 years	682	30.2
Always lived here	512	22.6
Total	2,261	100.0

Source: Wallman et al. (1987).

Making contact

The relative "newness" and lifestyles of affluent resident newcomers made them a difficult group to access. My initial contacts were made with those who were already linked with the Association of Island Communities, the churches and other organizations on the Island, and they in turn introduced me to their neighbours and friends on their own and other private developments around the Island. Business executives I interviewed in their professional capacity sometimes lived on the Island and were also helpful in providing contacts, but all of these networks were limited and rapidly exhausted so I had small clusters of at best three or four individuals on any particular development. None of these people corresponded with the images portrayed in the media. I mention this because at the outset I, too, seem to have been steeped in the imagery about Docklands and "yuppies" and perhaps this is why it never failed to surprise me how insightful, and often sensitive, most of those I interviewed were. "This stupid media mania for yuppie land", one newcomer said: "Whether yuppie land is further over into the Royal docks area I have no idea but I can assure you that on this Island I haven't come across anybody from the Isle of Dogs who's ever found the yuppie land, it doesn't exist."

I did interview a handful of people who fitted some part of the stereotype (although not all of them were young), working in the city and living life in the fast lane, whose attitudes were different, less insightful, less caring and sometimes, from my perspective, insensitive. Others I was told about, or who were suggested to me, never responded to my letters and calls. Consequently, I have little way of knowing how representative were those I interviewed. But in one sense this is unimportant because, whether they were representative or not, what the following account illustrates is that outside the media spotlight there was another side to affluent settlement on the Isle of Dogs in which people who moved there excited by the development and, as it gained momentum, the prospect of profit came, sometimes much to their surprise, to like the area and the established residents and to develop an appreciation of their struggles too.

Given its derogatory overtones, few newcomers described themselves as "yuppies". One said it was "an expression I loathe". "It annoyed me so much", another explained. "If I say to somebody 'I live on the Isle of Dogs' they say 'Oh you're a yuppie' and there's not really much point in countering that because if people want to be narrow-minded and accept what's been fed to them there's nothing you can do about it. It actually hurt. It hurts me. . . . I mean [such] gross generalizations." There are, she continued, "enormous differences not just from people broadly speaking who live on the other side of the road [in the council housing], [but also] people from this side of the road as well; they're all terribly different there's no homogeneity at all", but "the resentments have been so heightened" that "they just refuse to see us as, like, normal people. I mean I don't have any pretensions, pretty normal . . . In any other area I am a normal woman staying at home with her children. I just don't understand it. I think they actually have a false view of how much money everybody has."

"This is by no means yuppiedom, I mean you could hardly class it as that", a woman living on a very modest private development said, "but because it's like a new estate [the kids] . . . shout a bit of abuse at you down the street . . . The joke is they shout it to [my neighbour] who was born and bred in the East End!" Even established Island residents who built their own homes or were fortunate to move into new developments found themselves branded "yuppies", "which is not very nice", one explained "it's sort of like an insult really that you only get that sort of thing if you're a yuppie not if you're an Islander." "People . . . say 'You live in yuppieland'", another said, "They forget you've lived here forever."

"We call 'em yuppies because it's an easy thing to say", an Islander explained but, as another pointed out, defining exactly what constituted a "yuppie" was highly problematic:

> When they talk about yuppies I have a hard job to find out who they
> mean . . . I thought to meself well you know really if they judged people by
> how they dressed, you just look at my son, he works for the Bank of
> England as a computer programmer. Now he's an Islander born and bred
> but with his suit and his case and his car anybody without any knowledge

would say "Well he's a yuppie". So I think it's just a name tag. . . . I think if your approach is right I mean . . . they're people aren't they?

Given the problems of both perception and definition I have used the term "newcomer" to describe the affluent residents who moved to the Isle of Dogs in the 1980s and early 1990s. This chapter focuses on their attitudes towards the development and perceptions of existing residents. In many respects their relative affluence, expectations, experiences and lifestyles contrasted significantly with the predominantly poor working class population on the Island and in these respects they did indeed occupy *different* worlds. Yet, in spite of their differences, some newcomers came to identify strongly with the area and, as aspects of the commercial development threatened their quality of life too, joined with more established Island residents to fight a common "enemy".

A leap into the unknown

I once said to [a friend on a new development] "You have such a strange mix of neighbours . . . they're really Bohemian." . . . She said "Well we had to be. You had to be a bit strange if you like to move to a place that was so new and had such a bad reputation in the past. You didn't really know what you were moving to." . . . I suppose I felt a bit like that as well . . . [People used to say] "What do yer wanna move to the East End for . . . what are you doing down there?" . . . There's the old stigma attached to it. (Newcomer who moved onto the Isle of Dogs)

The newcomers who moved to the Isle of Dogs in the early stages of the residential development perceived themselves, rather like their business and LDDC contemporaries, as "pioneers" and experienced a similar sense of excitement about being part of a dynamic process of change. "There was a great pioneering spirit", a woman who moved there in 1985 explained: "Everybody felt that they were in on the beginning of something really very special" and "It was the sort of thing you can say to your kids in ten years' time, I was there right at the beginning." "I loved it from the moment I moved in", said another despite living on a "building site". There was a feeling "of great optimism" a woman who returned to Britain after working abroad said:

It was forging ahead. I'd been in Liverpool where my parents were living and all the companies were shutting down and making people redundant. We were getting up to three million unemployed in Britain and it really seemed doom and gloom. But then when I got here it all seemed to be moving forward and there was great promise ahead and they were going to bring it alive as far as the residential areas [were] concerned and this talk of a big financial centre, Canary Wharf. So I came looking forward to all that and was prepared to put up with the pile driving because it was progress and it was wonderful, it was uplifting and I felt lucky to be here at the beginning of all that development. I was very excited about everything that was happening.

"I was really really excited", another newcomer enthused:

> It was obvious to me that there was only gonna be one place and that's the Isle of Dogs . . . I saw drawings and pictures, artists impressions of what Canary Wharf was gonna look like . . . I loved it. I thought well this has gotta be it. I mean for God's sake where will you ever be where you've got all this wonderful open space, all this water, and . . . they're gonna build the City, they're gonna build Oxford Street, Bond Street everything is gonna be on 72 acres half a mile from where I live . . . I'm absolutely smacked out by it all. Every time I drive through it I think it's wonderful.

"It's brilliant", a property agent and resident agreed: "I think actually it's unique. Here you're gonna have this massive great office complex but where you can still live quite normally pretty close by. Almost countrified down some of these areas you know, you can hear the birds singing and it really feels quite quaint and to think you're fifteen minutes' walk from one of the biggest developments in the world. That really is something . . . These are massive changes from an area which previously you wouldn't have ventured into unless you were in a tank — it's phenomenal." "I was a bit tired of staid Victorian middle class suburbs", another newcomer said: "I thought it represented the future rather than the past and I was excited by that . . . [and] the idea of being near water in a new environment appealed to me."

Negative images of the East End seemed to evaporate in the excitement, as this newcomer described: "In a way I was sort of tunnel visioned . . . I so desperately wanted the job, I so desperately wanted the house. I saw the rest of the estate . . . saw the new offices in Millharbour and I thought this is so exciting. . . . I really want this." It was only that later "I thought what are you doing? You're moving into the heart of the East End, you've never lived in London before, you're not street-wise to the ways, you know, and it was a bit scary, but then again . . . I just thought it was all very exciting really."

Others needed more convincing, as a woman born in the East End recalled:

> We came over 'ere the very first time . . . years ago to a wedding . . . . I've never seen anythink like it. I was terrified . . . I said to 'im "I don't like it, I really don't, I can imagine this wedding's gonna end up with a big punch up and we're all gonna be shot and I really don't wanna come 'ere" . . . Actually it was one of the best weddin's I've been to . . . And then when he said . . . "Remember when we went to that wedding . . . ?" and I said "Yeah". He said: "There's all new developments going up over there shall we go over and have a look?" I went: "Not really no it's not the area I'd like to sort of go out on a Sunday drive" . . . We came over to see it and I said: "Oh it's really changed you know from what I remember", the wall was down, new flats up, . . . it had changed ever such a lot . . . and he sort of took over after that, he wanted to move. . . . I didn't know anything about the Island at all . . . so we moved on 'ere . . . completely blind.

The "newness" of the place (in terms of affluent settlement), the lack of facilities and uncertainty about the local welcome contributed to "camaraderie". As this newcomer explained: "It was very nice because it was not [like] moving into a very established area . . . If I'd gone to anywhere else in London it might not have been so openly friendly . . . I came very early [1985] . . . they sort of built houses, sold them, and with that money built two more. So people moved in very slowly . . . Because we were all new and arriving here you got to know everybody who was coming in." "I think that's one of the advantages of moving on to a new estate", another resident on a different development said. "You do get to know quite a lot of people very quickly." "I suppose of . . . the first twenty or thirty people that moved in we all got very very friendly very quickly." "We all had the same sort of problems getting gardens going, fending off the developer and getting him to do what he ought to be doing. . . . So we all banded together . . . common causes, to get this estate looking as nice as possible, as quickly as possible." "In a way that was a temporary friendship thing", another observed; a "couple of years later . . . people were actually getting settled in their own lives so didn't need to have everybody being involved."

Gans' (1967) study of an American suburban housing development had many similarities with the accounts of new Island residents who moved in early on in the residential development. Good community and neighbourly relations, Gans suggested, were founded on several factors: that the "new population must be culturally and emotionally 'open' [and] receptive to interaction with strangers"; with "sufficient social and cultural homogeneity . . . to enable strangers to talk to each other"; and that "they must face a common situation that places them all in the same boat" (p. 142). In another study Gans (1961:37) also highlighted the importance of neighbours being perceived to be "just like us". All of these factors characterized new, affluent settlement on the Isle of Dogs, and as I describe later, shared experiences, though initially focused on the individual residential developments themselves, for some newcomers very quickly extended beyond them and into the politics and conflicts of developing the Isle of Dogs itself.

Making a "killing"

The rapidity of Docklands' "success" in private residential housing in conjunction with external economic circumstances led to intense speculation that very quickly changed the character of the housing market. "Suddenly all the prices started going up", an early newcomer said, "and everybody was thinking oh gosh I've done a jolly good thing here." But those who moved in later were drawn to Docklands by the desire, as one newcomer put it, "to make a killing and move off". "I knew nothing, absolutely nothing about the development, or the place, or what was happening", a young professional man who bought a flat during the boom explained:

> I was playing football on a Sunday morning . . . and the opposition skipper was an old friend of mine and just before kick off he said to me . . . "You've got to come up to the Isle of Dogs. They're building some fantastic

flats . . . I'm working on a development that's going to make a fortune"
. . . At that time this place was heavily demanded . . . anything that was
built was being sold before it was completed . . . He managed to get the
sales woman . . . to put me on the top of the list because I think he was
having an affair with her. . . . I rang [her] up . . . [she] said ". . . You've got
to put . . . down a deposit . . . if you don't do that the place will be gone.
So I came up here. I had literally ten minutes to make a decision, wrote a
cheque there and then. And that's how come I moved to the Isle of Dogs.
And I never came back here . . . I mean I never revisited this place until
the day I picked up the keys which was something like nine months later.

Others described a similar story of buying on "impulse". "I'd never heard of London Docklands", a woman who got a job in the area explained. "We ran round the
area very fast cos I only had three days." Another bought her luxury flat on "the
toss of a coin" and it had "been on the market ever since . . . [at] a fixed price".
"It's not negotiable", she said, "and when it reaches that price I'll sell it and if it
doesn't I won't. . . . My neighbour below did the same thing. . . . He wants £30,000
profit, when he makes it he'll cash it. Until then he'll keep it. I dare say a lot
of the company-bought ones will do the same." A banker based his choice of the
Isle of Dogs on a belief that "Docklands was an area where you would be reasonably certain to outperform the market as a whole" (see Figures 5.1 and 5.2). "It was
virgin territory", he continued, and "because of the build-up situation [changing
working practices on the London Stock Exchange and financial derugulation]
and . . . the ridiculously long hours that the City works . . . I wanted to be over the
shop." In a luxury development with almost two hundred flats, a resident said: "An
awful lot of these apartments were sold speculatively for investment. Quite a few
were bought by companies. Quite a lot were bought specifically to site, wait for
Canary Wharf to be finished and sell, basically like myself." A significant proportion
of this investment came from overseas: "We seem to latch on to Docklands much
more than the English do", one such investor said.

 Given the motivations and lifestyle of these particular newcomers, few expressed
any interest in developing links either with new or established Island residents,
something a banker described as "a matter of complete indifference". When I
asked if he had met others living in the development when he moved in, he
replied:

No. Why? First of all if you go back nearly three years it was the top of
the speculators boom — hardly any of these properties were occupied . . .
There's still a high proportion of properties that are not occupied . . . The
social interplay is very limited: (a) the functional fact that half the place
is speculative, half of its empty and (b) the working hours . . . so your
chances of social interaction are low. The only time you ever meet people
is basically if some person blocks your car parking access . . . My biggest
interface with the people in this block of flats has been a running battle to
stop people dumping their rubbish in bin liners which means that the
store where we put our sacks was getting smelly and unpleasant . . . social
interaction is minimal . . . This was designed, and I suspect that I'm not
atypical in that sense, as a *pied à terre*.

Figure 5.1   Canary Wharf from a new development, Janet Foster

Figure 5.2   Gardens on a private development, Janet Foster

"There's absolutely no communication . . . at all", said a newcomer on a different development who had lived abroad for many years before moving to the Island, "but then . . . I feel as though I'm obeying the rules . . . the rule is you keep yourself to yourself, you don't mix and mingle." This was a source of considerable disappointment for a woman who felt so positive when she initially moved to one of the first developments:

> I thought [this development] will be a nice little community. But I found [we were] worse off because it was selfish young people moving in and they weren't very, not that they weren't interested perhaps — they just didn't have the time they were leaving at six-thirty to seven in the morning for their jobs in the City and coming back ten to ten-thirty at night. . . . I thought it was shame that there wasn't more of a cross section, older people, children and things. That's gradually come now, which makes it a more normal place to live.

This was not what an occupant of one of the luxury developments wanted: "I assumed that being here my immediate neighbours would be business-type people and not so many kids and families — that I'd dread."

The contrasting views of these two newcomers reflected an estate agent's experience. He believed that Docklands attracted some "people who don't mind the mix" but on the other hand "more people who've got so much money they'll never have to mix . . . they won't actually come into contact . . . they've got enough

Figure 5.3   This is a private estate, Janet Foster

money to keep them separate. They can go to all the posh restaurants and all the posh bars and all they do is get in their posh cars . . . they don't actually [mix]."

The private river and dockside developments were frequently perceived as fortresses by established residents. "I don't think they're ever going to mix with us, I just don't", an Islander said: "They are all their own little communities aren't they?" "If you think about it they go to work everyday they come back . . . what local do they see?", a newcomer said. "Their lives don't cross, an inevitable consequence", he felt, "of the way the development's been structured". (See Figure 5.3)

As the housing boom was very short lived some people who bought as investments found themselves trapped in unsuitable housing in an area they did not particularly like. A newcomer who hoped the profit from their property would enable them to move back to an expensive suburban area in South London said: "It was like the dream had fallen through. It was just becoming a nightmare." While a man who summed up his motivations for moving to the area as "profit and convenience" said:

> There was no point effectively in paying about £25–30,000 [extra] for a river view which I didn't anticipate using (a) because I wouldn't be here at weekends and (b) one was getting up at six o'clock in the morning and getting back at seven or eight at night you didn't see the damn river. Now the way things have broken subsequently I wish I had a river view and I wish I wasn't living here . . . It's a god awful place to live.

No Marks & Spencer

> Most people won't look anywhere where there isn't a cinema or a Marks &
> Spencer. [These are] the two most crucial [things] people look for when
> they move into an area. (A Docklands estate agent)

Whether they saw themselves as pioneers or profiteers, the lack of infrastructure
and facilities that hampered development in the 1970s caused considerable problems for newcomers well into the redevelopment. "It was a very dead place to be"
at the beginning, one recalled: "There was no wine bar round the corner, the pubs
are pretty grotty . . . there wasn't anywhere to just pop out to unless you wanted to
go really to the West End." Some "suffered an awful burden by coming here very
early", another said, although they "got the pick of positions . . . [and] the price
was right . . . you suffered in as much as you had four, five years of dirt and filth
and dust like you cannot believe [and] the infrastructure obviously was totally
inadequate." For many years, there were a few local shops and a single supermarket built in the early 1980s, described by one newcomer as a place with "no nice
twiddly bits there . . . Ask them for stem ginger and they go '*what?*'"

   "We don't have the basics for a nice area to happen", a resident explained
during the fieldwork (1990–1992), highlighting the absence of high-street chains
like Boots, W.H. Smith, and of course Marks & Spencer. There is "not the same
choices as you might have somewhere else", said another, a view exemplified by
the comments of this newcomer who said: "You did have to stop and think now I
need a present for someone, I need an item of clothing whatever, I'll make a trip
to the West End on Saturday . . . even decent birthday cards . . . You had to look a
few weeks ahead. Do I need anything from Body Shop over the next month or so,
let's get it while I'm here because there's nowhere else to get anything from here."
Mobility was the key to overcoming these difficulties, going to the shops in Surrey
Quays, across the river, and to the West End. Most established residents went to
Chrisp Street Market, in Poplar, on the bus and, after it was opened, the supermarket on the Isle of Dogs.

   "There are terrific drawbacks" living on the Isle of Dogs, a woman who moved
to the area said:

> There's facets of normal living which just aren't here yet . . . It's very chicken
> and egg the whole situation — nobody will come down here until there's
> enough population to support the turnover; on the other hand we're
> trying to get this poor population to come here before they've got any-
> thing to come to. [Newcomers are] sitting here like I am saying we know
> it's going to come we've just got to be patient.

   Given the industrial nature of the Island's past and the emphasis of the development on buildings rather than landscaping, many also found the physical environment poor. There is, "basically not a single blade of grass or planting anywhere",
a banker in the City said. "Now my bitch and my beef is that people who [are]

coming into this area . . . are going to be essentially middle class. They're not people in council estates who either don't care or think that the whole system is against them . . . But you know I was used to having a garden and some green; the place is just going to be a concrete desert."

"Everything's happened so quick that they haven't kept up with the essential things like schools and shops and the roads", another newcomer said, and some quickly discovered what the consequences of this, alongside the area's previously poor and working class character, really meant. Services like health and education did not correspond with their expectations and were sometimes inadequate and overburdened. A woman who moved to a new development with a baby just a few days old found that existing overstretched systems did not cater for newly arrived and more affluent residents on incomplete developments (where addresses were not even included in the catchment areas for services). She was left feeling neglected and resentful: "It was assumed we were the mean go-getters living in the new house in Docklands and the reactionary way that I was treated was almost taking on board that Thatcher philosophy of every man for himself. . . . It was all pretty miserable." Not surprisingly, her excitement quickly evaporated: "I just thought, what have I done?" Another newcomer went to register with a GP during her pregnancy:

> The local one's just down the road . . . I couldn't believe the filth and I lost my cool and I told that man "You call yourself a doctor, what's this?" . . . He was just so shocked because he's never expected anyone to walk into his clinic and speak to him in such a tone but I had to say it . . . The place is so dull and dreary and was so dirty and this is the place where sick people go . . . I've got a choice most of them don't and it's so sad.

Newcomers with young children often had serious reservations about educating them locally and many envisaged moving before their children were school age: "I wouldn't send [my daughter] to a local school", one said. "I'm not sure if I'm just not brave enough or if I'm too snooty or what. It's not the people, it's the standard." A newcomer committed to educating her children in the state system, "to fight from inside" as she put it, taught at one of the Island primary schools for a year and found herself "teaching nine-year-olds who couldn't read . . . I feel very sorry for some of the children who are at the school because they're being given such a lousy deal really, both from their home backgrounds and at school". Despite this, she felt it was "incredibly important" for her son "to have a good social mix". Another woman who had opted for a local school was frank about the problems there at the outset:

> The children were really cheeky; mums used to go up and pick on the teachers and have fights in the classroom with 'em. I mean the mums over 'ere they're quite hard . . . They've had to be I think. . . . the teachers were leaving at a rate of knots . . . in the end the school was only half staffed and children were being sent home like every other day. [My daughter] was in her second class in the infants she had about ten teachers in one term and they were just baby minders, they weren't learning a thing. The

171

kids were totally unruly . . . all the kids were just jumping on chairs, and
throwing books around and drawing all over the walls and teachers who
just couldn't control 'em . . . This was at primary age. . . . This was at six. . . .
you would normally see this sort of thing goin' on in secondary school
. . . She was swearing. The things they did was unbelievable . . . I actually
was gonna think about movin' her . . . but [another local school] were full
up . . . I thought, well, sod it if we can't move her then we've gotta fight. . . .
So I began to . . . ring divisional office . . . to say "Please you've gotta help
us out, you know, we're doin' our utmost, the kids do have to be taught,
don't ignore us we're really tryin' hard to resolve the situation, having a
word with the mums, don't go up and beat the teachers up cos that means
none of the teachers'll come here" and it was just really hard work over
there for a while and in the end we got the Inspectors into the school . . .
big report, all damnation of all kinds being written down . . . and eventually
[the head] ended up resigning.

The concerns of newcomers were justified. "Borough data suggests that attainment levels are lower in Tower Hamlets than in the rest of inner London. Poor
attainment at primary level means that children enter the secondary system without the basic skills needed to cope with its curriculum. In the London Reading
Test, which is taken by children in the final year of primary school, Tower Hamlets
pupils score lower than other London pupils, although performance is improving"
(London Borough of Tower Hamlets 1995a, quoted in Lupton 1997:47). Figures
for the two electoral wards covering the Isle of Dogs Neighbourhood (larger than
the geographical area that residents called the Isle of Dogs) reveal that almost 30
per cent of children in secondary education had "special educational needs" and
over 40 per cent of school aged children "do not have English as a first language"
(Lupton 1997:v–vi). One established Island resident who moved off the Island to a
relatively prosperous suburban area discovered that her son was two years behind
his peers there and required special tuition to improve his reading skills.

"We never dreamed . . . we would like it"

Whatever their original motivations for moving to the Isle of Dogs, some newcomers found a quality of life there they had not experienced previously and even
developed a sense of "belonging" derived from the area's geography and previous
white working class character. "When we moved 'ere," a newcomer said:

we wanted to make as much money out of this house as we possibly could.
We never dreamed in a million years that when we got here we would like
it . . . my husband's never liked anywhere we've been until we moved 'ere.
He said "I could live 'ere forever" and I could live 'ere forever. Don't
like the house very much. But I actually like the Islanders because they're
down to earth in a lot of ways, although you get a lot of back stabbing. . . .

They're very thoughtful people. I mean there's been times when I've badly need the 'elp and there's always been somebody who's [said] "Well I'll always pick [the kids] up from school." . . . I know I can ring a dozen people and they'll all say yeah [they'll help] . . . I never found that . . . where I used to live, didn't wanna know round there, look after number one.

"Most of the Islanders in some way or other are connected", she continued:

They're even related . . . People marry other Islanders . . . I . . . talked to someone in [the supermarket] couple a years back moaning about somebody . . . and I said "I know the lady's been there a lot longer than me . . . blah blah blah blah", but this lady said to me "You know she's my sister in law. I'll stop yer before you go any further." She said "Most of the people on this Island, you've gotta be careful who you talk to cos they're probably related to that person that you're talkin' about so you'll have to be careful what you say." So you can imagine after that I just sort of kept me opinions to myself a bit more.

"The later part of the twentieth century has rushed in on them", another newcomer observed. "I remember walking through the foot tunnel [from Greenwich], before I had the children, to the little cafe in Island Gardens and not believing that this place was in a complete time warp . . . you know still in the early sixties at the beginning of the 1980s . . . There's a lot of . . . local names cropping up again and again, a lot of in-breeding going on in the indigenous population." A newcomer who described his life before moving to the Island as a "normal British suburban background" in "a detached house in a snobby little part of the town" where "we just about knew our neighbour", came "to like the Isle of Dogs . . . there was a sort of quality here that I enjoyed", he said. "Now what I'm enjoying here . . . even as a new boy, is a little bit of London community life."

"People here are very into families", a newcomer from a wealthy overseas background said, "more than anybody else I've ever known". "What is nice about this area is you have the daughter next door and five doors away the mother will be living — they're all around here whereas most of the people I meet the kids live here but the parents live out in the country; here mum is just a few doors away." This pattern was confirmed in a survey (Wallman et al. 1987) of 5,965 people living on the Isle of Dogs in 1987 in which 70 per cent of women in the survey had relatives living in the area (10 per cent all of their relatives, 14 per cent about half, 11 per cent a quarter and 35 per cent a few), figures which were significantly higher than those of male respondents (see Table 5.3). "This seems to fit a pattern previously identified in East London of daughters getting married and then living near their mothers" (Wallman et al. 1987:36).

Despite the importance of kin, new and more established residents' attachments to the area were by no means limited to those with kin contacts or familiar networks of association (see Wallman et al. 1987). In fact "knowing many people locally rather than having local kin . . . [was] most significant" for the 80 per cent of the sample who said that they felt like "locals", a figure which included newcomers as well as those who had always lived on the Island, or those who had lived on the Island prior to the development (see Wallman et al. 1987:27; and see

Table 5.3 Proportion of respondents' relatives living locally, by gender[a]

	All of them		About half		About a quarter		A few		None		Total
Males	51	(6.4)	93	(11.6)	57	(7.1)	244	(30.4)	357	(44.5)	802
Females	139	(9.9)	190	(13.6)	153	(11.0)	481	(34.4)	434	(31.1)	1,397
Total	190	(8.6)	283	(12.9)	210	(9.5)	725	(33.0)	791	(36.0)	2,199
Percentage of all males and females	8.6		12.9		9.5		33.0		36.0		100.0

[a] Figures in parentheses are percentages.
Source: Wallman et al. (1987).

Table 5.4 How long on the Isle of Dogs by whether respondent feels a local[a]

	Not a local		Yes a local		Total	Percentage of overall total
Less than one year	187	(77.00)	56	(23.00)	243	(10.77)
More than 1, less than 5 years	205	(46.80)	233	(53.20)	438	(19.41)
5 to 10 years	66	(17.20)	317	(82.80)	383	(16.97)
More than 10 years	39	(5.70)	643	(94.30)	682	(30.22)
Always lived there	6	(1.20)	505	(98.80)	511	(22.64)
Total	503		1,754		2,257	(100.00)
Percentage of total	22.29		77.71		100.00	

[a] Figures in parentheses are percentages.
Source: Wallman et al. (1987).

Table 5.4). Furthermore the study indicated that 20 per cent of those living in the area who had no relatives there perceived themselves to be local.

Even on one of the private developments one newcomer said she felt that

> a lot happens here which won't ordinarily happen elsewhere ... It's so hard to believe something like this exists here ... I had my neighbour over this morning and she came in her towelling robe and I was still in mine ... I have friends who live [in Belgravia] who don't know their neighbours ... [and] I had to go dressed up to nip out and get the milk — bit of an exaggeration but that's the way it is ... My neighbour's always picking up my dry cleaning when I'm not around ... me going around for eggs and them coming around for whatever and I cannot see myself living any other way.

"London can be very amorphous sometimes, very alien", a businesswoman said. "It's better to live somewhere where there is a community ... that's got, say, a sense of identity ... and the Island is quite special like that."

Many newcomers seemed to develop an affinity for the area. "I tend to dissociate myself from the Docklands", one said. "I'm from the Isle of Dogs. I don't

Table 5.5 Total notifiable offences for 3 Area inner London Metropolitan Police divisions

Area	Total notifiable offences, 1997
Shoreditch	13,948
Stoke Newington	14,632
Limehouse	10,452
Whitechapel	13,308
Forest Gate	13,616
Plaistow	14,101

Source: Metropolitan Police Statistics 1997

want to know about Beckton, Wapping, and Stratford is naff in the extreme. Isle of Dogs is different definitely. I feel quite a part of the Isle of Dogs." "I love having water all round me. You feel like you're living on an Island", an overseas resident said. "You don't feel like you're living in London . . . and the area's very friendly." "This is a small piece of New York", a young affluent man said, mirroring advertising images used in selling Docklands, without "all the crime".

The Island was also seen to have some advantages over other inner-London city areas, and major cities elsewhere. "The Island's quite safe I think", a male European resident said, echoing the beliefs of many newcomers. Women, especially, commented on their feelings of safety: "My friends said I was mad to come down here and live as a single woman and I must be careful at night", one recalled. Yet despite their fears, "I have always felt safe walking round here." "The Isle of Dogs . . . feels so quiet and so gentle", said another:

> I'm talking about the residential areas not the commercial space — it felt completely unthreatening . . . subsequent to moving in I have discovered from the police that the Isle of Dogs is that part of London that has the least . . . street crime that exists. And so I found it completely unthreatening. I'm quite happy about walking home at night in some way more happy then I was in [a provincial town].

Certainly, in comparison with other inner-London areas, Limehouse division, which includes the Isle of Dogs, had the lowest rate of total notifiable offences of the inner-East London police divisions (Metropolitan Police Statistics 1997, see Table 5.5), a trend which has been sustained over a ten year period (Metropolitan Police Statistics 1987, in Foster 1987). As Table 5.6 indicates, street crime on the three beats that cover the Isle of Dogs (this time defined as the geographic boundary of the Isle of Dogs, and not the local authority neighbourhood) was low compared with many other inner-London divisions across the Metropolitan Police area. In the six months between June and December 1996, for example, only 4 such offences were reported in the three beat areas. In 1997 there were less than 500 recorded street crimes on Limehouse Division, compared to more than double that number in some other inner-London divisions in the East End and higher levels still in other parts of the Metropolitan police area (Metropolitan Police Statistics, 1997).

Despite overall lower recorded crime rates, the one group for whom the area felt far from safe was Bengali residents, who were also newcomers to the area in

Table 5.6 Major crime in the Isle of Dogs Sector, June to December 1996 (number of recorded offences)

	Beat 1	Beat 2	Beat 3	All
Theft from motor vehicles	27	48	28	103
Theft of motor vehicles	8	9	1	18
Taking and driving away	12	11	4	27
Criminal damage to motor vehicles	18	18	5	41
Residential burglary	18	22	5	45
Non-residential burglary	14	11	23	48
Robbery	1	3	0	4
Theft from person	0	2	6	8

Source: Metropolitan Police, Limehouse Division.

the late 1980s and early 1990s. They, as I describe in Chapter 7, had very different feelings about the Isle of Dogs and for some it not only felt unsafe, it was unsafe.

The local welcome

The arrival of significant numbers of affluent residents to a largely homogenous working class area was not likely to go unnoticed, especially when the contrasts in lifestyle and relative affluence were often marked. The isolation and insularity that had shaped the Island's history were just as apparent to many newcomers as they had been to other "outsiders" who moved to the area in the decades before the development. "I've never come across anywhere quite like this area," a newcomer said, "it's got its roots in very strong traditions . . . it's a very powerful impression that you get of this place." "The Island has almost been like an island", another new resident explained:

> Normally even in places like Clapham or Hackney or wherever there's always been a cross section of people and here it was all working class, all dockers, all that sort of thing, there wasn't, you know, the middle class . . . I don't like the class system but, you know [on the Island], there wasn't sort of anything different. They were all the same . . . that's why there's the problem. Not because people have come in but because they were all the same before.

One newcomer described an amusing incident that highlighted the extremities of some Island parochialism:

> I had a cab driver the other day . . . and he was supposed to take me to Covent Garden and we came to Bank and he said "Where's Covent Garden?

I've never been there." I said "What!" . . . He said "I've been on the Island all my life, never been outside the Island. I don't know London" and he was driving a cab from a local cab company . . . first day in his job. . . . It was quite a lot of fun and I told him the way out there and I had to show him a map how to get home again!

Despite rather simplistic media portrayals of the beleaguered white working class (often referred to as "the locals") and their opposition to "the yuppies", there was no single response to newcomers when they arrived, just as there was no simple divide between established residents and newcomers in defining themselves as "local". As Suttles (1972:13) suggested, "It is" only "in their foreign relations that communities . . . have to settle on an identity and a set of boundaries which oversimplify their reality", and it was these stereotypes that were reflected in media coverage. Consequently, while some established residents like this Islander felt resentful — "They took away what was ours originally and made it available to yuppies and just forced us into a corner" — others were relatively unperturbed: "I'm not against it. I think it's good to have this mix." However, both existing and new residents anticipated trouble at the outset and this expectation was fuelled to some extent by the activities of hard-line activists and some anarchists, commonly said to be part of an organization called "Class War".

Class War, with a membership of approximately one thousand, called for resistance, and in their campaign material "urge[d] local people to mug a yuppie, scratch BMW cars and make life as unpleasant as possible for the affluent incomers" (Short 1989:185). According to one of their number, although the popular perception was that they were "just into violence", in fact: "We have our own political theory. We do not call ourselves anarchists any longer. Yes, we want to overthrow capitalism and if that has to be violent then so be it. We are interested in community politics for the working class to stick up for itself" (Lashmar and Harris 1988, quoted in Short 1989:187).

This brand of politics had few sympathizers on the Isle of Dogs. "We had a bunch of anarchists came here and they were trying to generate a lot of hostility towards what they called the 'yuppies'", a community activist explained, "and all kinds of slogans was appearing." "When I first came here it was 'shoot the yuppies, burn Asda' on a placard" a newcomer said. "There used to be regularly, . . . 11 o'clock after the pubs: 'Yuppie, yuppie, out, out, out.' That hasn't happened recently but it did happen. . . . there was a lot of antipathy against the yuppie element — against the incoming people, which you can understand" (part quoted in Foster 1992:177). Another said:

There was serious hostility in the early days because of the rage of what the LDDC had . . . put on top of everybody and of course we had the really nasty, hard [left] . . . If you remember it was in the days of Militant, whatever you call the revolutionary workers wotsit party and all that gang . . . they were operating down here like there was no tomorrow and — (name of a particular activist) got absolutely hacked off with the way they [locals] were being manipulated . . . They're all traditional, old-fashioned Labour voters but the rubbish that was coming out of what I call the doctrinaire socialist brigade was driving them all bananas. In the end it was the Island people themselves that tipped out the trouble makers. . . . It's actually not been as huge an issue as Class War would like to have made it.

"When I first moved in" another newcomer recalled "Islanders . . . were actually being politicized from the outside . . . There was infiltration going on from the extreme left to actually be as nasty as possible to people moving on to the Island. . . . When I tried to speak to the local Labour [Party] branch you could see that they weren't prepared to listen; their view on me was that I should get off the Island as quickly as possible. It was actually quite threatening. They didn't want us here."

I mentioned in Chapter 2 that Docklands was a symbolic site for militant and anti-government protest, and therefore attracted a number of outside organizations into the area who did not command grass roots support from the residential population itself; a situation that combined with the Island's long reputation for dislike of political interference and a strong independent activist movement. Consequently, there were important differences between the overt opposition to newcomers epitomized by the activities of Class War and the far more commonly held and powerful underlying feelings towards newcomers that many established residents experienced but that frequently did not manifest themselves in direct hostility. "Although you say things, you wouldn't do things, the likes of us wouldn't", an established Island couple said, "but there are people who will have a go at 'em." Yet, as another local observed: "The system is not their fault, what's happening is not their fault", which highlighted the need to distinguish between newcomers as people and the processes that bought them to the Island, a factor that was important in minimizing conflict. "My main thought was, well it aint their fault", one local said:

> It's the LDDC, they created the problem. If I had £200,000 to spend on a house I'd have bought an house on the river so it aint their fault for buying an house you know what I mean. It's the LDDC for creating it. But there was a lot of [needle]. The trouble is, you see, you get people like Class War then who come down here and they started daubing "Yuppies out" and "Mug a yuppie" and they don't even live round here for Christ's sake. I actually had a row with a bloke outside the [pub] . . . he give me one of his leaflets and I said to him "But why are you blaming the yuppies like?" I said "If you had the money you'd buy an house here." . . . I said to him you should be challenging your efforts against the LDDC, they've created it." He wouldn't have it. No we'll have a go at the yuppies, you know, scraping a two bob down the side of [their] car. Well I couldn't see the point of that.

"When they saw they weren't getting any support from the community," another established resident said, "they packed their bags and went back to Hackney."

Despite the lack of support for Class War, there was an underlying sense of unease on both sides, described by one new resident as an "atmosphere" in which "there was this dark expectation of hostility". "I don't think it actually amounted to very much," he continued, "a few windows being broken mainly in unoccupied properties and car crime" and as another experienced, youngsters "shout[ing] a bit of abuse at you down the street". Nevertheless there was, as one newcomer described it, a palpable "friction". "The . . . people who came here on the few private estates were . . . cold shouldered — they weren't given a welcome by any stretch of the imagination", an established resident said: "They got a bit of a frosty reception really. No one ever mixed with 'em." "We was here first if you like, know what I mean?" (quoted in Foster 1992:177). This feeling was made all too apparent to this woman:

> When we first moved in, a friend and I went to . . . a [local] group . . . and we were asked where we lived . . . That was frowned on immediately . . . They knew I was a newcomer and I didn't feel very welcome there at all . . . They . . . didn't make you feel relaxed or at home or anything. It was a real . . . family-run affair and you really felt it was very cliquey and hard to get into. We went two or three times, we tried, you know, we persevered but it didn't work for us at all . . . But that's the only unwelcome experience . . . everything else has been absolutely fine.

Another woman described the discomfort she felt going to the family planning clinic: "They're quite aggressive, some of the East End ladies, and they can be sort of a bit forceful and they seemed to resent the fact that I was in the queue . . . I was never comfortable there . . . I used to think I was going to get shoved or hit or something and some of them were quite nasty . . . but that's the only time that I've felt it." "They're not a very welcoming community," was the initial verdict of one newcomer who found established residents "rather suspicious and even resentful." "I just had to keep a low profile", she said, and resolved that "if this doesn't go away . . . I shall move. . . . It didn't prove that bad because I'm still here."

These feelings of discomfort also extended to newcomers of similar social class background, including those who came from other parts of the East End. "Because I'm a Cockney doesn't make me an Islander", said a former council tenant who bought a house on a new development with his redundancy money:

> I came from Bethnal Green . . . if you move from Bethnal Green to Poplar, Poplar to Stepney and Stepney to the Isle of Dogs they're all different people they've all got different ways strangely enough, although there's four or five mile in it . . . These Islanders are very cliquey . . . I'm all right with 'em . . . [but] they did resent these people coming in . . . why I don't know because it shot the price of their houses up . . . So we've done them a favour and anybody who bought a council flat here, they was in the money.

Another person with East End roots said:

> When I first moved 'ere . . . me accent was slightly different . . . I think I was a bit posher when I moved 'ere than I am now and . . . I dressed differently as well to everyone over 'ere . . . and I think they thought oh she's just one of them who's just moved on the Island. . . . I obviously wasn't an Islander, I stuck out like a sore thumb . . . I got [my daughter] into one of the little play groups over the road and I was totally ignored,

completely ignored. And I thought if only they could speak to me and know I'm an East Ender . . . And then one day someone who was pregnant at the play school there liked this [carry cot] thing that I took [the baby] up in one day. She said "Where did you get that?" So I said "Down Roman Road." She went "Oh Roman Road", as if well how do you know Roman Road if you come from out there somewhere. I said "Well I was born in the East End". "Oh . . ." and they all wanted to know then, I was a novelty then, they wanted to know where I'd been and what I'd been doing.

Despite the obvious monetary or visual differences between newcomers to the Island and their more established counterparts, Elias & Scotson (1964:xi, 1994:75) describe very similar experiences in a community study of "a relatively old settlement" and two that later developed "around it" in which residents moving into cheap privately rented accommodation were similarly "cold shouldered". Furthermore, having been ascribed lower status because they were perceived to be "outsiders", it was "even more difficult for the newcomers to take an interest in their new community and to break down the barriers of their initial isolation" (Elias & Scotson 1994:75).

However, a newcomer who moved to the area in 1985, and over time developed a relationship with more established residents, suggested that what she had initially interpreted as an unwelcome reception was not quite what it appeared:

There was a lot of graffiti about the yuppies and when I went into [the supermarket] I felt that they [local people] weren't being particularly friendly. I realized afterwards that it wasn't that they were being unfriendly, they just didn't recognize me. They were just used to knowing everybody on the Island . . . after about six months to a year then I started recognizing them, we used to recognize each other at the cash desks or whatever and they'd be very very friendly. . . once we started to mix with other people, walking the dog or sitting in the park chatting to people you realize that there isn't this friction at all.

In the complex nature of human emotions, those responsible for "cold shouldering" new residents also criticized them for making little effort to mix with local people and resented the expectation of hostility, as this discussion between two women both born on the Island aptly demonstrated. "The thing what I don't like is when people say, my dentist she said to me, 'People have been very nice to me.' So I said to her 'Well we're not aliens — we're only people — just because we've always lived here don't mean we're going to be any different to you.'" "They think we're natives or somethink", her mother said. Yet accompanying these feelings was also an intense dislike of the processes of change that locals recognized influenced their behaviour towards newcomers. "You get very [angry] you feel like sayin' 'But I live here. I've got rights. I've always lived here' . . . It's terrible, you get really nasty."

Moving into an established working class community required a degree of sensitivity and accommodation. "It wasn't just the Islanders," a woman who moved to the area said, "it was newcomers as well . . . They were quite snooty and you actually started thinking to yerself well p'rhaps the Islanders have got a point . . . they weren't making an effort either." An example was provided by a woman who had lived on the Island since the 1970s, who described a dispute between her neighbour and a newcomer:

> She's typical real old gor-blimey East Ender and I come down the stairs, she said to me "'Ere Marge", she says "what do you think of them fucking yuppies?" . . . so she says "Them yuppies", she said, "Yankee yuppy as well, he was . . . knocked on my door". He objected to opening his door, he didn't pay all the money for his property to open his back door and be greeted with a view of Brenda's washing on her line. I wouldn't repeat to you what she said but she said "Since then I make fucking sure I put dry washing on my line every day." He told her "There's a laundromat over the square, over the shopping precinct, you should use that."

"I think the people who come unstuck", a new resident said, "are . . . newcomers who . . . I wouldn't call them yuppies, but the wives go on all the time about how the husband, . . . "He'll be landing now" . . . and you're supposed to ask, . . . "where's he landing?" . . . and how much money they've got: grossly insensitive in an area where many people had little or no disposable income and would never even be able to get on the housing ladder, let alone lead a jet-set life. "I get fed up with it so the Islanders must really get fed up with it", this newcomer continued:

> And you go to the clinic, you know, and the babies are little and you hear mothers sitting there saying "Oh of course we've got a specialist in Harley Street". And I feel like saying "Well what are you doing here then?", you know, bugger off and I'm sure the Islanders must think, you know, this is terrible. . . . There's a number of them have second homes . . . and they talk about it and I think that's a bit tactless cos there's people here that don't have a home or they're living with their mother because they can't afford a home and it's a bit tactless to sit there, you know, going on about it.

Another woman had very similar experiences: "Some of them . . . the only thing they can talk about is when they're changing their car and what exotic holiday they're going on and basically flaunting their wealth . . . there really are people like that . . . My next door neighbour but one, she's like that . . . renting their villa in the Italian lakes, which sounds lovely but it all comes out at really inappropriate times."

"Most of the locals who live here, they're quite welcoming you know. That is because of the way I am to them", another newcomer said. "I know somebody who'll tell you just the opposite . . . They are a bit wary of you when you first come in and if you're friendly they open up, they are themselves. Now when you're a little bit snooty I guess they give you the full load." She went on to describe one newcomer who got a taste of the "full load" when she was "thumped" by a local woman for asking her not to swear "in front of her children". She then "got her by the hair and said 'We don't want you here' ". Ted Johns made an interesting observation that seemed particularly pertinent to these events. Newcomers, he said, were accustomed to having "verbal disagreements" but the "problem is when you come

Docklands

into a working class area like this . . . if the working class person feels that they can't compete in the verbal, because they don't know what to say, and they can't bring their mind, or their tongue to the right word, and get frustrated, that's [clenching his fist] what comes next." "What was atrocious", a newcomer said, continuing her account of the events, "was that all the other women stood round from the school and didn't lift a finger." Another new resident said about this incident: "When I heard . . . I was like what! . . . but I had to laugh because I know. . . . [the newcomer involved], I know the kind of hoity toity voice she puts on and you just don't do that . . . If you're nice they're nice back . . . nobody went to her rescue because they knew the kind of person she is." Suttles (1972:97–8) in a study of an area that underwent development in the United States, suggests that new-comers "take the view that the people who have been hurt by the area's clearance have a right to be suspicious and even violent". He continues: "This leaves the new residents in a peculiar position: they sometimes seem to be siding with other segments of the population but at the same time are willing to accept the aliena-tion and avoidance which persist between the older and poorer residents and themselves." This situation seems to relate closely to events on the Isle of Dogs.

"I don't find any ill feeling from the local people. There is a lot of talk that they're fed up with the yuppies or the 'dinks' [double income no kids] who are moving in. But any of them that have have brought it on themselves", another newcomer said. "An example, I was in the local pub the other night and two guys walked in and sort of said [in very plummy voices] 'Landlord, two gin and tonics' and of course that doesn't go [laughs] . . . so they were turfed out." Newcomers, it seemed, recognized and accepted that it was they, not established residents, who must accommodate and those who got into difficulties "brought it on themselves" through insensitivity. "As far as the Islanders are concerned, I'm new", a newcomer said, "and the response I get . . . is positive. It's not gosh you yuppies moving in and taking our houses . . . I'm sure that happens [but] I think the vast groundswell of opinion is you're taken for the person you are, particularly if you're prepared to contribute" (in Foster 1992:177). Another agreed, "I think if you make an effort they're prepared to meet you half way." This process of accommodation is dis-cussed later in this chapter and again in Chapter 6.

Although overt conflict was unusual, resentment often simmered just beneath the surface and focused on particular symbolic issues, especially the river, which in some places could only be accessed by walking through new developments. "They think they own the . . . waterfront", an Islander complained: "They try to bar us . . . they do try and stop you going in there if they can." Suspicion that locals were being denied access to the river was fuelled by rumour:

> I know for a fact that, somebody told me once that there's some kids playing . . . down where the slip way [is] . . . [and a resident] said to them "You've no business down here", and there happened to be . . . [a local person] there that . . . overheard this and he just said to him "I shouldn't say that too loudly because . . . you could find your car floating in there." . . . This is the type of thing [that] can happen. (Elderly Island woman)

On another riverside development, "They've got a beach . . . and it's lovely round there in the summer", a woman who moved to the Island in the 1970s

182

explained, but "they've got security . . . to keep the kids off of there." Furthermore, on another development with dock access, she continued, "kids that have been born down 'ere go round to swim in the docks and they're turfed out by 'em." On the riverside development residents felt compelled to employ a security guard because of the problems they experienced. "If anything, the real hostility comes from the adjacent council estate", a newcomer on the development said, because "a percentage of them just let their kids run amok . . . Residents are terrified". "They don't like the ugliness . . . they feel threatened . . . by the yobbos that come over the wall and make trouble for which we have to employ a very expensive security guard . . . They're worried and there's really a right to be cos [of] the crime figures on this estate . . . they'll mess up cars with no trouble at all and they're absolute pests." As if to confirm the difficulties the following appeared in the *East London Advertiser*: "I was delighted the other day when sitting with my younger sister on the Isle of Dogs and saw some youngsters ripping up newly planted trees and using them to attack yuppie homes. Hopefully some young people locally will still have some fight in them and will repel these new Eastenders by making life unbearable for them." (Kane 1987 in Short 1989:187). So there was some foundation to the fears.

A managing agent for two of the private developments said that when residents initially moved in they "did have a crime problem . . . let's say 30 per cent, yes, in the first six months to a year we've had either our car damaged, stolen or our house burgled. If you asked that question now [1992] I would say 3 per cent, 5 per cent maximum, but that must be less than the rest of London." Several newcomers described their own experiences of victimization or those of their fellow residents. One Island woman on a new development where renting was common said: "Every house around here's been burgled", except her own, while another said: "We've been burgled quite a few times . . . I guess people get frustrated . . . In a way I can understand it. This area has been so bad and all of a sudden people with a bit of money have moved in." Unfortunately, the police do not retain local crime figures dating back to the 1980s and early 1990s. For the last six months of 1996, motor vehicle crimes (theft of, theft from, criminal damage and taking and driving away) were by far the most prevalent recorded offences in the major categories of crime on the Isle of Dogs, as they are elsewhere (Metropolitan Police Statistics, see Table 5.6). Although the police figures provided did not differentiate between social and owner-occupied housing, given what is known about patterns of offences and their concentration in the poorest areas (see, for example, the British Crime Survey, Mirrlees-Black et al. 1996, Hope 1996), it is likely that a relatively small proportion of the overall residential burglaries on the three Isle of Dogs police beats occurred in the private residential areas.

On the private development with its adjacent beach where residents felt the need to employ security, some newcomers had a very different view of how tenants on the neighbouring council estate felt about river access:

> The people I . . . meet [from the council estate next door] did say that it was an awful lot nicer living next to our estate than it had been a derelict warehouse and they rather appreciated access to the river cos we've got a sort of shingle beach here and the slipway and the pond feature that they all thought it was rather nice and attractive to sort of walk round and a pleasant place to have next door.

Conflict between new and more established residents was sometimes avoided by not discussing the development. A Danish man from a luxury flat who played football with some locals said: "They don't seem to want to talk . . . about it [the development], not to us at least. . . . They see us as yuppies and we see them as real East Enders . . . But there's no problem." "You don't class them in with the Docklands, do you know what I mean?", a parishioner at one of the Island churches said of the senior commercial and LDDC figures who sat along side her on a Sunday morning in church. "When I see [an executive in a leading company] I speak to her and say hello because she's just a person, she's not [her name and the company she works for] she's just [her]."

Despite the civility of these encounters (which shaped newcomers' perceptions of locals attitudes and left them with the impression that "if they didn't like the development they wouldn't have liked us"), underlying feelings and resentments about the development and newcomers remained, as a new resident who had close links with Islanders discovered:

> It would come up in conversation about "Oh it's not right what they're doing . . . they just put these yuppie flats up that we can never afford." So I said "Yes but what they're doing is something that you can't stop it's just progress. So you have to sort of learn to live with it as comfortably as you can . . . instead of fighting against it" and then all of a sudden I was an alien again, you could tell by the look on their face.

Although many newcomers were unaware of established residents' resentments, these were often a topic of conversation with people who did not live in the area. For example, a local church featured in a television programme revealed that "some of the congregation were very resentful", a newcomer recalled. "They feel they've been invaded", emotions of which she had previously been largely unaware.

Newcomers who only had superficial contacts with established residents sustained a rather rosy picture of relationships: "When I first came here there was a lot of bad press about the yuppies", one of these explained ". . . and I started chatting to biddies at [the supermarket] which still have overtones of the original place and I started chatting to ladies at the cash desk queues and things like that and everyone I spoke to was delighted — 'Oh goody, oh goody, you should have seen it before. We never had transport like this before, never had a supermarket.' I haven't had one local tell me that they resent the new people coming — not one."

Those with more interaction, though, recognized the negativity. "There are some women who say 'I'm glad this is happening' but by and large I don't think they're happy about it", said a new resident whose child was in the local school. "You see, of course, it points up . . . what I suppose is missing in their lives as well a bit, doesn't it?" Newcomers who sought to develop links with more established residents, and sensed their resentment, trod very carefully and were often self-conscious about their "differentness". For example, a woman whose East End roots exposed her more directly to locals' views was left feeling caught between two worlds:

> I got quite friendly with a girl [at the school]. She said to me "Where do yer live?" I said "Oh just over the back", I couldn't tell her where I lived. . . . I'd heard all this, "Oh this wasn't built for us and the Arena's only put

here for the yuppies and the Docklands Railway was only built for the yuppies" . . . and I felt they were being really stupid . . . I said "That's a really silly attitude to have if it's there you use it, I can't see why is it for one and not for anybody else." "Well they didn't bovver with us before the houses went up" and once I sat down and spoke to 'em and tried to put myself in their position I actually found I was hiding where I lived. I wouldn't tell 'em. I said "Just over the back, just past the school" and they all thought I meant over there [the council flats]. "Oh well where? You must live near me then, where, where, where abouts, what flats?"

When she finally told them:

"I don't live in the flats. I live on one of the new developments . . . then I got comments like "Oh your husband must be a crook then, you know, cos if an East Ender can afford one of these sort of things, must be a crook, he must be bent." I thought "Bloody cheek", you know. And then after about six months I started to get annoyed and then I was really standing up. I said "Look you know . . . I've had four, five jobs . . . working through the night as well to get out of the dump that we were in when we started off." I said "I've left having kids 'til late to get somethink behind us and we weren't just gonna stay where we were forever, we wanted to get on, we wanted to move on". . . . [One day] I was walking down, just past the school . . . and two women were coming towards me and [my newcomer friend] had the speech goin' at the time. She's very well spoken . . . , very nice voice and I think they must have thought well what's she doin' with her then? I went into school to pick up [my daughter] in the afternoon and they said to me: "Who's that lady you were walkin' along with?" So I said "Oh a lady who lives opposite me." "See you're hangin' out with all the yuppies" and I said "No. You don't understand just because she's got a nice voice and she looks different from everybody else doesn't mean to say [she's a yuppie]." I said "Actually she's got a heart of gold that girl" . . . They sort of took it for what it was [on face value].

This woman's experience speaks volumes about the power of the "yuppie" image and the barriers that it put in the way of social interaction, and how sensitive those who mixed with established Island people were to avoid being labelled in this way, perhaps because of the potential for conflict, perhaps because of the resentment it might invoke, or perhaps because they themselves at some level felt guilty that while they had prospered others had more restricted opportunities that were not being addressed.

Positive "vibes"

Although many of the diverse group of people embraced within the term "newcomer" recognized that locals might resent their arrival, they frequently found negative attitudes towards the development difficult to comprehend:

I don't feel that the development that's going on is intrinsically wrong. I think there are things that have got to be improved about it and a lot more public consultation has to happen, but to rubbish everything, you know, is not my style at all. But how to persuade people who are so entrenched in their viewpoint that it wasn't done for them and therefore it was wrong, I just don't know how to handle it.

"I've seen some prejudice from people who almost defeat their own arguments", another newcomer said. "On the one hand they're saying 'We've got nothing down here' but on the other hand they're against anything happening. Well you can't have it both ways. I think that is sort of blind prejudice that whatever anybody does if it's new it's bad and I'm afraid I just switch off I can't cope with that sort of negative attitude." Like other newcomers, she could not understand why some of the Islanders resisted changes from which they might benefit: "I can't understand why they complain about getting more facilities here. . . . as far as the shops and the cinemas and that sort of thing, I think that can only be good . . . It's not so many years ago you couldn't pay people to live down here . . . they just wouldn't come because it had such a bad reputation and to me anything that can enhance that has got to be good."

Whatever their background, newcomers tended to have an individualistic perspective towards the development and the opportunities it offered. For example, one newcomer immediately thought: "It's got to be good from their point of view because their property's worth more." Another's account of his meeting with two Islanders in a night club in the "very, very early days of all this development" aptly demonstrated the individualism:

I said to these two girls "Well . . . what do you think of it all?" So one of 'em said, "I think it's fuckin' disgustin' bleedin' people coming here doing this that and the other". . . . I said to the other one "What do you think?" She said . . . "I've never known anything like it . . . We live in a council house me and my husband", she said, ". . . we've bought three flats and an house", she said, "and that is it mate we are gonna make a fortune" and I thought well all right bleedin' good luck to you mate . . . and that to me was the difference in mentalities. One could see a golden opportunity for cashing in and taking advantage of a situation. All the other one could do was stand there and gnash her teeth and think "Oh these people are comin' on", forget that they're bringing roads and bus services and things that they've never had and employment.

I discuss the issues about individualism more in Chapter 8 and the extent to which the development itself and the opportunities for a few to "cash in" were changing the nature of "community" on the Island. But the comments above also indicate how diverse the views about the development and its benefits were.

"What's happened here has been brilliant for everybody, everybody here . . . apart from . . . the people it's affected in terms of actually they live . . . [but] I've never met any of those people", an advertising executive who lived on the Island said:

I think the major dissenters are the ones . . . that . . . don't really want to work, they won't make an effort to make their lives better . . . And if you talk to the old Islanders I talk to, blokes that are in their forties and fifties . . . those people will turn round and say to you there are an awful lot of scum bags and those are the words they use, . . . which I fortunately have never met, that live in this Island . . . that will do everything in their power, everything to bring this, the whole community down. . . . They fight, they don't work . . . you hear terrible trouble [in one pub] with people getting their heads cracked open on a Friday night and the blokes that are doing it are unemployed. They go in there, they drink six pints of Special Brew, get out of their heads and then . . . knock somebody else's head in and they're the sort of people I think are major dissenters.

Another newcomer continued in a similar vein:

One of the things that I was told is that the East End apparently is a very poor area and so people are jobless and these new buildings . . . [offer] a lot of opportunities for them to work, but I was also told that they're uneducated or not very well educated so the kinds of jobs they can go for is maybe receptionist, office cleaner or something like that. You know how often those people are advertising for locals and nobody bothers? . . . Every three months I'm putting out ads for cleaning ladies. They come they work for a couple of months, they take the cash and they go off on their holidays. They're just not interested. They just don't want to work. I don't understand it they seem to be going on holidays all the time to Spain and Portugal . . . how the hell do they afford it, half of them are on the dole and they're so proud of saying "I'm on the dole" and I think why don't you work?

Newcomers often had very little understanding about the conditions and circumstances in which the bulk of the poor Island population lived. One woman said: "There's a lot of poor people, I mean really poor, and I didn't realize until we went to church how poor some of these people were and what trouble they were having . . . There's a couple of people there that are very poor and they just don't seem to be in the system, they don't seem to be getting help they probably ought to be getting." But this did not seem to contradict her belief that many Islanders "don't want to work . . . I think those who do want to work have got on and have got their own home or done a self-build and what have you. And those who expect to be given it are still how they were before but more resentful because they see other people getting on." The impact of the development on the poor Island population is discussed in Chapter 8 and was far more complex than newcomers' simplistic views suggested.

Windows on to "other" worlds

Newcomers and established residents often came from very different worlds: the former lived in relatively self-contained new developments in which the bulk of the population worked and invested little in the area; the latter relatively poor, diverse and fragmented working-class population predominantly living in council housing and many of whom were unemployed. These two worlds sometimes came together in the Island churches, community organizations, pubs, schools, shops and play groups, the shared spaces on the Isle of Dogs.

The accounts that follow provide an insight into the different worlds of six newcomers and the links they forged with individuals from the more established Island community. All of these occurred in places where the opportunity existed for relationships to gradually evolve (breaking down mutual stereotypes), and were facilitated by common purposes (community groups) or by shared faith (which traversed the "new/old divide", as all were part of the Christian community), by some unifying factor (in the play group and schools where women of all backgrounds had their children in common), or simply by the fact that both sides were receptive to building new relationships. All offered windows onto other worlds and access to "broader communities of interest".

Building bridges: the vicar's view

My predecessor had been here ten years . . . He had been very up-front in opposing the sorts of changes that were taking place and exercising . . . I think considerable leadership in the local community . . . He was particularly good at being a voice . . . [and] also a support for the people in the Association of Island Communities. . . . The line that they took was, of course, change had to take place but not this sort of change. It's not the sort of change that is good for the local people. Now there came a point at which it became fruitless to say that. . . . Somebody new needed to come.

Part of my briefing was . . . about being able to engage with the changes taking place. . . . When I came . . . there were about thirty to thirty-five people in Church on a Sunday morning and they were all old Islanders, well . . . nearly all old Islanders. It's an incredibly warm, friendly congregation, but it was also quite cliquey, close — difficult to get into as a visitor. And, so I think there was a bit of a history of people coming once or twice and then feeling as though they weren't quite part of it and going away again. And what's happened now is I think newcomers are here in such presence on the Island that the Church congregation has grown to about sixty and it's a mix of people.

It still feels to me as though it's the sort of church that locals would feel at home in, and basically they run it, but there are lots of other people who come as well who tend to be not so regular and not so reliable, but who pass in and out and do contribute a great deal and, you know, they tend to be people on fixed contracts working in London for three or four years or something, so they're never really going to settle down and contribute to the life of the Church in the way that some Islanders have, but at least there's a mix of people. And I think it is one of the few places I know where that mix has happened and that's quite good. . . .

[When I came] it was quite polarized and it felt like I'd moved in on the beginning of people trying to say "Come on we've got to make something of this together" . . . And I think this process was happening but . . . I think that there was still a sense of things being pretty polarized. I think if you scratch the surface you'll find that now . . . But people are making more of an effort to find common ground.

"Mums out and about": an affluent newcomer

I know more people here than I've known anywhere else, ever . . . There are a lot of mums here . . . you meet them in all sort of ways. I mean, when Charlotte was a baby and you go along to the clinic you're introduced to the mums through the health visitor because she happens to know someone who lives in your street for instance. I met a couple like that. [A neighbour] has introduced me to a few. You just meet them walking around and bump into people. They're very friendly. Again, people on the estate, because they're new to the area as well, they know you're a new mum in the area so they start chatting to you, and the Toy Library, there's a lot of mums there. So they're easy to meet, mums, when you've got a kid.

Sharon is one of the mums I've met and she's lived here all her life and her mum lives here and her mum's mum lived here all her life and she's as local as they come and she's a wonderful person, she's lovely. She appreciates the change that is going on and she's benefited from it. . . . So she's great there's no resentment there at all . . . I'm aware there is resentment but I suppose you just avoid it. I mean, I don't take Charlotte to the play group in the . . . estate, for instance, because I wouldn't feel comfortable there. That's my problem, it's not their problem. But I wouldn't feel comfortable there. A friend of mine lives round the corner, Jane, and she's from (the East end) and she went there one day and she just came away and never went again because she felt so ostracized. "You don't belong to this estate", "why are you here", you know. . . .

The Toy Library [where children's toys and equipment are loaned for a small fee] is the best thing, that's brilliant. That welcomes everybody and its very friendly a lot of young mums go to that. There's probably fifty/fifty local people, well no it's divided into three actually. There's local people, new people and then there's nannies . . . but they don't quite fit into the same slot. But that works very well with new and old people. A really good mix. But then . . . you're there because of your child, that's the common interest. It sort of stops at that really. . . . I've met sort of four or five mums that I meet and I can call whenever I feel like it or I feel they can call on me . . . it's great . . . and they're really genuine . . . like they know I'm about to go into labour . . . and they say "Well phone, you know, if you want Charlotte taken care of, phone." And it's really nice, a lovely feeling . . . they're from all walks of life. . . .

There are so many young mums with kids . . . very much wrapped up in their own communities around here. I was thinking about it last week when we had Charlotte's birthday party. I went down to the church and I borrowed the tables and chairs from the toy library, took them back the next day and my dad came with me and . . . I felt like I was in a village just driving down to the church dropping everything off, coming back, you know, hello to people. It was all very nice. I mean it has that friendly feel about it. It has an intimate feel about living here, definitely. . . . It's great the community we have here . . . Doug's often amazed at how many people I know who'll say hello to me when we're walking down the street.

I hear such weird comments sometimes like, I went to Sharon's little boy's . . . party, I came away, I actually was in tears that night . . . I came away and Doug (my husband) came home and I really just broke down and said I really feel as if I don't belong. This is the only time it's ever happened and it was really a one-off and it hasn't happened again and it was because of Sharon's uncle. He was there all day and he was drinking beer all day long and he was burping all day long at a kid's party. And I thought it's not necessarily a reflection of him being a local person, this is just his character. But I came away thinking, you know, I really felt like I didn't belong because that just would never have happened at any of my friend's parties. It just wouldn't have done.

And then about a week later Sharon was saying to me "Oh, my mum thought you looked like Princess Diana, she thought you were so glamorous". And I thought how ridiculous I'd gone along in just something like a track suit, something very casual but obviously she'd highlighted me . . . as being something a little bit different . . . So it worked both ways. I noticed something in them and they noticed something in me. It was really odd. But it was just that chap who . . . put me off. And she's a lovely girl, Sharon, and her little boy is really sweet, but never in a million years would we as a couple mix with them as a couple . . . there's no common ground at all . . .

There is a mix but it's all probably, maybe it's more superficial than I'm thinking it is, you know. May be there isn't much depth to it. But that's generally, as a sort of young mum, there's a lot of mixing going on. I don't need to have these people as my friends socially in the evening cos I have my own social life anyway. And I don't sort of feel depressed that I'm seeing them just because we've got the kids in common because there's a few now that I've met that I'm really happy seeing the women as themselves, not necessarily as Jack's mum or Amelia's mum or whatever, so it works quite well.

"They all help each other": an affluent resident with "East End" roots

I was shocked when I first come here. . . . I s'ppose I'd been here about nine months to a year, one day someone said to me "My car's broke down, are you doin' anythink this morning?" and I said "No not really." So she said "You wouldn't mind running me down to Stratford would yer." I thought bloody cheek, all the way to Stratford so I said "No alright then." I thought bloody cheek, but I'll take her anyway.

I went down to Stratford, did her a bit of shopping and she said to me "You wouldn't, on the way back, pull into the garage to see if my car's done?" and I said "Yeah OK", thinking I'm gonna be bloody out all day with 'er, I'm not gonna get anythink done indoors, you know. Called round her car was done so she said to me "I don't know my way 'ome from here" . . . so she said "You wouldn't wait for me and show me the way home?" I said "Yeah OK", brought her all the way back and she said "I really appreciate that, thanks very much." . . . I didn't realize that they all did this for each other . . . she didn't feel she had a cheek asking me . . . the Islanders will all 'elp each other out. If I said to 'er "Would you run me to East Ham, I haven't got a car", she'd say "Yeah" and we've become quite close friends . . .

When I was pregnant . . . I was really ill, I was just so ill I couldn't get out of the bed. It was just the worst experience of my life. Barby picked Emily up probably for the nine months, took her to school, picked up of a night bought her home. I bought her a big bunch of flowers and I kept sayin' to her "I'm sorry." She said "Please don't keep sayin' sorry for goodness sake. Don't worry about it. I'll come every morning I'll pick her up, you don't wanna worry about it, really." . . . I've got into the habit meself now of doin' it like if someone says "Oh I've got a hospital appointment", I say "Well is it early cos I'll take the kids to school for yer if you want" and I go and pick 'em up and we drop each other's kids round to dance classes and "Oh well if you can't make it I'll pick her up and take her home, give her a bit of tea and drop her off later." But when I first came here I'd never had that experience before . . . I didn't realize the whole of the island was doin' it for each other.

They don't mind what they do for each other. It's just such a tight close-knit thing with them. . . . They don't mind, they'll do it . . . because there'll always be a time when that person can help you back anyway . . . All the Islanders are like it, their mums were like it, their dads were like it and their grandparents were like it. . . .

"Us wifey types": a woman on the luxury development

I assumed that being here my immediate neighbours would be business-type people and not so many kids and families, that I'd dread and that would really affect me, whereas the people here you can take them or leave them as you find them. So it really doesn't make any difference to me what they're like. They're diverse. They're all sorts.

I used to know the lady next door that way very well, she moved fairly recently. I don't know who's gone in there now. . . . I'm a little bit picky in more than the hellos and how is it going? . . . My buddy is on the second floor who I phone every morning and say "Well I'm doing this, this and this, what are you doing?" OK we'll rearrange it and we'll do that together. We spend 50 per cent of our daytime together and that's most days. Lunch every day. Then there's another friend she's my neighbour immediately above . . . I actually met her husband in the foyer, "My wife's here, she's at home all day and, you know, give her a ring sometime." . . . Then in [the adjacent apartment blocks] I've got another friend, she's two weeks here, two weeks in France, go for luncheon occasionally . . . I don't see very much of her really but sort of enough that the next time her husband disappears abroad she sort of gets cheesed off she gets in touch.

There is a [sense of community] with us wifey types. But I don't think there is particularly for the nine to fivers. I don't think many know each other . . . when you're rushing to work, back from work you don't sort of chat. Whereas if you're swanning off to [the supermarket] you sort of bump into somebody and say "Do you want to share a cab down there" or whatever and you're probably a bit more laid back. Now we've got something for all the ladies here who are the wives of, sorted out a luncheon club [and set up a list of tennis partners] . . . so that people are only isolated if they choose to be.

In search of the "authentic" East End: John, the executive

The day I moved in, it was a Saturday, I picked up the keys and I didn't even come into the flat, it was a very strange thing for me to do but I picked up the keys . . . , I didn't even come in here I drove straight back down to — to play football . . . that evening I came back up here, I came in, stayed here about ten minutes and went down the pub . . . and thought I must be mad cos . . . it was awful. . . . I had this romantic image of the East End really because my family are originally from — and although that's south of the river it's . . . [a] very similar mentality . . . and I went into the . . . [pub] and I thought I've been living in a fool's paradise. What have I done?

To me it's a social necessity going to the pub . . . I can't even remember a Saturday going by since I've been 18 where I haven't been in the pub for half an hour, or a bar or something . . . Like half the week I will go down the pub and have a beer 'cos I like the atmosphere. Men talking about stupid things like football — I've always liked that.

After realizing . . . [that the first pub] was not necessarily my social vehicle, I decided to go to [another pub] cos that had been done up recently. Now that's inverted snobbery really. It was just simply, and also insecurity. There was a pub that had obviously been done up to welcome the influx of the new breed of Islander. I knew I could feel quite secure there, I didn't feel as if there was ever any threat [but] I didn't particularly enjoy [it]. The atmosphere wasn't bad but wasn't really, I still hadn't found that East End atmosphere of people telling jokes all the time and lots of humour, lots of banter, lots of deep discussions about the meaning of things. I didn't find that in there and I realized that actually the reason for that was obviously it was purely cosmetic. This thing had been created for the future rather than something that represented or referred to past or present actuality. So I kind of stopped going there as well cos it didn't give me anything.

Then I let the place for [a few] months . . . Then the . . . pub which I use now was refurbished. Now I used to walk past it, and again this is insecurity, I walked past it and I was scared to go in there cos . . . there was an estate sort of surrounding it that looked pretty awful, but why I should think that I don't know why, but . . . I thought I can't go in there . . . so I didn't. Then one day . . . [we] were walking past on a Sunday. It had been refurbished. Even then I was nervous. Scared of walking in. . . . We did go in there. We met the guy who runs it . . . [He] has become a really good friend and that's where I found the thing I was originally looking for. That's where I found all the East End humour, the warmth, the generosity.

The generosity of these people who have got, relatively speaking, very little is just unbelievable. I go in there and they will not let me buy a drink and I've got more money than half a dozen of them put together and they won't let me buy a drink and of course I insist, knowing that it all becomes a bit of a healthy joke. And I love it. I just absolutely love it . . . It's like everything that I thought about the East End is represented to me in that pub. The humour the warmth, the generosity. I really like it . . . I've got to know a crowd [in there] that are all old Islanders. They've all been here since, well, their fathers and their grandfathers before them . . . So I'm involved quite strongly with that crowd of people.

Worlds apart: an overseas perspective

Apart from maybe a handful . . . I don't know anybody else on this estate and I've lived here for four years now, but the East Enders are just the opposite. They're always there with a smile. I can understand now why the East Enders are called the "salt of the earth"; because they will go out of their way if they like you. They really will and I guess the others on [my estate] aren't like that. One up for the East Enders you know!

My very first friend I made here was a true and true East Ender. She used to work in the — shop . . . and I used to go like every day cos I was so bored. Oh God. I used to pop into . . . [the shop] and go and see Susie and I started to get to know her, you know, and it's a different world for her. I remember the first thing she said to me when she came in here and she looked at everything and she said "Phillipa, it's a different world" and at first I couldn't understand that because this is not much to me but to them it's everything, everything they want. The little things that I would probably [not even notice] to them it's mega big.

[Susie and I] got along very well and we became close. She used to come here and I went to her house — well she didn't invite me to her house until much later. I guess she was embarrassed ... so we used to meet in the —— shop all the time. ... I used to talk to her, she was a divorced mother and she was working a few hours so I said to her "How do you manage with two kids and just working three hours?" and she said: "Oh, you know, I'm on the dole" ... and yet she went on holidays and I couldn't believe that. It's not so much, you know, that my husband is being taxed ... so that people can live off him, it's not so much that, I felt she was wasting herself.

So I gave her those lectures and talks and this that and I said "Look you're wasting yourself, why can't you get a full-time job?" and I'm happy to say a few years after that she's now [a] manager, she works full-time off the island, she's got married and she's a lot happier ... My son is 10 months old. I've had a nanny because I was travelling a lot and I feel really bad because I don't have a job now ... I've got nothing to justify the nanny but the thing is we entertain an awful lot, either three or four nights a week we are entertaining either having dinner parties at home or out to the theatre ... so at least that way it's better than getting baby sitters, but even then I feel guilty that I should be doing something during the day. I just can't understand how people can just sit there and wait for the next holiday!

Through my husband's work the kind of people I meet are the very reserved English people who are always trying to pretend they are something they are not, or they're so reserved they're not open and they're not friendly ... They'd rather live in SW1 or SW3 ... and have Belgravia stamped [on their address]. A lot of them also have houses in the country, so come Friday they shoot off there ... great for shooting didn't you know! That's how the other half live I suppose ... It's funny saying that because I remember Loraine [Susie's sister in law] coming here and my husband gave me this beautiful birthday present. I remember Loraine looking at it and saying "Oh Phillipa, this is another world!" and I used to go to these places with all these people in their houses and think Christ it's another world isn't it!

East Enders are just so different. Susie is very open, very friendly, she's always there and I can be myself with her and she's introduced me to her boyfriend's sister and we became great buddies me and Loraine. ... They are really genuine people ... I remember when they came for my son's christening. I invited the whole lot, the mum, dad, Susie, her husband, Loraine and her boyfriend and we had the christening at Westminster Abbey and had tea at the Ritz and they had never been there. The christening was in October and they all bought their Christmas outfits two months ahead because they didn't want to let me down you see. They said, "We don't want to let you down." They wore their hats, yeah they spoke with an East End accent, big deal, but that's how they are.

What really upset me was nobody outside of my mother, myself and my husband spoke to them at the Christening, nobody. I just don't understand why, because they spoke with an East End accent? ... It became embarrassing for me because nobody spoke to them and these people claim to be well educated and well bred, is this how a well educated, well bred person behaves? Let me tell you they can learn a lot from this lot. ...They're great, that's one reason why we live here, we've looked at other places we just don't want to move because at the end of the day they'll always be around and they're not the type who'll call everyday on the phone or call up "Oh Phillipa *darling,* we *must* meet for dinner. We *must* get together." There's no bullshit. ...

Another common myth they [the East Enders have] is that we get things easy. It's not true. My husband's out of the house by 6.15 because it takes him about 13 minutes to get into work and he's at work every day at 6.30 and he doesn't come home 'til 7 o'clock or 8 o'clock. Three or four times a week we don't come home until about 11 or midnight because we are entertaining. For me it starts at 6 or 7 in the evening for him he's been on the trot since 6.30 in the morning right up 'til midnight. Now people don't know things like that. Oh yeah, they say "Where did you go for dinner?" and you say "Savoy" and they've only ever heard of these places and they think it's wonderful but it's a lot of hard work ... We have a box at [a famous race track] and entertain clients there ... They work bloody hard. ... When you drive around in your car and they say wow how can anyone spend that kind of money on a car. Robert bought a Mercedes and they all came round to look at it wow ... and this [car] was worth more than the houses they live in ... OK when they say things like that I feel bad, but my husband works really hard for that.

Bill: I'm not a "newby"

I regard myself as sort of returning, rather than a newby ... My parents are Cockneys, both born in the sound of Bow Bells and [I] liked the idea of getting back into the scene, getting back into where bits of London were happening.

Soon as I moved into my allotment, the lady on the allotment next to me said "Gawd, don't you talk posh!" That's I think a classic and normal response ... say it to yer face and you address it and you sort it. ... Now they've dragged me onto their committee cos they think I'll be able to talk with the LDDC to get some funds we very much need, and that's right because those are areas that I have some skills at and can do, just as they've got a whole range of other skills and that's what communities are all about.

Fighting a common enemy

We all do it, don't we? We all categorize each other without thinking because you live there, because you dress in a certain way, you go there and stay there and because I live here and I do this, I stay over there. Now when it happens that somebody like me from there has to meet somebody from over there . . . they suddenly realize . . . we are all the same. (New resident)

Few newcomers, attracted to the Isle of Dogs and the promise of a bright future in Docklands, ever imagined that negative aspects of the development might generate in them the kind of anger that their more established counterparts expressed towards the LDDC and developers, or that they might one day establish "common" interests with them. Those who arrived early expected disruption, but it was all part of the excitement. "I think with a place like this you've gotta have vision", an investor and resident explained:

I said when I came here which was [nineteen] eighty-five . . . I didn't think you would see any significant things happenin' for ten years. I don't mean that you wouldn't see buildings going up and things like that, but I didn't really think you would see a proper integrated community with everything that a community needs — shops, cinemas, theatres restaurants . . . and the infrastructure that goes with it. Because you look at what they were trying to achieve . . . the sheer scale of the whole damn thing; I thought ten years would be a reasonable period.

Yet newcomers quickly discovered that they, like the residents who had lived on the Island before they arrived, were not the primary consideration of the Corporation once, as one newcomer put it, "the commercial enemy took over", and that their views as residents would be ignored just as their poor counterparts were. "The LDDC just appear to be out for what they can get", an exasperated newcomer said, a comment that could just have easily have been said by an established Islander. "They want to get people living down here and when they've got them here they really don't give a damn about what the conditions are like" (in Foster 1992:178), said another. "I'd really like to bang their heads into a brick wall. The contempt with which they treat people is appalling. I can understand why there was so much angst by the people on the Island about the LDDC." The comments of one newcomer spoke volumes about the unaccustomed sense of powerlessness they experienced as commercial interests gained primacy: "It was very difficult to find somebody who was prepared to listen and you felt actually would take into account how you felt" (Foster 1992:178). "They began to see what we'd been arguing about all over those years", Ted Johns said, and started to "suffer from it with . . . pressure on the schools, and all this sort of business, and the roads . . . continually being dug up . . . [and] the developers come 'ere, build right next door to them, block out their light."

"They've not been treated any different", he continued, "because the attitude that guides the regeneration is not about people . . . [it's] all about profit . . . They're in the way as well as the working-class community." As the reality began to sink in, the "positive atmosphere" became more fragile and the disadvantages of "pioneering" became increasingly evident, as a woman returning from an extended trip abroad found to her cost:

> They [LDDC] allowed *that* [a block of flats] to be built outside my balcony, which used to be a river view . . . [it] was no more just like that. I came back and blow me there was two storeys of a building. I particularly checked when I first bought that the planning permission for there was single-storey light industrial.

Unlike the Island couple who had an office block built at the bottom of their back garden, she was not reticent about approaching the LDDC and asked: " 'What the hell do you think you're doing?' But the attitude was [it's] EZ (Enterprise Zone) . . . How much joy did I get out of it — *none*, [I got] nowhere." Surprisingly, she was not bitter merely philosophical: "If you're sort of gonna be all pioneering and brave and new front and say no to bureaucracy you have to live with the down side as well. . . . [But] I wasn't wildly impressed with the LDDC."

Other new residents were not so accepting and were exasperated with the Corporation and their activities. "I was actually very pro it [LDDC] in the beginning", a newcomer explained:

> As time has gone on I haven't been so pro-it. I mean, one gets the feeling really just how much have they really taken into account the feelings of the local people and feelings of most of the residents now, whether you're an old or a new resident — how much do they take into account or how much are they just interested in the business community and the financial side of things? I think perhaps they could have done more for the community as a whole than they have done. I feel our needs haven't been catered for.

Although they began as unlikely coconspirators, some new and old Island residents came to recognize their common interests. "The new people that are moving in are having the same feelings as the original residents", a newcomer observed. "The things that affect the older locals, they affect the new people as well . . . so we've got the same complaints really." "It turned out we had a lot in common", an Islander said talking about his friendship with a newcomer: "You know, fight the enemy sort of thing." "We're all in the same boat", said another, a comment that an activist fully endorsed: "Their struggle is the same as ours."

As the development gathered momentum and newcomers realized they were not immune from its effects, a small group joined forces with established residents to fight aspects of the development that threatened their quality of life. "I didn't really get to know anybody . . . on the [council] estate", a newcomer said, "until . . . we had a big campaign to stop the scrap yard operating on the other side of the river." "We got to know all the neighbourhood because everybody was suffering the same as we were." It had "a machine . . . capable of pulverizing, I forget how many cars a minute, with explosive force. . . . It was just incredibly noisy and . . . just ruined . . . the quietness of this estate and the adjoining estates . . . I think that was the common sympathy . . . so we decided that we would contact other estates in the area and find out . . . if we could do something together." After a court battle, "a planning appeal and a public enquiry . . . it got stopped."

"Human nature being what it is, you only tend to get involved in things when there is a need", another newcomer said, and the turning point for one was "when they started to dump things on us that affected my living at Clippers Quay, then I got quite angry and joined local groups and discovered what else was going on". Those who had been battling for many years not unnaturally questioned the motives of affluent residents who got involved: "A lot of these people . . . only come when they want something done . . . All they're interested in is their problem they're not interested in anybody else's problem", an Islander said. One newcomer was open about the role of self-interest in her initial contact with community groups over the scrap yard. When I asked her whether locals were cynical about the group's motives, she said: "Absolutely . . . they were quite right. I said yes, fair enough, but they . . . also thought that they were just going to see me for one day . . . I said 'I know this is the ostensible reason why we're coming in and this is giving us a reason for coming in, but we actually want to join anyway so . . . regardless as to whether . . . you can help us in this issue, we're in' and I've been there ever since." "You can't just sit back and moan and criticize and not do anything about it", another newcomer said. "All right it's probably the negative things that started me off, you know, I wanted to do my bit to complain against the Arena and this type of thing, but then you get positive things as well . . . I'm beginning to feel that I belong because if you do go meetings and you do find out what's going on otherwise you're a bit out on a limb."

Whatever their motivations, contacts between new and established residents over issues related to the development in which they had common interests served to break down barriers. "I go to these meetings and no one offends me, although they've had a go about [our development] on various things", one newcomer explained, "and I've been lucky I've been there to say 'Well that is not the case' where they would perhaps have thought it to be if we hadn't been taking an interest." On one of the private developments involved in the scrap yard campaign a newcomer discovered that the adjacent council estate, which gave "the impression" of being "a rather hostile place" was quite different: "Some of the flats are lovely . . . [and] I suppose generally speaking people I met . . . were exactly the same as me. They commuted to work in the City and wanted to come back for some peace and quiet in the evening."

The siting of the London Arena (a large multipurpose space for concerts, exhibitions, etc.), described by one newcomer as a "common evil", led to concerted opposition and "united new residents and old residents" alike. "I just wish the LDDC would think things through before they act like putting the Arena here", a newcomer involved in the campaign against it said. "The Island can't cope with the number of cars coming on to it normally, it can barely cope without 12,000 extra people to get on and off the Island — crazy. Again it's the ooh that'll be a money maker attitude, you know. Damn the poor people who live right next to it. I don't think they ever thought about the associated problems that went with that kind of venue."

In the rest of this section I have drawn heavily on the account of one newcomer and her experiences of opposing the Arena. The account demonstrates how the contrasting worlds of newcomers and more established Island residents, their differing stereotypes and life experiences, influenced their approach. Given the sensitivity of local people's feelings, one of the private developments most affected by it was initially cautious.

We were told . . . that if we didn't have the car park [adjacent to their estate] . . . [the Arena] would fail and we would cause the locals to lose the Arena. That was the tack they tried to take . . . We were treading very carefully because we do feel there's a barrier . . . We were very careful that it didn't look as though we were trying to stop something for the locals. We went round all the other new developments and said "Do you realize what's happening? The local population are automatically opposing it but do you realise the impact it's going to have on us? We won't be able to get off the Island. It's going to be noisy it's going to be hellish" and they said "Oh well it's not really going to affect us." Then you tell them it doesn't really matter where you are on the Island, it's going to attract a lot of extra traffic and, as they soon discovered, a number of other problems: "We've got videos of people not just urinating but doing the works on our door-steps and being very drunk and coming through. [For a while] I thought I just can't bear all this and I thought I would move and I think a lot of people did. Then we started going to the local meetings we found it wasn't just that they [the locals] were against the car park, they didn't want the Arena anyway . . . Thank goodness we got talking. I met . . . [an Islander involved in opposing the Arena too], she wasn't terribly convinced that [we] were genuine in [our] objections. When she got to know me and she knew how against it I was she said "But you've got a representative on this liaison group." I said "A what?", she said, "Yes . . . there's this man and he says he's the . . . representative" . . . We looked into it and this chap . . . had nothing to do with the residents committee. They . . . used to turn round and say "What do you think?" and he'd say "I don't think they'll mind" and they'd [local tenants] say "You traitors" . . . We sent this chap a solicitor's letter saying under no circumstances was he to represent us. That was trying to turn the old ones against the new ones. We all think good came out of it, as it happens.

Not all new residents on her development approved of the opposition:

To be fair there were people who didn't like what we were doing, the bad publicity because they felt the value of their properties would be going down and certainly . . . a lot of us didn't think that was important — the quality of life was more important — to build up a community. It was very short-sighted [of the others] because that in the long term would ease selling your house . . . so whether it was from a selfish point of view as far as the value of your property's concerned or whether a genuine desire to help develop the community in the long run, it's a common aim so we were able to persuade a lot of the new residents that in fact it was in their interests to help the community. They needed quite a bit of education and the Arena did motivate people into writing letters and complaints which I don't think they would have bothered to have done.

"I've always been a bit of an activist" another newcomer involved in the campaign said, but felt she was not representative of most of her contemporaries: "The apathy is just unbelievable." As with their "local" counterparts, those willing to campaign actively were very small in number, but it was said that other newcomers shared their concerns even if they were unwilling to act upon them. "There's terrific good will for the Island here even if it's not expressed in very practical terms", a representative from one of the private developments said. "Everybody's awfully pleased that I'm involved . . . [and] that I'm around putting in [our development's] point of view when necessary." The geography of the area also helped to generate an affinity. "People do think they're on an Island and they actually see the boundary of the river, anything within that is of interest to them," a property agent explained, "whereas in London you could be in Kensington and something could be happening at Westminster and you couldn't give a damn about it. Maybe if it was down your street you might. But on the Island people do tend to . . . consider the whole little area."

Newcomers often felt that their professional expertise gave them an edge in their dealings with developers and the Corporation: "[I'm] not playing down what the locals have done because they have not given up, especially the hard core, but I do think it was a bit of a shock for the LDDC when they got these people who weren't intimidated", a newcomer said:

> There are people here who felt no hesitation to knock on the private homes of, say, the chairman of the Planning Committee, the chairman of the LDDC. They knew how to go about things I think that's more what it was. The locals had put a lot of time and energy into it, but I think that people living here had contact through to politicians and more influential people that original residents hadn't . . . We contacted newspapers, TV and there were a lot of programmes and they gave both sides of what was going on. (Part quoted in Foster 1992:178)

"I think they were a little bit luckier", an Islander also involved in the campaign said, "because they had solicitors and lawyers and things like that on their estate and that's a big help isn't it? And also they had the money . . . to be able to do the paper work." The "weapons" newcomers brought to "the struggles" were different a community activist said: "Our . . . recourse . . . is actually to go out on the street . . . they would pick up a telephone and say 'I want to talk to so and so in the LDDC' and they go 'Do you realize who you're talking to . . . I'm an architect' . . . or . . . whatever they might have been": status made it easier to mediate with those in positions of power and authority.

Comparing the differences between her views on the development and the negativity of many established Island residents with whom she came into contact, one newcomer observed:

> I'm very very lucky I've been born where confidence is [inbred] . . . you just assume that you're confident . . . that natural air of authority. . . . I find it consequently very difficult to understand people always seeing the bad side of everything and never the positive . . . I see bad things constantly but I also see good things and I think that has something to do with the background . . . What really depresses me about this area in terms of the Island people [is that] the dark side is always the side that wins.

But these newcomers were not seasoned campaigners and some negativity, they soon discovered, was not misplaced: "I've never been terribly outspoken or revolutionary or anything", one active newcomer said, "but you just can't help but get angry about what they've been doing. You feel that you've got right on your side and it's just a matter of making sure that they know you're determined to get things done." But she also admitted: "I was terribly naive thinking that if we went through the formal channels, we were given time to object, we had very legitimate complaints and we offered alternatives, what's the problem, everything would change, we'd made our point and they would listen. And the locals said 'You're wasting your time!'" (in Foster 1992:178). They were right. "The LDDC are very clever", she continued:

> The locals did warn us of this, that they would keep refusing permission for the Arena car park or put them on trial and then put them on another trial, put them on another trial until it wore us down and in the end we just couldn't get the support of people to go to meetings and writing letters and that did happen.

The commercial enemy proved bigger than the combined opposition of both newcomers and established residents. However, for new residents the first-hand experiences of the development's negative aspects led them to reappraise their view of local opposition. "The awful thing is, I don't think we even thought about the old residents", one newcomer admitted:

> Cos it was very easy to live in a place like this without even mixing or without coming across the older residents. I think all along I felt and understood that it was inevitable that there were going to be people who'd been bought up here who resented these new people moving in and at that stage making money very quickly on properties which they couldn't necessarily afford themselves. These properties had been built for new people to come in rather than catering for their needs, which perhaps should have come first. I mean you can understand both sides of it.

"Getting to know the local people I . . . found they were so unhappy about every-thing and I thought they were a bit naive to begin with, I must be honest":

> I thought well you had nothing here before, now you're going to have a railway line, you've got Asda [a supermarket] and you're going to have new shops and there's nothing for the young people here and I did think perhaps they were just automatically being anti any change. But then I started to experience it myself. We'd only been here a few years and I have great sympathy for them now because it's just the way things are gone about. I do think they have a right to have a say because they've been here so long, which perhaps I didn't think they had at the beginning. I thought they should just be grateful that anyone was coming down here and what they would get as a result. . . . I think it's quite disgraceful the way they've been promised things and lied to. (Part quoted in Foster 1992:177)

"When you see the way that the LDDC's treated people it's appalling", another said. "They really don't give a damn about their [local people's] feelings and how things change their life . . . I certainly think they [LDDC] should have consulted more fully with local people who were originally here, I really do. I don't like being walked all over, why should they?" "They just can't see any advantages coming with all the disadvantages" explained a third newcomer.

> That's what I think's unfair. I do think that they could have adapted a bit, a bit of give, but they won't now because they just see anything that happens as against them. The railway line — they were promised it would be silent and it's not . . . They have their community people at the LDDC but I can't see how they could be fair — they'd be going against everything that [the Corporation's] putting forward. I can't see they've [LDDC] given the community anything for compensation for what's happened — that was very blind of them . . . as all the new houses went up, there should have been other things.

"I can understand now how frightened the Islanders must have been", another newcomer with close contacts with local people said. "Their whole life was threatened, I think with all the changes. . . . it was a very bad time . . . when we first moved 'ere, now I look back . . . I think I probably would have been as scared as everybody else . . . It's not knowing, I think, was the worst thing for them. Not knowing what was gonna become of them."

New frustrations and fragile liaisons

Newcomers who forged links with established residents not only came to sympathize with local people and the way they had been treated but were always mindful that there were sensitivities that they should avoid. This frequently led to a deference in their dealings with established Island residents, aptly demonstrated during a meeting about a riverside development that attracted both new and established residents:

> One of the private residents was speaking and . . . was obviously quite nervous, 'cos you're actually in a local community centre . . . and had previously heard . . . chaps getting very irate with this planning bloke, you know, where they've earned their living on the Wharf for 30 and 50 years . . . And there were a lot of Islanders there and . . . at the end she said ". . . I've only just moved 'ere" so she was actually making a point — don't think I'm being pushy, or patronising or anything else — but she actually felt uneasy and an old fellow shouted at the back he said "But you live 'ere" and that is coming from an old Islander, you know, you are 'ere, you live 'ere, you don't actually have to say I've just arrived. What they wouldn't like her to say was "I am an Islander", you know, but you don't actually have to apologize for living 'ere and that's what she was doing but I understood why she was doing it and I felt sorry for her. [Established Island resident]

The ambivalent emotions that established Island residents felt towards new-comers strongly influenced the feelings of those who got involved in community groups like the AIC. "I feel very much the new boy," one newcomer explained, "don't want to trespass." This was "perhaps partly their making", he continued. "That's unfair in the sense that I've never ever had any one other than greet me and welcome me." But this experience was readily appreciated by a woman who had been involved in community politics for almost twenty years. Some newcomers did want to get involved but, she said, "They're not allowed to . . . they've never been welcomed." "They're welcomed from the chair" but underneath the message was "piss off we don't want you, we don't want you knowing what we know, how dare you, you might wanna take over." This feeling left one newcomer in the local church feeling that her contribution was not valued:

> Whenever I made a suggestion . . . nobody took any notice of me . . . They were quite happy for me to take my turn to do the flowers and do every-body's else turn when they were on holiday. [But] they didn't want me to alter anything. They wanted everything to stay the same as it was. I was just going to make helpful suggestions . . . they just didn't want to know. And I feel I shouldn't press it because it isn't really my, I don't think it's really my church.

Another newcomer was very frustrated:

> You can see so much that clearly needs to be done and you have the energy to do it but your energy is not required. I don't like that a bit. You're not really wanted, your input is not valued, you're not seen as an important resource, you're not seen as anything. You're excluded. Now it just so happens that that doesn't bother me. I know to expect it but as a newcomer I'm frustrated when I think that there's a lot of phenomenal talent on this place and nobody knows about it.

Despite fears that they might take over, newcomers themselves had a very different agenda. "I wouldn't like to sort of bulldoze my way in . . . I feel I'm very much a newcomer really", one said. "I think people resent newcomers coming in and taking over . . . and I certainly wouldn't want to do that because then you are creating two communities the new and the old and I think we should be breaking that down." Another said: "I come from a background . . . where I did time in my local village as a member of the Women's Institute and various things and . . . I had to learn to be one of them . . . Now here is very different from a country village, you move in slower funnily enough because the apathy here just blows your mind. When you think about it you've got half a dozen, you can count on the fingers of one hand, the people who actually do anything." But "the thing that I'm finding so hard to cope with" she continued "is what the hell they're bloody aiming at. What is their real direction?" — issues that are discussed in Chapter 6.

In his study of the American surburban housing development, Gans' (1967:141) observations closely mirror what happened on the Isle of Dogs:

> What brought a community into being in Levittown . . . was . . . a complex process of external initiative and subsequent internal transformation that

produced organizations and institutions which reflected the backgrounds and interests of the majority of the population. The new Levittowners had not thought much about their future community life before they arrived, but once they had settled down, they chose to enter the community being founded for them and then to alter it to meet their requirements . . . people went into and reshaped the organisations *on the basis of needs that developed in the situation in which they found themselves in Levittown.* (Gans 1967:141, emphasis in original)

On the Isle of Dogs, newcomers were coming to an established place, even if the housing developments they occupied were new. The situation in which they found themselves was one where accommodating and becoming absorbed into established community groups allowed them to make their own contribution and change from within, yet at the same time without changing the dominant tenor of an organization like the Association of Island Communities, for example.

Newcomers would "find their niche and join in", one activist said, but he did not "see . . . people . . . capable of bridging the gap (between old and new)." "I think at the moment there are a number of people that can stand astride those stratas . . . There are camps, if you like, of self-interest." But whether they would ultimately "pull together in the same direction" was more questionable. However, as another commented, "They are people . . . and if you're gonna have any chance at all of surviving as a community then you've gotta come together."

Conclusion

They're [newcomers] now accepted as a matter of fact. I think it's a sign of the times. . . . People have come to recognize . . . it's here, and let's make the best of it. (Community activist)

The gap is closing . . . I don't think there's us and them so much anymore. (Newcomer)

This chapter has outlined the experiences of newcomers who moved to the Isle of Dogs, their attitudes towards the development and motivations for moving there. Their accounts were diverse and challenged many of the stereotyped notions of "yuppies" that prevailed in the media coverage of Docklands in the middle to late 1980s. Despite a belief that trouble was inevitable and that poverty and affluence could not coexist, there was little trouble and much accommodation on both sides. "Most of the people I know have got no objections to the private developments providing they don't push the local community out to do it", an Islander said, while another woman who had also lived on the Isle of Dogs all her life said: "I . . . like having the new people as well. It's nice. It's refreshing." Some established residents even came to sympathize with newcomers caught in the recession and the collapse of house prices in the late 1980s. (One report suggested that 99 per cent of first-time buyers across Tower Hamlets who purchased property in the last three months of 1988 had negative equity, a higher figure than anywhere else in London (Churches Standing Committee for Docklands 1995 in Lupton 1997:24).) As this Islander said:

> There's a bit of sympathy for 'em, actually, because with the big crash a lot of 'em had to lose their homes and had to sell up and you know the prices of houses have actually come down where they were shooting up. But they're still out of our reach, miles out of our reach. I think we've just come to accept the fact that [they're here], we've accepted it . . . and they're living in their world and we're living in our world.

But as the individual accounts (pp. 188–95) also revealed, some worlds were not as far apart as they at first appeared. Despite the evident contrasts in lifestyles and expectations, friendships developed between individuals of contrasting class and income, and the intensity of the commercial development led to a "common struggle" against a "common enemy", though the numbers actively involved in campaigning were very small.

Newcomers, despite the talk of hostility at the outset, did not in the end represent enough of a threat, because those who wanted to be were absorbed and came to value the positive sides of the Island and its community life, and those who did not largely kept themselves to themselves. "The yuppies haven't had any impact socially", a man who moved in to the very first private development on the Island in the 1970s said. "The explanation tends to be that they go somewhere else at the weekends or they stay in their houses as much as anything. They're not really interested in the Island. They just want a place to live." Certainly their lifestyles were largely private and self-contained, forming, as one newcomer put it, "a different kind of community within the community". "A lot of people say to me", one said, "that they'd rather 'yuppies' to the sort of hooligans [and] the criminal element that were coming down here."

Some newcomers, however, were interested in the Island and, much to their surprise, discovered that they liked living there. They found a sense of place and community, not in the "glitsy" new Docklands city, but in the remaining parts of white working class life on the Island. Despite the fact that the numbers of Island residents with very established roots in the area declined progressively through-out the process of development, many residents from whatever walk of life, rich or poor, white or Bengali, still felt the symbolic power of the established white working class population and the similarities between newcomers' accounts and those of the Islanders and established residents outlined in Chapter 1 were marked. "Sometimes I'm amazed how I can't go to the shops without having to stop two or three times", one newcomer said, "but it's nice you know because I come from a village and it's sort of like that, you know, people are friendly if you make an effort, if you say hello and chat to them". "Although I'm the newcomer in the sense that I've moved on to a new estate from an area outside Docklands," another said, "I don't feel like an outsider, I never have." "I've got so much time for them [the Islanders], I really have", the young advertising executive said. "I love the place. . . . it's the people." "I've made this my home", said another. "This is not a passing through stage for me. This house isn't a passing thing. I've made this into a proper home."

Those newcomers who got involved sought to fit into existing community structures, exhibiting a deference to the established residents rather than openly challenging them, while others simply commuted in and out to work, rarely making contact with the wider community. Whichever course they chose, they were rarely perceived to have "overstepped the mark", as this elderly Islander said:

> It's just their lot, it's just their gain as long as they don't tread on your toes; as long as they don't take something off you that belongs to you, why worry, let them get on with it. It's only when they overstep the mark that you would have reason to and I don't think they've overstepped the mark.

Although they were in no way responsible, it was the arrival of Bengali families, frequently forced to move onto the Isle of Dogs against their will, who were perceived to have "overstepped the mark" because they were, unlike the affluent newcomers, in direct competition, occupying the same territory and social housing. However, as Elias (1976:xxx) argued, most importantly they were also less powerful and the focus on "racial" or "ethnic" issues, he said, "singles out for attention what is peripheral to these relationships (e.g. differences of skin colour) and turns the eye away from what is central (e.g. differences in power ratio and the exclusion of a power-inferior group from positions with a higher power potential)".

Sadly, the strategy worked and the British National Party for a brief period succeeded on the Isle of Dogs where Class War failed. "The locals who are into trouble, they're into trouble with the racist issue and they don't bother anybody else", a newcomer observed, "and that really is happening . . . That's probably why we've been left alone." In fact in the debate concerning access to council housing for the white working class, newcomers found themselves cast into a neutral and even benign role, aptly illustrated in the prejudiced comments of this Islander:

> I think I would prefer to welcome them [affluent newcomers] here than the other way [the Asians]. Those sorts of people [affluent newcomers] are the people would look after their property, would keep everywhere nice. But the other way around, you don't know who you're gettin'. [Asians] come and they're slums in no time you know.

Elias (1976:xxvii) suggests that "established groups with a great power margin" like that enjoyed to some degree by the white working class on the Isle of Dogs, especially those with extensive links in the area, "tend to experience their outsider groups . . . as not particularly clean . . . Almost everywhere members of established groups and, even more, those of groups aspiring to form the establishment, take pride in being cleaner literally and figuratively than the outsiders" because they have greater investment in it. This might begin to explain the automatic assumptions in the quote above that housing occupied by Asian families would be "slums in no time" but such assertions, and the stereotypes which underpinned them are, of course, at their most basic level, simply racist.

The plight of the Bengali families who moved to the Isle of Dogs is described in Chapter 7, and the battle over public housing that ensued as housing policies changed was a very bitter one with some very unpleasant overtones indeed. There was little indication of accommodation in relation to the race and public housing issues but, as I have described here, and as I move onto in the following chapter, the criticisms of the development raised by both established and new residents and those of developers that I mentioned in the last chapter eventually forced both government and the LDDC to take a different approach and in so doing moved into a phase where for a short while the community took a higher profile.

Note

[1] To retain anonimity pseudonyms have been used and certain details have been changed in the interview extracts.

Chapter Six

"A slice of the cake"

The Corporation didn't change overnight. It changed because of what people had said outside because of what lots of local people had said, local authority members and politicians, and it changed because there were lots of people internally saying we ought to hear what people are saying, what they're saying is perfectly sensible. (Senior executive, LDDC Community Division)

Docklands isn't just development is it, because regeneration isn't just development . . . It's creating new communities. It's getting the new community integrated into the old community and getting the old community wanting to integrate with the new community — that's regeneration. It's saying to both there are benefits of the development. (Business journalist)

The power to change things

The anger expressed by local people about the development was anticipated by the LDDC from the outset. The disillusionment of newcomers and the business community was not. Yet, as the accounts in Chapters 4 and 5 demonstrated, the LDDC found itself under fire from all directions as the Docklands dream became discredited. What was initially perceived as the unjustified complaints of a few politicized activists soon began to extend into more powerful circles where views could not so readily be dismissed, especially when they were accompanied by sustained and negative media coverage and parliamentary select committee reports criticizing the LDDC, as this community representative said, "for its lack of social commitment, lack of training and targeting employment to local people".

Although a representative from one of the leading campaign groups said they had been "a thorn in the side" of the LDDC, with "enquiries to challenge them all the time", they recognized that "the major factor" in altering government and Corporation policy was the intervention of the business community, especially developers, described by one businessman "as a powerful force" with the ear of government ministers. However, the vociferous opposition of a minority did draw attention to the damaging potential that high profile opposition might have and this, alongside other problems "caused great consternation in the upper echelons of power", an LDDC employee said, not least because some of the most disruptive aspects of the development, including much of Canary Wharf, and the major transport infrastructure were still being planned at this stage or had only just begun. "People aren't complaining at the end of five years", a businessman observed, "they're complaining after the *first* five years when they've got potentially another five and maybe another five after that." "Business doesn't thrive on conflict with the local population", another local businessman said. "For heaven's sake, they form part of it", and it was this which led some businesses to perceive "the need" to get involved with the community, something that might not have occurred, as this journalist remarked, if the community had "just laid back and waved their legs in the air".

While the Corporation were roundly criticized for their neglect of social regeneration, those within the organization perceived much of the criticism as unfair. "The original concept" for the LDDC "was a fairly simple one" said one of its executives:

> If you look at our constitution, what do we have — a bit of planning authority, so we can say yes or no to anyone who wants to build developments here, we're an acquisition authority very much in the initial stages . . . we're not even a public transport authority, we're not a highway authority, we're not a housing authority so the whole idea was to initiate change, act as a catalyst and then really ebb away.

"We felt quite frustrated", he continued. "There we were being knocked on the head by select committees and yet we didn't have the money . . . it wasn't until 1987 when Margaret Thatcher was being criticized in terms of her attitude to the inner city . . . and she was told to soften up, that we were able to actually start getting money."

It was apparent, as I mentioned in Chapter 3, that behind the scenes the LDDC had recognized the need for investment, especially in social housing, as early as 1985 and that while, as Reg Ward argued, there was no change in the Corporation's remit officially, they "created an illicit and limited freedom on social and community support" in a hostile political climate both centrally and locally. However, as one of his closest allies said:

> Reg handled the Board and the D of E [Department of Environment] and ministers appallingly badly. He felt, I think, mainly contempt for them and made no attempt to hide that. He ignored their instructions. He did what he wanted to do . . . and there came a point when they wouldn't put up with that any longer. Now in a sense Reg's demise was his own fault . . . My belief is that they should have been big enough to say "Yeah the guy's a pain but look what he's achieving." But that would have been asking a lot of them and they weren't that big.

During Christopher Benson's Chairmanship of the Corporation (1984–88), he too expressed concerns about the residential population. Described by the Chair of the Docklands Forum as a "Victorian gentleman" (*The Times* 19 April 1985),

Benson said he was surprised at the depth of "hopelessness around at that time". "The turning point, or the revelation that we had to be more considerate of local people", he said, came on a visit to Southwark where "agitators" had

> actually shown me where they were being deprived, where they've been shut out, where they were not being considered . . . and drilled into me, the need actually, to come back and look at our role relative to local people and we started trying to put out feelers . . . to bridge . . . with the local people as much as one could, and getting rejected pretty much by that time because they were suspicious. (Sir Christopher Benson)

Consequently, the Corporation and the Conservatives, now into their third term in office, were shifting towards a more interventionist housing strategy (LDDC 1998:20) prior to the critical House of Commons Employment Committee and Public Accounts Committee findings. Once these reports were published, however, they were forced to act:

> Government ministers cannot easily ignore criticism from House of Commons committees and it was obviously important that Docklands met with local as well as international approval. The need for greater cooperation was obvious. Housing and community improvements were both areas where greater emphasis and spending should help to knit new and existing development together and create a better political and media climate. (LDDC 1998:20)

The recognition, as one LDDC employee put it, that more should be done "to make regeneration work for the people who live there" led, for a brief period, to community issues taking a higher profile and indeed it was even suggested that it became "flavour of the month". The Corporation entered what one board member referred to as "a consolidating period", designed according to a colleague "to bring law and order in to our affairs rather than more vision." "That's when Michael Honey came in with his local government experience [and] a slower pace, tidying up administratively . . . this tremendous effort". Honey, a former local authority chief executive in Richmond upon Thames, who had also at one stage in his career worked as a city planner for the Boston Redevelopment Authority, immediately set to work on an altogether wider vision of regeneration in which addressing community need had an integral part in the regeneration process. As one of his staff explained:

> It wasn't until about 1982/83 that we appointed a lone single community officer to handle community liaison whereas with Michael Honey . . . he came on board in May 1988, by the autumn of that year we had a new community division, with a chief officer, with a fairly impressive budget and appropriate staffing resources. . . . Here was a sort of 60-strong department. So we geared ourselves up in organizational terms to be much more positive, much more proactive, in dealing with the community.

The change of chief executive and political climate in central government also offered the opportunity, according to Sir Christopher Benson, for the Department of the Environment "to begin to understand what it was that we are seeking to do in Docklands" because regeneration is not "just building things and selling land" and "I think they (DOE) actually felt that's all we were about." "By being very open with them", he continued, "and taking them almost into a partnership as it were, one got an enormous amount of support."

This chapter describes the changes within the LDDC as a result of the "softening" political climate and the way in which conflict between the LDDC, developers and local people was replaced by conciliation (see Foster 1992), at least for a short while. The difficulty was that raising the profile of social regeneration coincided with a "dead housing market, the recession and administration for Canary Wharf, by then a potent symbol of Docklands"; and while relationships improved with the centre, "the road back to departmental trust and the winning of public approval and local acceptance was uphill" (LDDC 1998:22). As in other chapters, the focus of the discussion is on the Isle of Dogs, not the broader area covered by the LDDC's remit.

Social regeneration: a very different approach

> The most instructive thing you can do when you first come to the Isle of Dogs is walk round the Enterprise Zone and then step one foot outside it and feel the difference, the quality of the environment immediately changes, not just because it moves straight from office block to residential but because you're immediately in a very typically run-down, poor, urban local authority housing [area] but with . . . their view and their lives dominated by the monoliths on the other sides of the road. (LDDC, Community Division employee) (See Figures 6.1 and 6.2)

The LDDC's community division established by Michael Honey, and headed by Elizabeth Filkin (recruited from her post as chief executive of the National Association of Citizens' Advice Bureaux), began with a critical view of the Corporation's approach in its early years and had aspirations for an altogether more expansive notion of regeneration than their predecessors with, as a senior executive believed, "the opportunity of cracking urban deprivation . . . both in the . . . arbitrarily determined little bit called the UDA [Urban Development Area]" and also "in surrounding neighbourhoods". The hope was that by the time the Corporation "finished its job" adequate public services would be in place "across education, housing, health, social services, emergency services and the . . . voluntary sector"; that the Corporation should "secure benefit for local people [which] was quantified and real"; with "other arms of the Corporation . . . geared towards trying to consult better . . . to ensure that people's voices were heard". A very ambitious agenda indeed.

The Community Division, staffed predominantly by people from local authority backgrounds and attracted by the opportunity, as one of the team said, of "having money to spend" at a time when local government was starved of funds, not only sympathized with local people's feelings about the development and their exclusion from it, but were firmly committed to consultation and partnership

Figure 6.1 The Enterprise Zone, Janet Foster

Figure 6.2 What lies outside the Enterprise Zone, Janet Foster

with the local authorities because without it, they believed, the development would not succeed. "People were also attracted", a member of the division said, "by the changes that were going on and the possibility of being at the beginning of something and actually planning it, planning services in a better way." The emphasis on planning, partnership and consultation was far more reminiscent of a local authority agenda than the anti-planning dynamic development model of the Ward era, and there was a real need for it.

The merits of social inclusion in the regeneration of Docklands were self-evident for the community division: "Most people . . . recognize that for all of the monolithic city that's being created over there, to succeed they need a skilled local workforce, and [one] that actually participates in that", a member of the team said. "They also need a local resident community", she continued, "divided as it will be between private residential areas and local authority residential areas where there won't be the gaping gulf of access that there is at the moment." Local people's anger and disillusion was also readily understood. "You've only got to work here for a short while before you start seeing the Island as the Island and the rest of the world as the Rest of the World, it's a really strange thing", another of the team said:

> If I feel that after four months working here, if I'd lived here all my life God knows how strong the feeling of identity and pride of place that I would feel and I think that's what you find amongst the people here, they have a very solid history and sense of identity. They don't like what's happening because they feel threatened, they feel, . . . they won't be here in ten years' time. They really do need to be made to feel secure and they actually need to be made to feel that their kids will be here and that their kids will be working here in this lot and actually have access to economic well-being [and] working skills as well.

If the social infrastructure had been put in place earlier in the programme, said another, it "would have been the key way of demonstrating to local people" that the Corporation was "serious about benefiting them". Although this analysis was undoubtedly right, perhaps because they joined at a time when central Government were committed to more social intervention, they, like other critics did not appreciate the constraints placed upon the LDDC originally. "Housing is the most obvious manifestation of where money could have been spent more imaginatively at a much earlier stage so that . . . from day one local people would have had their immediate lives, even if not their economic base, but their immediate environment improved", a senior executive said. Instead "for seven years it was talked about getting some new build low-rent social housing on the Isle of Dogs that was in a prime site". It was not until the early 1990s that Mast House Terrace, a riverside site with almost two hundred low-rent housing association dwellings, was built — a delay that undoubtedly made people feel "very disillusioned". Indeed it did, but the mood centrally and locally was very different by the time of Honey's arrival. It was difficult for anyone coming into Docklands in the late 1980s to believe that there were no "knowns" about the commercial or residential development at the outset, and what a struggle changing the image and attracting inward investment had been. The politics of partnership, even if the LDDC had wanted to pursue them, would have been almost impossible given the history and conflict over Docklands development in the decade before the LDDC's arrival and the conflicting positions of the Conservative government on the one hand and the Labour controlled local authorities on other.

By 1989 the scene was transformed. "The Government had made clear that the LDDC must deliver housing for local people at prices they could afford within a balanced housing programme" and "so emphasis was switched", given the recession, "from owner occupation to social housing." The LDDC became "a social housing development agency within the area. Acting as a financial catalyst it would help fund the local authorities and housing associations to build new homes for rent and contribute towards the costs of modernizing existing estates internally instead of simply improving the external environment" (LDDC 1998:20–21). As estimates for refurbishment of existing housing were put at £220 million, the emphasis was placed on new build housing, which a Board report said "would demonstrate that it [the LDDC] has been able to promote access to housing for local people", with a budget of £50 million, of which £30 million would go to a new build programme, and £20 million to refurbishment (LDDC 1998:21), a drop in the ocean compared with the estimates of what was required but a welcome one nevertheless.

The LDDC's social housing strategy also extended into the three borough areas as a whole, not simply those that fell within the Urban Development Area (Hillman 1997:45) — a very important gesture because housing need, though bad in the UDC area, was equally bad and sometimes worse in areas outside it. However, this raised new conflicts, as I describe in Chapter 7, because "in an area of acute housing need with increasing homelessness, the first priority must be . . . to increase numbers of available dwellings, however poor the living conditions of those already housed" (LDDC 1998:21). These new homes became the focus of a very bitter and public battle between Island residents who felt that after years of inconvenience and little benefit at last they might gain something tangible, a new home. The Community division and the local councillors, working on a borough-wide needs-based system had other ideas.

"Battles from within": physical versus social regeneration

There will always be a tension between the community side of regeneration and the people who just want to build roads and that's reflected within the Corporation, within the Board itself . . . Although we're supposed to be a nonpolitical organization, we have a very high-profile political position and everything we do reflects in local politics and national politics and whilst it is true that this current [Conservative] government sees this as a very important experiment, they'd never allow the LDDC to flounder, they also are concerned to make sure that it doesn't achieve its objective at all costs, it achieves its objects within the parameters that the government want to set down. (LDDC Community Division Employee)

Although there was a degree of consensus about the need for investment in the community, and powerful voices in government and business wanted an end to embarrassing and negative reports and publicity, the Corporation's primary task was still perceived to be physical regeneration. This left the community division in a difficult position, frequently treading a tightrope in which, as one member of the team said, it was "difficult to keep the priorities as the priorities".

The narrow remit of the Corporation's initial approach was based on a firm belief, as one former Board member said, that "health, education and housing rested either with central government or the Boroughs" and, as an LDDC executive said, on a "naive assumption" that these other public sector agencies "would be able to keep up with" the Corporation. Given the political and financial climate in the public sector at the time with statutory agencies, starved of cash, often ideologically opposed to the Corporation, and excluded from the regeneration process, this was not very realistic.

By the time of Honey's arrival, the level of criticism, and the need to get major transport infrastructure in place, made partnership, especially with the local authorities, a necessity. Ironically, therefore, a central government committed to reducing state intervention became embroiled, through the Corporation, in developing partnerships with local government and other agencies, funding a wide range of "social regeneration" projects that were normally the realm of other public services. A member of the community division explained the difficulties:

> We are not a statutory authority. We do not have social services functions, health functions or any of those things. We can only work in conjunction with our statutory colleagues . . . Those public authorities are . . . dramatically under-resourced and unable therefore to deliver a good quality of service sufficient to meet the needs of the population. The degree to which we are actually able to pour and pump money into those services is a tricky political balancing act: how far will the DOE go? . . . Trying to get health schemes through . . . is a major headache 'cos . . . they're sort of saying "Well what precisely has the Health Authority contributed to this project?" and the classic one . . . in trying to get resources into schools is "Well why aren't [the education authority] paying for this?" . . . So that's the level of frustration, what exactly can you do, how can we actually prop up a very ailing public sector, how can we really assist?

The commitment to these "softer" aspects of development was not always popular either. "The Community Division is slightly at odds with the rest of the Corporation", one of the team said, while another described their relationship with other parts of the organization as "delicate":

> There [are] still a lot of people in the corporation who have been used to the Corporation being about economic and building regeneration not about the kind of social objectives that people like myself and my colleagues bring into the team. . . . The divisions that are here to reclaim the land and sell it see that as the most important function of the Corporation; those who are here to build the roads think that's the most important part. They are generally people who have come from a commercial background and view things in that property orientated way and we are having to re-educate internally.

"A lot of the . . . community division or the *New City* [as it was later renamed] . . . are seen as loony lefties within a very conservative organization", a local authority officer said. "They're marginalized . . . They're not part of the 'suits', which is what you have to be to get on in that organization." Nevertheless, the division had "an enormous impact", one of the team observed, bringing in women managers "in a way that had not been seen elsewhere in the Corporation"; and, because of the background of those working within it, it also raised awareness about other equal-opportunities issues too. "Coming to work here from an inner city urban local authority where . . . key equal opportunities issues . . . had been hammered home for some years now, where policies are not only on paper but actively ingrained in people's minds, it's a bit like stepping back in time", one observed.

So the division's critical view, differing experience and social commitment did not rest easily with other parts of the Corporation despite a recognition that their role was necessary. "You can't have a viable urban city when half the population feels completely disenfranchised . . . any planner will tell you that regardless of their social or political objectives", one of the team said. However, some could not "emotionally believe it", said another, "that without getting the community and the boroughs on our side we could not have built those roads".

Consequently, ongoing power struggles and conflicts developed within the LDDC that were not helped by the less than smooth transition between Reg Ward leaving as chief executive in 1987 and the appointment of a retired army officer, Major General Rougier, who was only in post for three weeks before Michael Honey was appointed in 1988. Many of Reg Ward's core team left the Corporation shortly after his departure, some of them expressing concerns about the political machinations that had led to his leaving. "It was criminal", one of his closest allies said,

> Reg is very special in that he is truly innovative, a truly creative force . . . He's covered in warts, but at the heart of him there's a truly original mind and there are very few of those left. . . . You could argue . . . that Reg's loss at the time he went was not a terrible loss for the Isle of Dogs, Limehouse or Surrey Docks . . . he'd made his mark on that and the technicians would finish it off, but the Royals hadn't been started and it needed Reg to get that started.

When Honey was appointed, he brought in a new team and some "Wardites" who remained struggled to adjust to their new environment, as one senior executive explained:

> [Honey] certainly inherited a lot of problems — never claim that Docklands development is easy — but one thing he did inherit internally, he inherited a team that was full of enthusiasm and zeal and ambition. OK that was a little dented because of the pain of Reg's departure but that could easily have been restored. Instead of which he converted it. They've been involved in, as far as I can see, two and half years of internal bickering, office politics. He's forced out of the Corporation some of the most able people to replace them with his own clones. (LDDC executive)

"It was always quite difficult", a "Honeyite" senior manager recalled "the old guard . . . had become rather defensive about why other things weren't done" and this combined with professional differences between the "people-type professions" (local authority, education, recreation) versus "the bricks and mortar-type pro- fessions" (engineers/property surveyors) with their "different professional back- grounds" and "values" which "inevitably led to quite a lot of tension". This situation was exacerbated by competition over resources as Michael Honey's announcement of £50 million investment for community projects inevitably resulted in less being available funds for other parts of the organization.

Edging towards a pragmatic response: the local authority gets involved

> The local population and the local authorities have moved from a position on their side of armed hostility to unarmed neutrality. (LDDC Executive)

The need for partnership with the local authorities was self-evident from the Corporation's point of view, because the bottom line was that the Limehouse Link road could not have been built without their consent. Similarly, partnership was also desirable in the Royal docks where plans were underway for a major new swathe of commercial and residential development (LDDC 1998:18) to avoid the kinds of difficulties which had emerged on the Isle of Dogs. There was a new mood in the local authorities too, who now saw an opportunity for community gain from the development. Prior to Honey's arrival, negotiations began with Newham and Tower Hamlets, in which "the development momentum and major infrastructure" were "used to persuade Government" (Reg Ward, written communication, December 1997) to sanction "accords" intended to enhance "the quality of life for many local people including the provision of new social housing" in return for permitting the Limehouse link and development in the Royals (LDDC 1998:18).

Although the notion of partnership was straightforward in principle, it was not unproblematic in practice, especially in Tower Hamlets. Apart from a brief period at the outset, there was no local authority representation on the Board until a Liberal council were elected in 1986. (Southwark and Newham did not elect members to sit on the Board until the early 1990s). However, to complicate matters, under the Liberals' decentralized structure, seven semi-autonomous neigh- bourhoods were created and the Isle of Dogs was one of two that remained under Labour control. "The Labour councillors are on the whole fairly far left and don't seem able to work with the Liberal council in any constructive relationship at all", a professional working in the area said:

> So we're stranded within the borough. And they also can't work with the LDDC who are an unelected quango imposed upon us who aren't doing things for local people. I can see the line but actually it's very destructive not to work with them. So it means that you've got no inputs into the real decisions that are being made for the community. And I think that leaves people here feeling particularly . . . isolated.

"The whole local neighbourhood/central frictions" operated "all the time", a member of the Community Division explained, and illustrated the difficulties with an example:

> We produced some social housing leaflets . . . in consultation with each of the boroughs and the neighbourhoods . . . Wapping when they received their draft copy thought it was wonderful and ordered 10,000. Isle of Dogs Neighbourhood took one look at it and said they thought the whole thing was patronizing and arrogant and the LDDC should not be involved and the chief executive . . . issued instructions [that] these leaflets are not to be seen in their Area Offices. That's the same local authority, over the same leaflet and we consulted centrally but we didn't consult with the particular individuals.

A report in *The Independent* captured the tone of the local Labour administration:

> At present, the Isle of Dogs (Neighbourhood) council is run by Labour from Jack Dash House, a luxurious post-modernist centre leased to Tower Hamlets at a peppercorn rent by the London Docklands Development Corporation. It was named by Labour councillors after the Communist docker who lived in Millwall and whose constant wildcat strikes contributed to the destruction of the Port of London Authority a generation ago — which says a lot about the sort of Labour Party Millwall boasted. (*The Independent* 29 March 1994:17)

The "political in-fighting" that characterized the relationship between the central Liberal administration and the Labour neighbourhood was further compounded by conflicts at a local level. "The politics was very complicated", a former LDDC executive explained:

> The relationships between the neighbourhood committees and the local residents' association were horrendous, particularly on the Island where I think it was personal and the AIC and the Isle of Dogs Neighbourhood Committee were absolute daggers drawn. . . . The internecine warfare was fierce. Add the LDDC to that and it's not surprising the result was pretty horrible in one way or another.

"If (councillors) saw there was political capital or political point scoring to be made at the expense of the Liberal Democrat town hall, or central government, or the LDDC . . . then that was a high priority", another explained. "Never mind the issue, never mind necessarily the rights or the wrongs or the benefits of the particular issue. Go for political capital . . . rather than what makes the most sense for our community." Not surprisingly this resulted in intense frustration. "We've lost a lot of ground because of their views and because of their political machinations" a community activist said. "They are the people who really have the power, some of 'em, not all of 'em, but they have more than we have, they can pull certain strings that we can't. So we didn't ever, never had that support." Eventually the discontent and frustration reached the stage where two prominent AIC members "stood against" the Labour councillors in the local elections. "We done very well", Peter Wade said; "we got the highest vote in London as Independents and I failed by 102" to get elected. Not surprisingly, this episode did nothing to improve the relationship between certain sections of the Island community and the elected Labour councillors.

Despite the ensuing conflict on the Isle of Dogs between local councillors and the central Liberal administration, and between these councillors and some sections of the community, the partnership between Tower Hamlets and the LDDC marked a "new phase" a senior executive said in which the "accords" led to "a joint understanding about how we were going to proceed in new and urban development and its impact on the community".

A golden opportunity?

Not only was local authority consent needed to sanction the Docklands Highway which government, developers and the Corporation wanted, to ensure the viability of the development, but in order to build it, poor, run down and increasingly scarce council accommodation needed to be demolished to make way for it, leaving over four hundred displaced households requiring new housing. Given the extent of opposition to the Canary Wharf development, there were real fears that the road might, as this senior LDDC figure said, become the focus for "serious unrest". The company responsible for building the road, acutely aware of how politically sensitive it was, for "the first time in the history of the construction industry" recruited a public relations officer to liaise with the community affected.

Local Labour councillors, not surprisingly, were fundamentally opposed to the road and saw it as yet another example of their powerlessness to influence political decisions. The LDDC, one said, "basically want to smash the working class community by driving a road through it." Others viewed it as another method of diminishing council housing stock. Their Liberal counterparts, however, took the pragmatic course and in return for granting permission for the road gained a "package of measures to compensate or mitigate the impact of the highway" (DCC 1990:70) that involved rehousing the households whose homes needed to be demolished and £30 million to be invested in education, training and other "social, economic and community schemes" (DCC 1990:70). This deal, signed in 1988, became known as the Social Accord and was one of the few serious benefits in terms of social regeneration to emanate from the new spirit of partnership.

The "problem" of rehousing those affected by the road was undertaken by a joint task force with local authority and LDDC personnel. Given the sensitivities, "tenants had a very unique offer made to them", one of those involved in the task force explained:

> We agreed to offer . . . various options for rehousing; one of which would be . . . brand new homes on a private sector development, with a Housing Association landlord, but with "the right-to-buy" retained. Tenants were also offered two other sites, . . . just outside the Urban Development Area

(UDA) boundaries and . . . a choice of other local authority re-lets, housing association re-lets. They could go and find a property they liked and we would get the Housing Association to buy it and . . . rent it to them and they were also offered a discount which was equivalent to the number of years' right-to-buy they had [or their discount] in cash if they wanted to go and buy elsewhere privately.

At last it seemed that out of the problems caused by the development a few of the poorer people in the locality might significantly benefit in terms of an enhanced quality of life and access to good housing. "In 1988, this major exercise was expected to cost £47.5 million net allowing for the sale of surplus land and homes. In the event, the total, according to a subsequent National Audit Office report, was more than £100 million" (LDDC 1998:19).

A community division employee suggested that as a result of this major decanta-tion process:

> The majority of those people are benefiting . . . in terms of the improve-ment of their quality of life. Their homes are . . . modern, . . . well appointed and provide them with a much better living standard than they had before. . . . We've been very generous in terms of the way we've treated hidden households. [They've] . . . been given their own homes, they didn't have any choice [like those] who had the legal tenancies, but sons and daughters who were registered as homeless at home, if you like, have been given individual tenancies as a reward. Now for a lot of those young people they would have had years of waiting and probably they'd have been allo-cated something completely gruesome, and what they've been allocated is close to their relatives and a brand new place and carrying with it the right-to-buy if they want to do that in the future.

This optimistic view of the tenants' opportunities was not shared by Labour coun-cillors. "The people . . . didn't have a choice whether they benefited or not", a councillor said, and expressed concern that the transfer of council tenancies to housing association did not offer the same rights and protection for tenants. The LDDC were accused of buying tenants off.

At one level it is difficult to see how an improved quality of life and better housing could be anything other than beneficial. Indeed, for some of those moved to the new estate on the Island, their lives were transformed. An elderly woman who moved in to a house with her husband and sister said it was "like a dream come true". Her home was a "palace" and she saw herself as the "lady of the manor". The poor status of the 1930s estate due for demolition meant that some of those who had received the poorest housing allocation in the past, among them a significant proportion of Bengali households, at last had access to good housing. One Bengali woman described her feelings about the new estate:

> We were really excited . . . My parents . . . were really happy when they saw it. They said it was really nice and everything. . . . We thought ooh it's going to be famous here, everyone talked about it . . . brand new houses. . . . Docklands, we thought it's going to be a special place and everything.

We were very excited. We told all our relatives and everything that we were getting a lovely place.

Despite the benefits, there were mixed emotions. "Some people were angry, some people were frightened", a former resident recalled. "There was a lot of anti-feeling . . . more than might be made out by certain agencies like the LDDC who just saw it as a wonderful opportunity for everyone to move." "There was a lot of old people", she continued, "who had lived all their lives on that estate and to them it was disastrous . . . they just didn't want to move . . . Even though they might be moving into a brand new house, a nice new house, it didn't compensate for the change, they couldn't tolerate the change." Others experienced difficulties exercising choice — a concept with which many were totally unaccustomed. "They really had no choice", a community development worker said, about whether or not to move, their only choice was where they would go:

And I think people forget how much that can affect people. All right, they've got lovely houses but it wasn't their decision — the only choice they had was where to go and they don't necessarily know if they've made the right decision. A lot of them didn't want to move down here . . . some of 'em did and now they're down here they don't like it. Some of 'em didn't want to move and now they're down here they do like it.

There were also "interest groups encouraging certain people to go to certain places for certain reasons", she continued: "There were Bangladeshi people with power in the community actively discouraging people from moving down here . . . [because] the more that the community splits the more the power goes. . . . [In] the white community . . . there's a strong feeling among some of the people . . . that they were used, that they weren't given the options of moving in the way that was best for them." These dilemmas were reflected in this woman's experience:

For me personally it was good because I had a one-bedroomed flat and I had a baby coming . . . but it wasn't my first choice to come here. It's just that there were personal things going on at the time and I didn't have time to sort myself out and this was the easiest option, just to say we'll come down here. . . . I wanted to move somewhere else but it would have involved more work and I had to find the house . . . do the research . . . A lot of people didn't know about being able to buy a house. . . . A lot of people are wanting to move off already. If I'd known what'd be like I wouldn't have moved here.

Those involved in rehousing tenants were confident that the move was in their best interests:

Although certain people will tell you that there was unhappiness about the way it's been handled. I think the majority of people cannot believe how lucky they are being moved from very, very run down [19]30s blocks, which were pretty unforgiving environment, infested with cockroaches, very poor conditions, to brand new homes. All the chance for mobility wherever they wanted in the country. . . . For some people it's been able to mean a move nearer family and friends outside the borough.

David Widgery, an East End GP, had a very different interpretation of these events, describing the unhappiness of some of those with whom he came into contact. He suggested that people were "bribed to go", citing examples of a man with "bone cancer" who went "to Ireland" where he would "not be able to get proper hospital treatment", a "local community organizer bundled off to Essex with a part-paid mortgage" and "many families shunted down to . . . a soul-less fake Georgian estate which the LDDC bought off the peg to house its displaced persons . . . Rather than a plan which fits the people, the LDDC have got rid of the people who don't fit the plan" (Widgery 1991:156/226).

When I asked a key figure in the company responsible for building the road whether there was a lot of resistance, he replied: "What I encounter is a lot of hostility towards the LDDC because they didn't put double glazing in; they haven't done all the improvements that they promised the local people. They haven't re-housed a lot of people that are supposed to have been decanted because of the road going through."

What was underestimated, both by those involved in moving the tenants and by those who envied them their new homes, were the stresses and uncertainties associated with a forced move. "To the LDDC I think it was logistical exercise", a community worker said. "It's just getting things out of one place into another. They're just things that have to be moved cos they're in the way and they're not thought of in terms of people." Yet people require support and services and insufficient thought was given to these matters, resulting in a number of problems many of which could have been anticipated in advance, as this leading figure in the tenants' association, on the Island estate, where many were moved to, explained:

> There are two major . . . problems. . . . One is the caretaking which isn't sorted out . . . so [the estate's] never been cleaned properly . . . The TA [tenants' association] opted for one residential caretaker [but] . . . there isn't a property allowed for that. . . . Not all the tenants can be bothered to tidy up after themselves but it's . . . the administration that's really to blame . . . that's led to unhygienic bin areas and there are rats in two houses now. . . . It just seems like it could go downhill really fast. . . . The other problem is the ceiling heating in the flats and apartments. People . . . are getting like £400 bills [per quarter]. . . . Repairs have been a problem because everything is so designer wonderful that there are no spare parts. It's all meant to look good and not be practical at all. . . . In the mansion blocks . . . there's a service charge . . . because of lift maintenance and things like that . . . the service charge is £15 a week . . . so if you have got a one-bedroom flat and you are a young person, single on a low-paid wage then you are paying as much as someone in a three bedroom house. . . . they can't afford to pay their rent, service charge and these huge electric bills.

Unfortunately, many people, largely unaware of the problems these residents confronted, were quick to judge those who seemed to have been given a golden opportunity only to have squandered it with alarming speed. A senior figure in the LDDC said "It distresses me, that those who have moved to — have already got the place looking an absolute shambles, you know, gosh, . . . we are one of the dirtiest nations imaginable, and those people there, their garbage is all over the place, the place is looking [dreadful]."

Others were a little more understanding. "When we first heard about this decant business", one local said,

> we went there and we were just absolutely astounded at these wonderful houses, you know, with integral garages . . . wonderful flats with balconies that went round the corners, and I thought this is all right and then before [long] . . . there's dilapidated vans all over the place, you know, burnt out cars [rubbish, broken windows] there's dogs roaming the whole place — no thank you very much. But it's because people weren't given the support, they weren't given the back-up service they need. In a way people almost need counselling on as big a thing as losing their homes . . . But . . . nobody cares, we'll move them, that's it, end of story.

"This whole saga", as a community worker commented,

> is really sad when you analyse it. If it had been that the people who built [name of estate] and the local authority and the LDDC had actually planned things to happen like this then it might have been quite nice but everybody knows that it only happened by accident. There was a slump in the housing market, [the company] couldn't sell the houses, so the ideal solution was to flog it off. It's really nice that the people here have got nice homes out of it but it's by accident. It's not by design.

Whatever the failings of the rehousing programme, some lessons had clearly been learnt and people did not sit in front of bulldozers as some had predicted. That in large measure was due to the community division's work and partnership with the local authority, without which it is highly likely that further conflict and confrontation, instead of conciliation and accommodation, would have occurred. For a while, too, the changing mood of the development and its more inclusive agenda also had an impact on the ground, as the discussion in the next section demonstrates.

An unusual alliance: property *and* people

> The community doesn't seem to be as important but in the long term it is because if you have a discontented community then you're gonna have vandalism and you're gonna have dissolution and you're gonna bring your people down for business lunches and they're gonna see people behaving in ways that they don't like very much . . . There is no reason why local people should not have benefited as much as business people from local

development. . . . If they really want the Docklands to take off then every-body who's involved in the Docklands whether they live or work there has to be reasonably content. So I mean the perception that you can fob the community off with anything is really short sighted. (Community worker)

We've all got the same interests at heart, we all want the area to be success-ful and we all want business to flourish down there and want the community to flourish. (Businessman, Isle of Dogs)

Given the potential for serious opposition, it was not only those responsible for the Docklands Highway who sought to protect their interests by building bridges with the community. With considerably more resources at their disposal, and a differ-ent philosophy and approach to business, Olympia and York pursued a strategy for the Canary Wharf development in which community issues became integral to their approach. "If you're going to be successful in property development", one of their senior executives said,

you somehow have to do three seemingly contradictory things well, and if you do them all well, then they mutually reinforce each other. . . . (1) you have to produce a competitive product . . . (2) you have to produce something that is aesthetically and qualitatively good, and . . . (3) you have to do it in a way that's sensitive to what's going on in the community . . . frequently development goes off pitch if it tends to over emphasize one of those to the exclusion of the others or two of them to the exclusion of the others. (Quoted in Foster 1992:175)

"The people . . . we . . . replaced had some difficulty in the local community", he continued, because they "felt that their interests were being ignored." Con-sequently from the outset Olympia and York recognized the need "to keep the lines of communication into the local community open if we were not going to have difficulty" (Foster 1992:175).

The company's approach of building to "own and manage" resulted in "a totally different attitude to the nature of its developments and also to the sur-rounding area", one of their executives explained, because "it is going to be part of that community". "We bring 90 per cent of the [construction] material by water . . . as opposed to by road", one of the company's executives said:

Every other developer . . . has bought it all by road . . . it has cost us more to bring it by barge but it has eliminated a lot of the congestion. . . . That is an example of a decision that was driven in some considerable meas-ure by a community concern as well as a realization that . . . if there was already a traffic problem . . . if we added several thousand trucks a week to the existing congestion . . . we would simply shoot ourselves in the foot. There were commercial considerations as well as community considera-tions. In the dynamic that I've described . . . they interact with each other and this is an example of the way something that's good for the commer-cial side is also good for the community. That kind of thinking was there.

"Community relations is not something that half a dozen people in Olympia and York do and nobody else does", another executive explained, "it's something that really imbues the whole planning and thinking of the development". This was an approach that was strongly influenced by the principles of the Jewish brothers who owned the company and their "personal history".

"Immediately we noticed that we're talking to different people", a community representative said, and in an extraordinarily shrewd move Olympia and York reached out directly to the heart of the opposition by recruiting Peter Wade, a key local activist, who had been at the forefront of the campaign against the initial Canary Wharf scheme and was chair of the AIC at the time. At a stroke this "divided the community right down the middle". "He was the instigator of 'No Canary Wharf'" a fellow activist said:

> He wound this community up to a demonstration and I mean *the* community. There was hundreds and hundreds of people . . . He led the way. . . . It was a feeling of total faith in the AIC by the community . . . people were lining the streets . . . and you thought there's a man that's dedicated, there's a man that'll never turn his back on the community.

"I just couldn't understand why he done it", she continued:

> People say like, you know, Canary Wharf took a leader of the community away but if that community leader didn't wanna go then he wouldn't have gone . . . All the things he's been involved in. One can do it for the gain and one can do it for the community and all the time now people are saying he only did it for the gain, for recognition, like I'm the big I am and if you take me the community falls apart. Well, it has in that respect because they've [AIC] never been able to pick up a strong chair [since].

Feelings ran so deep in some circles that "slogans" were painted outside his home, on the community resource centre and "alongside his grandson's grave" (Foster 1992:175). He was branded a "traitor". "Those who I've clashed with over the years, and certainly on a political front, would scream poacher turned gamekeeper", Peter Wade said: "It was a bit of a rough time" but "a lot of people in the community . . . supported me." Although, as a former LDDC executive commented, "from the outside it was seen as . . . Peter selling out . . . not at all . . . [It was] another way of Peter doing what they all do — try to get the best they can for their community. And if they can do that by releasing bees they'll do it by releasing bees; if they can do it by taking the job on the inside they'll do that."

Pragmatism certainly played an important part in Peter Wade's controversial decision to join Olympia and York. "A project this size that actually gets Maggie [Margaret Thatcher] down 'ere personal", he said, ". . . you can't do nothing about it . . . you can't influence it. We've tried everything, do what you like, you're not gonna [stop it]." "Therefore you've got to start thinking about your children and your grandchildren. The community are gonna have to come up through this . . . so therefore you've gotta . . . bite your tongue and try to find somewhere within that that's gonna have some benefit." What local people needed, he said, was employment and training: "If you give 'em the weapons they'll win the war, but if you don't they've got no bloody chance, and the weapons they needed were the new skills . . . training, . . . education . . . That's why I came to work for Olympia and York." "Obviously there's a terrible mismatch in this area of skills", he continued. "We have marine and heavy engineering related . . . you've got construction and then . . . financial development", he told Olympia and York, "so they agreed to set up a training scheme, . . . for construction trades . . . in partnership . . . [with] the borough, . . . LDDC and CITB [Construction Industry Training Board]."

Although the alliance between the community and the might of a company like Olympia and York was inevitably an unequal one, there were gains for both sides. "It might be enlightenment but it's enlightened self-interest", an executive said. "We want to make sure that we've got local people that we can employ on Canary Wharf and so the reason that we target training and educate is because the skills that we need aren't there at the moment and we want to help put them in to the local community because we're gonna need staff. We want to employ local staff to do it. It's very commercial, that's very much self interest."

Initially Olympia and York, like the Corporation before them, had a rough ride with the community groups they consulted:

When we first got involved . . . in July [19]87 . . . we started getting into a whole series of meetings asking to go and talk to people and to listen to people. . . . the first of all those meetings were pretty acrimonious and what we were really hearing was people saying in various ways we've heard all this before, we've had promise, after promise, after promise over the last few years since the docks shut . . . why should you be any different why should we believe you? Our local council has been taken away from us, the jobs that we had have been taken away. We keep hearing rhetoric and nothing to show for it . . . There was nothing we could do to disprove this feeling that we were just like everybody else. We were just making promises because it takes time to do that.

"The hostility", the executive continued, "was a symptom of mistrust, of alienation, of a feeling of having been let down", aptly expressed by an activist who attended one of these meetings. "They were telling us that they were gonna put money into education and help out this and help out that. . . . [We] were looking at each other saying [cynically] 'They're just gonna change our lives on the Isle of Dogs!'" They "started talking about the workforce . . . I said 'You're not gonna get 'em from here . . . We're gonna do, like, the making of the beds in your hotel, clearing somebody's shit up down the toilet, delivering meals, room service, receptionist . . . that's all you want from us.'"

But over a relatively short period attitudes towards Olympia and York changed. "After the first year people were probably beginning to believe us and now I think we are viewed differently from others", a significant change given the extent of antagonism which greeted the initial proposals for this development. "They're eight years too late" was the verdict of one of the AIC Committee. "They're terrific, you know, with their community relations, with their setting up of job schemes, training schemes for local people", a newcomer commented. "I don't think they could be expected to do more to be honest. I think they're doing a very very good job out of a very difficult situation." A head teacher whose school benefited from

Figure 6.3 Canary Wharf, Janet Foster

a twinning arrangement with Olympia and York said: "They're very perceptive people and the right kind of people in their community relations department. People that others feel comfortable with." He continued:

> We were listing a couple of meetings ago what had emerged as positive things from that twinning arrangement and we got over twenty-six things that had occurred just in a fairly short space of time . . . we regard it as very important indeed. . . . and I don't mean in financial terms necessarily but in all sorts of support . . . I'll give you . . . an example . . . We're in the middle of persuading the local authority to do some internal modifications to this building. O&Y have leased one of their employees . . . to do a design brief for that, . . . free of charge . . . So we benefit considerably from that, we couldn't afford her . . . she's a very expensive lady who does a brilliant job.

A senior Olympia and York executive also observed a gradual, but tangible, shifting sense of place about the Canary Wharf development (see Figures 6.3 and 6.4):

> When you're messing around with a place as fundamentally as we have done, you run the risk of offending people and so I think that to a certain extent there has been an evolution of the reaction of people to it. I mean the initial reaction . . . not just to Canary Wharf but all this development

Figure 6.4   Canary Wharf, Janet Foster

was X and as people have lived with it and accepted it . . . we've moved in to Y. I wouldn't necessarily want to articulate the difference in X and Y, but I mean people are more accepting of what's going on today than they were five years ago . . . there would be a range of opinion but certainly there are positive voices of people saying this is good . . . and this is in some part a result of our trying not to offend people unnecessarily and stick their noses [in it] . . . so we try to respect the fact that there is a community there and not to do a great deal that sort of made life more difficult.

Others had discerned these changes too, but were not always comfortable about their consequences. Although, as a local priest said, Olympia and York were "at least benevolent towards the local community and . . . intent on making sure that some good things happen [and] to that extent it seems . . . possible to be positive", the underlying trends were more worrying. "We've moved into something which feels like benevolent feudalism", he explained. "When you've got a developer the size of O&Y you've got no electoral input into the decisions they're making and you're dependent on their good will for any gain that the community gets, I think actually there's an uneasy relationship there." He used an example to demonstrate his concerns:

One of the most symbolic events was the concert they organized at Island Gardens . . . It was a nice thing for Canary Wharf to have done. It cost them a lot of money ... It was a big publicity thing for them. But there were also some nice touches, like they organized a concert in Island Gardens and they gave tickets for the concert free through local community groups. So all the pensioners got in who wanted to be there ... and people in community groups — churches, schools — they got in, and the people who couldn't get tickets were the newcomers because they hadn't joined community groups. And so there were all the old Islanders in the stand and people in green wellies and yellow anoraks standing around the side. And that felt to me right. You know, I think Canary Wharf really got that right! But there was a very, very interesting moment. It just so happened it was Robert John's birthday. Robert John's the chairman of O&Y and, this was announced and ... we were all expected to sing Happy Birthday ... And, you know, I came away saying to one of my colleagues, "I'm sure we're going to be invited up to the Hall to sing carols at Christmas." You know it just felt ... medieval in the relationship between the new squire and his people. And of course we weren't invited to the Hall at Christmas to sing carols because it's a Jewish firm. But they did have a party to say thank you to all the people who'd been helpful to them in the local community which just happened to be in the week before Christmas. They're kind, they're nice, and I'm actually very fond of some of the people who I know who work for them, but the underlying relationship's quite worrying.

"Don't be kidded by what they say about ... they want to be good neighbours and all that ... I'm not so sure that ... at the end of the day that the community was going to gain anything particularly out of Canary Wharf", Ted Johns said. "We aint gonna get very much off 'em. But what we're going to get is a bit more than the others have talked about. At least ... they were putting their money sort of where their mouth was." This ambivalence was reflected in the comments of another active community worker who said: "Big business are actually overwhelming. They actually deny local opportunity ... big money coming in just buys out." This feeling is mirrored in Fainstein's (1994) comments about another development built by Olympia and York:

Battery Park City ... is a carefully co-ordinated total environment, with substantial financial benefits disseminated to the general population. Its principal private developer went beyond meeting its contractual obligations and provided the public with significant amenities. The authority has also used its funds to make the city more appealing. Nevertheless, except in some of its park planning, it has involved no public participation whatsoever. Its commercial and residential structures are reserved for the wealthiest corporations and families in the United States, although any well-behaved individual can use its public facilities on a daily basis ... The public sector could hypothetically have borrowed the funds it put into capital expenditure for Battery Park City for some other, more socially beneficial purpose. The taxes foregone might have been directed toward another, more productive endeavour that would have produced more jobs for the neediest and less spatial segregation. Within the constraints of New York's political and economic situation, however, it is hard to imagine an alternative strategy for the site that would have resulted in a more desirable outcome or that, in the absence of any development there, commensurate resources would have been mobilized for a more socially beneficial cause. (Fainstein 1994:184–5)

"It's sad . . . seeing the almost insignificant gain the community got on the building of Canary Wharf, which is the biggest office development in Europe", said a local councillor, who saw even the location of the training centre in the former Poplar swimming baths as indicative of the power of large Corporations over the most vulnerable:

[The training centre] that was the only disabled pool in East London . . . it was one of their highspots of the week and to see people like that just sort of brushed aside . . . because of commercial interest. Canary Wharf wanted it so, you know, it didn't matter who was using it, that was it, it went to Canary Wharf, to be used as a training centre. You . . . think of the resources that Canary Wharf . . . have . . . it could have built a training centre . . . just because this happened to be close. I think it was very sad that they could just sort of brush people aside and people with such obvious social needs. (Labour councillor)

Others saw Peter Wade's appointment to Olympia and York as "very symbolic" of the way in which "the local community . . . had to change gear in its attitude to what's happening", as this man explained:

What I have observed is that they have fought for what they believed in, fought well and hard but have been outmanned, out-resourced, out-gunned, everything in terms of the fight that they've tried to take on. They've won the odd little battle but the result of the war was inevitable and the way that a number of them have resolved that is actually . . . to defect and there are odd little illustrations of that . . . That's not just a simple accommodation, that's radical conversion and you can identify in other people and other ways the sort of sense of accommodation that's gone on . . . The issues haven't changed but what's happening is that people are increasingly accommodating themselves to saying don't raise the questions — it hurts too much to keep fighting and people are exhausted . . . The direction and the nature of what is going on continues and it's a case perhaps of adapt or die and yet implicit in that . . . process of adapting you're also dying.

Whatever the wider ramifications, Olympia and York until their failing financial fortunes under intense recessionary pressures, were viewed in a very different light from the Corporation or other companies in the area and their community input played a significant role in this. One Olympia and York executive suggested these positive feelings were also reflected in local people's attitudes to businesses moving on to the Wharf: "They want to welcome them, they want to involve them. They don't want the tenants to alienate themselves. So some of our tenant companies have got involved in community events, are looking for it and wanting to employ local people when they've got vacancies".

Other businesses in less politically sensitive areas also employed community relations staff, arguing that their involvement made sound financial sense and for some companies was part of their overall corporate strategy. "In [our] annual general report you've five main aims", an executive from a prominent company said:

> The fifth one is to contribute to the community in which it trades, to put something back in to the community. And I think it put back last year about £12.5 million, something like that, and it's also a member of the 1 per cent club . . . This reflects on my job, it means that the community are as important to me in my job as the developer, as the business customer. I like the job because I'm very much a community man.

Another man whose company was committed to securing some community advantage explained their reasoning. "The Isle of Dogs, if it is going to be a success in commercial terms, can't have cheek by jowl areas that really are suffering, you know; it's gotta be seen to be, not only seen, it has got to be caring in terms of its ability and willingness to listen and react." He continued:

> A lot of the managers who came down here for their companies in 85, 86, 87 were never ever told by their companies "You will go down and do this". They naturally seemed to think that it needed to be done and we never ever got together and said, you know, "This is our sort of cynical corporate approach", you know, chucking glass beads at the natives . . . that never came into it, it just seemed to be a natural thing.

Companies like "British Telecom, British Gas, Olympia and York, Barclays and a few others" who took a higher profile in community issues were not in direct competition with one another, which as this businessman explained, contrasted with their more typical involvement in community affairs. "Predominantly we work in competition with each other", he said, ". . . Barclays would try and do one over on Nat West or Midland or . . . Lloyds . . . British Telecom would want to do one over on Mercury." However, in Docklands, "the fact that there were five or six, no say ten, key managers who were very keen to do something for their own companies but were not competing against parallel industries meant that it was being done very genuinely".

But there was a quid pro quo. "To get down to basic business", a community relations man from a major company explained,

> the way that ideally you should work business is that you create a brand awareness at a very early age. If it's by inviting the schools to come and see all the [things we do here] that has an effect . . . if it's supporting a local fete, a local event that has an effect. Companies the size of ours realize the importance of that. Whether it's the right way of doing things, whether it's morally right, I could argue for another hour but it's a fact of life and it

seems to me down here that it is giving something back to the community.
(Part quoted in Foster 1992:176)

A man in a similar post in a different company developed an almost identical
argument in which community giving was linked to "product awareness":

> You are using your involvement in the community to get publicity to make
> people aware of your existence so that when they think about . . . financial
> matters hopefully they'll think about [name of bank]. That's the whole
> object of it, it is public relations linked to community giving and commun-
> ity involvement is clearly designed to plough the field, so to speak, to
> make people aware of the bank's existence and that is quite clearly the
> commercial edge. That is what PR is really all about, making people aware
> of the companies so that when they think about a cigarette or a bottle of
> scotch they think about Bells or Benson & Hedges or whatever.

Although self-interest was the motivation, it did not in itself conflict with the
possibility of community gain. In fact, some suggested that the expectation of
community giving created by companies like Olympia and York, British Gas and
British Telecom would make others likely to contribute too:

> Originally it was developers, okay, being cynical it was probably "Okay we
> will be seen to be proactive to the community like Balfour Beatty", you
> know, they make no bones about it. They've got [a community man] . . . he
> does his PR bit because they are constantly being barraged by local people
> who are fed up with the noise and the dust. . . . now you're getting the end
> users coming in . . . saying "Well we have got this money available and we
> want to get involved in the community".

The bottom line, as one community worker said, was that businesses "ought to
care about people for their own sake, but even if they don't they ought to have the
business sense to see that it's enlightened self-interest to care about people. . . . It's
worth spending money on them and their environment in order to get the results
you want". It seems that, in the conflict-ridden intensity of development on the Isle
of Dogs, this was a lesson that had been learnt.

Changing tactics and growing pragmatism

Even if the social programmes that companies and the Corporation enacted were
a pragmatic response to a highly politicized, potentially volatile situation driven
by "enlightened self-interest", the approach did have some tangible benefits and
changed views about the role of community groups in the consultation process.
The fact that certain parts of the Corporation and business community visibly
demonstrated a commitment to giving local people a better deal, combined with

the realization that there was little that could be done to prevent the development occurring, resulted in a change in tactics not only by individuals like Peter Wade but by a number of community groups, with a movement away from conflict and opposition to increasing pragmatism and accommodation.

The clearest example of this "different way" was provided by the Docklands Forum, the umbrella organization representing community groups across the Docklands area, and the consultants involved with the Docklands Light Railway (and later the Jubilee line extension). "There's been a meeting in the middle somewhere", one of the consultants said, from "very hostile" beginnings. "They more or less denounced us from start to finish", a colleague explained, for

> being a tool of capitalist enterprise . . . everything was wrong — planning, investment levels . . . It was all pretty unsatisfactory. But when the Beckton extension Bill surfaced from about that time onwards . . . they changed their tactics to using ways and means, you know, petitioning the Bill, things like that, and on the City extension they petitioned . . . and an undertaking they were given was that the railway would meet regularly with them . . . So they got plugged into the establishment as it were and things have evened up since then . . . So the attitude has changed for the better. They still don't like a lot of what's going on but they've gone about things [differently] and they're now regarded really rather seriously.

"We're actually quite good friends with them", another consultant said. "We have a very good working relationship . . . everybody's learnt a lot from those days and here we are in the middle of the Jubilee Line . . . and we, for the first time, are regarded as the engine room of the project. That's never been said to us before. [Previously] we've been the cul-de-sac . . . they'll go get us out of that little jam that we're in, but now we're the focus of attention."

"I think we've all kind of changed our visions over the period", a Docklands Forum representative said, "For instance . . . the Jubilee Line extension":

> Once upon a time people were saying this is outrageous, you know, the serving of developments vis-à-vis communities, and basically linking up major development sites with major London terminus, commuting termini, it's disgraceful. And you can't plan like that. The GLC would never have had it. But now . . . we . . . say well we very much welcome this line it's a public transport improvement, perhaps it could go via the Greenwich peninsula . . . We've all changed I suppose . . . we've all become . . . let's say pragmatic about what Thatcherism has done to the general way we plan environments.

A change of tactics was also required in other areas of community politics, as Ted Johns eloquently described:

> By 1985/86 I think, it's a bit like any general, you can't keep your troops in the front line all the time. You've got to take 'em out and I think a lot of people just sort of got sick and tired of it in that sense. You know, they

said, "well we've been over the top now so many times" you know, like the [19]14/18 war, . . . where . . . the war started and where it ended, in fact in France, you can actually see . . . they've got a stone on one side of the road which says this is where the war started and just down the road you can actually see where the front line finished when the war ended in 1918. That's where it ended it moved about sixty yards up the road. And they lost three million men or whatever it was. Crazy.

When I asked whether any benefit had come from moving such a short distance, he replied:

I think there is always a feeling that you did achieve something, and I think that there was victories, you know. But one of the things that I was saying in 1980 . . . was pointing out the danger that in Docklands gener-ally, that groups would sort of disappear in their own manor, you know, and what we needed to sort of do was act collectively.

But the pattern had been one of fragmentation, competition and growing individualism and acceptance because, as a former councillor explained: "You can still wind people up . . . but what do you do when it's here? I mean Canary Wharf's not gonna go away."

"Eyeball to eyeball across the table"

We can sit and talk about the nostalgic things and the good things and the funny times we've had protesting and demonstrating . . . we've had some great times. I mean really a community sticking together in what we believed in. [But] I think there's a different way now. (Community activist)

The changing role of the LDDC and its community brief, the activities of Olympia and York and a few other companies in community relations, changing tactics and a growing pragmatism on the part of many community groups led to concili-ation in which the battles were replaced by conversations. Olympia and York, the LDDC and a handful of other companies held regular meetings with groups like the Docklands Forum and the Association of Island Communities (AIC) to discuss aspects of the development that affected the local community. "Quite frankly", one AIC member said, "instead of [being] an agitator organization we've become a consultative organization." The AIC

used to be constantly agitating with the LDDC because of the lack of plan-ning, the lack of opportunity, the lack of information, the lack of consulta-tion, and so on. There were constant battles. That has now changed because the LDDC and the AIC have got together and decided to have regular con-sultation meetings . . . we're able to put forward our views on what's taking place. We're able to get up-to-date information on what the LDDC is doing

> ... I feel that you can achieve a lot more eyeball to eyeball across a table
> than you can by throwing bricks and bats.

"I don't believe the relationship with the LDDC is a single unitary thing, either it
works or it doesn't", a newcomer involved in the organization said. "It's gotta be a
continuing relationship with new people on both sides coming in and entering the
dialogue with some levels of agreement, some levels of disagreement ... but the
first thing's gotta be communication."

Initially the community division played a very important role in this, as an AIC
executive member said:

> I think our best relationship is with the community relations division of
> the LDDC — no question about that. In the past the LDDC looked upon us,
> if you like, as a nuisance body that had to be tolerated but not given too
> much information. Now they've begun to recognize that we are a force to
> be reckoned with, we represent most of the community and tenants groups
> on the Island. Many people coming in and buying private dwellings have
> formed themselves into residents' associations and have joined the AIC and
> we can claim to really represent a wide cross section of the community.

These changes were also reflected in other organizations. "We have a more, fruitful
relationship", a Docklands Forum representative said: "We're not daggers drawn
constantly. Obviously we are very very critical still of what they're doing" but

> in terms of information, in terms of perhaps getting things done, you
> know, minor things around the edges but at least having some kind of
> input ... And our relationships with them have of course improved con-
> siderably, ten-fold since their change of direction because we're talking on
> a common basis now whereas before we were talking very much, at each
> other. And there was no common vision. Now that's coming to light. But
> we're still highly critical of them. In terms of something like transport ...
> and in their social strategy.

There were concerns, however, about the extent to which entering a dialogue
with businesses and the LDDC would actually benefit the community: "The AIC was
always a radical campaigning group that did not feel like it wanted to apologize to
the LDDC, Tower Hamlets, developers, whoever", one of its members said:

> You know, if something needed to be said, it ought to be said and it was
> said. At times you get the feeling now that the AIC wouldn't say boo to a
> goose. It's not necessarily about confrontation, what it's about is saying
> right is right and wrong is wrong and if unfortunately ... there is a lot of
> wrong then you're going to be in this business that you call confrontation.
> But then that's the nature of the game.

While recognizing the need for change, others felt there were dangers in
moving on to less familiar territory, fearing they lacked the experience to be able

to negotiate effectively with those in powerful organizations. "Ted himself says that since we've become arguing across a table sort of mob rather than have a protest march, the community, feeling has gone", one AIC member commented.

> Loads of people used to join in on a march because it was a bit of fun apart from anything else . . . the day he let the sheep go, t'rrific that was and it gets into the paper and you think yeah I'll go along to that next one, certainly sounds worthwhile. Gradually it's going as it got to a little nucleus of people like the AIC arguing the toss across a boardroom table and publishing reports and that sort of thing. People didn't go too much for it . . . it's gotta to be an active thing

Ted Johns felt that it was not negotiating that was the problem but the way it was handled:

> I've always accepted what you do, you look at what the enemy has got ahead of you, you say yes OK we'll come along and talk but while you're [doing that you're] manoeuvring, how to outflank, that's what you do. But you don't go along and pretend to wear their uniform and that's what was beginning to happen, in fact did happen . . . In a sense we were in there in almost in a privileged position . . . we were gaining knowledge of [the organization but we were not] using those meetings to get information, pass information on and then in a sense to radicalize in a broader sense that particular group that may be affected or disaffected by that and then actually bring them together . . . That is where we go forward, that is what community is about.

As the community took a higher profile, some local people began to feel optimistic that they might gain something from the development after all and that the "enlightened self-interest" of some companies and the work of the Corporation resulted in the lines between "us" and "them" becoming far more blurred. A woman highly critical of the development and any potential gain at the outset later said: "People . . . on the Island are beginning to see that there needs to be a balance, between community groups and business and there can be a very good relationship, a very profitable relationship to be enjoyed . . . There is an element . . . [who] feel that business is trying to take everything away from them . . . Whereas in fact business just wants to enjoy a fruitful relationship with the community."

If the Corporation and business image had once been uncaring, the dedication of a few key "committed . . . people working to get things out for the local community" changed it, a local vicar said. "Nobody's going to say right from the beginning 'we love the LDDC, we think you're brilliant'", one of the community division said,

> but I think people are watching and waiting and if they see that there are benefits then they're prepared to take part. . . . I think . . . that memories are quite short in terms of the bad feeling there was around about how it was imposed on the area, I don't hear that anymore, I don't hear people saying, you know, it's undemocratic, it doesn't represent the people's views,

you don't hear that so much. Now the arguments are where's our slice of the cake? Okay, we've accepted the cake's here and now we want more, we don't want the money going on X, Y and Z, we want the money on our homes. So there is a shift in people's attitudes.

In 1990, two years after the community division was established, 49 per cent of respondents on the Isle of Dogs reported having "confidence in the LDDC", compared to 36 per cent in 1988 (LDDC, 1991:120). Although Isle of Dogs residents remained the most cynical about whether the development would benefit local people, there had been a notable shift in the number who felt that the developments had benefited them (LDDC 1991:101–3) (see Table 6.1). 53 per cent of Isle of Dogs residents felt local people had benefited to "a great" or "to some" extent through the LDDC's activities (LDDC 1991:142) although there was still a considerable number (41 per cent) who felt there had been little or no benefit (LDDC 1991:140–42).

"One of the things we wanted to do . . . was to improve . . . people's perceptions of the LDDC and what we were doing in the area", another member of the Community Division said:

We came out much, much improved on that score in the . . . MORI survey . . . I think it is partly the people have felt that we have been more accessible but our programmes have been more geared towards their needs and we have been able to listen to what some of those needs are, . . . some of the schemes we'd long hoped to do came on to site, . . . and I think all of that counts. . . . People's expectations were previously very cynical and then they began to think "Good heavens, well actually something's happening".

Although Nigel Broackes was not in favour of the social regeneration approach, his analysis of the situation in 1981 was, by the early 1990s, shared by this community activist. "I would hate to think what it would have been like had the LDDC not come along quite frankly", he admitted. "The LDDC took a long time to get its act together but having got its act together I think we're beginning to see the emerging of a new city and I think it's gonna be good not for us but certainly good for our children and grandchildren and I think what's happened is to be welcomed rather than otherwise." Another said:

I think that the LDDC has at last come round to accepting a lot of the criticisms that were made of its first five years were valid and the fact they've increased their community budget and are putting so much effort into training initiatives to community development and actually begun to say housing, education or health they're issues that we've got to address . . . they're issues that we've got to address because we can pull things together in a way that others can't.

"I think there has been a change of perception", a former councillor explained: "The fact that the Corporation are there now, the fact that people are turning towards thinking about maybe we can work with [them] is beginning to change people's perceptions." Nevertheless, "I think the gap is still very large."

Table 6.1 The proportion of residents who claim they have benefited from certain new developments (by Docks area)*

(Figures are %)	Total		Royal Docks		Surrey Docks		Isle of Dogs		Wapping/ Limehouse	
	88	90	88	90	88	90	88	90	88	90
Refurbishment of existing council estates	41	54	37	39	54	58	37	57	42	67
Leisure/recreation/ sports facilities	35	53	39	66	41	48	22	48	50	48
Environmental improvements	49	49	53	45	65	57	34	49	50	43
Shops/pubs	47	49	44	31	58	76	46	48	44	40
New housing to rent from council or housing association	40	46	54	49	44	53	24	39	37	42
New housing to buy	43	45	61	55	55	52	22	38	35	30
New industry	42	39	42	32	46	47	37	42	48	34
New office developments	37	39	31	26	39	45	39	44	47	40
New housing to rent privately	30	39	37	41	42	47	16	36	27	31
Community and social facilities	27	37	32	44	30	37	19	33	30	34

← = The last ones coming under development

* 1988 figures refer to developments over the previous 5 years
1990 figures refer to developments over the previous 2 years
Source: Report of the London Docklands survey (1991). Reproduced with the permission of CNT as owners of LDDC archives. © Commission for the New Towns

Too much of a good thing?

Although the new direction pursued by the Corporation and some businesses led
to a changing mood, reaped benefits in terms of raised resident satisfaction, generated the beginnings of a feeling that local people might become part of the development agenda, and alleviated much of the previous conflict, the path that Michael
Honey carved out for the Corporation was one that very rapidly the finances and
politicians would no longer sustain.

For those wedded to a notion of physical regeneration, with minimal social
massaging to pacify the locals, Honey's more socially oriented course became
increasingly problematic. Memories were short. The conflict had subsided and the
importance of development that involved people as well as property now seemed
like an unnecessary and expensive luxury.

One source suggested that "a number of civil servants . . . in partnership with the
old guard in the organisation" worked to "do what they could to cut out everything
but the bricks and mortar investments", and they had some powerful allies who
asked questions about whose interests the LDDC was serving. Certainly not, it seemed,
some sections of the business community who became increasingly disgruntled,
complaining that the Corporation spent disproportionate amounts of time and
money on the community at the expense of their primary regeneration remit.
"People ought to be grateful for what they've had here", a businessman said:

> Bloody sight more than anywhere else in the country's had. £58 million
> they're spending on local social projects. . . . the LDDC devote a dispropor-
> tionate amount of their time and resources in buttering up the local
> community . . . So I don't see any particular need for conflict maybe conflict
> between the business community and the LDDC over things they've done
> or haven't done.

Despite the impressive headline figures, the reality of what was on offer was
somewhat different, as this Community Division employee pointed out:

> The LDDC has gone for a high publicity profile . . . we announced in the
> press . . . £50 million for the Community Division. If you're going to go
> announcing those sorts of figures then of course people are going to start
> saying "Wow, that's big money" Unfortunately . . . of the £50 million for
> Community, £20 million will just go on one new build housing scheme
> . . . and another big chunk to the City Technology College, which is, sur-
> prise, surprise, promoted by Docklands.

"Not only has the corporation not continued to make progress, in the . . . years
that Honey was there it has gone backwards", a businessman lamented. "They've
spent their money in the wrong way, they've treated developers in the wrong way,
they've just made an absolute pig's ear of it." One developer described Honey as "a
social messiah". The LDDC are "worse than many of the local authorities", another
said. "There has been so much damage done by the Honey regime, I mean the

developers are daggers [drawn]. . . . It's terrible, absolutely terrible the situation . . . There has been a sea-change in Docklands from the positive to the negative, it's absolutely scandalous."

A former employee who used to defend the nondemocratic role of the LDDC explained, "I used to say 'Look, emergency procedures for an emergency situation' — the place, it was dead, it was derelict, all the money's come from central government . . . and it was working and I was personally quite happy with the nondemocratic element of it." That was, until the Honey regime: "The problem is when a nondemocratic institution is not working, how do you correct it?" The LDDC had become a "politically motivated organization rather than an entrepreneurial organization", said another businessman, at a time when it needed "to take the lead". "Whether you're talking to the business community or the real community you find that view is universal", a former employee said. "The LDDC is in the way — developers think the LDDC is in the way. It's got no supporters really anywhere." Honey "created this enormous bureaucratic monster", a local journalist said, "that did nothing. It just talked". Instead of promoting development the LDDC was seen to be inhibiting it.

While Honey's approach certainly contrasted with that of his predecessor, it was both understandable and legitimate for the LDDC to act with less haste and more consideration of long-term issues. After all, the impact of the development would be felt for many decades to come and it was the broader strategic questions about Docklands and its future (in which community investment made sense) that had not been adequately debated at the outset of the development or during its most frenetic stages prior to Honey's arrival. A member of the Board argued it was not Honey's approach but the political climate that made social regeneration difficult to pursue: "We have increasingly recognized these social programmes as being an important part of long term secure, stable regeneration", he said, but continued:

> That has been a source of some concern . . . to ministers politically. They had had what they saw as the awful lesson with the GLC [Greater London Council, once the elected local government for London] pouring, what they thought was lots of money . . . to support . . . causes . . . which ministers certainly didn't think were particularly worthy causes, . . . and I think the Department of Environment felt that we were going to pursue this sort of minority groups support down the road that the GLC had gone . . . because it was wanting to support disadvantaged groups of one sort or another.

"Michael Honey made so many mistakes", a long time LDDC employee said: "The biggest one was doing what Reg had done, upsetting the DOE [Department of the Environment]." "In all fairness to him," she continued, "his brief . . . was to concentrate more on the community and how the development was affecting them . . . [But] he really went overboard . . . we were doing all sorts of things that really should have been done by the local authority." "Here was a short-life organization with no formal locus getting involved in very sensitive . . . long-term projects", another employee explained, "like dealing with ethnic minorities, people whose English was a second language, the long-term unemployed" when these areas were the responsibility of the "prime authorities".

In his defence, a businessman involved in community relations said that what the LDDC put "into place" under Honey was "far more professional than in the past" and identified areas where investment was needed that had been previously ignored. "People at the bottom of the Island", where the established Island power base was centred, "are well resourced and very aware of what their needs are", a local authority officer said, yet "at the top of the Isle of Dogs [neighbourhood area which included parts of Poplar] . . . you've got like a third-world country", where coincidentally, she continued, "all the major building works are going on, where all the major amount of rehousing of Bangladeshi people is, where the deprivation is the greatest" and which was not reflected in funding. "A lot of the resources get focused down here and nothing happens [up there]." A situation the Community Division sought to rectify.

Sadly, the opportunity for spending significant sums on these groups was superseded by events. "They just about started to get it going", a council officer said, "when the bubble burst, property values collapsed and they found themselves . . . unable to support the programme that they wanted to put in place." The "net resources for developing both the roads and the community side had to be revised downwards" and placed the LDDC under severe financial scrutiny.

"If we'd had . . . more available finance when we needed it", a senior Board member said,

> we could have done things dramatically different. . . . Like every government-funded body . . . Treasury cut-off means that worthwhile programmes have to be deferred. If we could've had the power to borrow, . . . it would [have given] us that insulation so we didn't occasionally have to go dead slow because money was running out. . . . and as events have moved . . . the massive slump in the South East of England has meant that if we could've finished off more things earlier we would've been well ahead of the slump.

"They were bound to feel the pain from that," a developer said, "but I don't think they can hide behind that . . . Michael Honey was a disaster."

If external circumstances were problematic with the developers at "daggers drawn", with little to show in the way of physical regeneration and dwindling revenue to support an ambitious programme, internally the LDDC was also fraught with conflict. At the outset "we were a very small hand-picked team completely committed" an executive said. The organization grew in Honey's era to "over 450 people on the payroll", working in an environment that had become rife, as a journalist observed, with "empire building" and "back stabbing".

From the outset the estimates for the road and rail infrastructure were, a senior executive said, "grossly understated", and when their true cost was evident they "immediately put pressure on the budget". "The Department had a throttle hold on the organization from then on" he continued "and once it had they were not going to let go. They used the finance to . . . control the organization in a extremely detailed fashion." It was not simply the expenditure on transport infrastructure that caused difficulties. With the recession came dramatic falls in receipts from the sale of land. In 1989/90 anticipated receipts had been £130 million, but actual receipts fell to just £10 million (LDDC 1998:22).

Alongside financial constraints and conflict over how resources should be spent came frustrating bureaucracy. "The organization's (got) bureaucratic controls coming out of its ears", an LDDC employee said, "both DOE to the Corporation and within the Corporation itself and that reflects a change in management style. We're being overmanaged now . . . The room for initiative isn't [there]." The frustrations were acute, as this woman, working in the Community Division at the time, described: "It takes months to get things through our procedures, to go to the DOE, to come back again", she said,

> and my credability is seriously trampled on by procedures such as, recently, having negotiated with the borough some quick spend projects, because we had some money left over . . . Got it through our executive committee, I wrote to the chief administrator, "You can have the money, sign the contract now, they're on site", and a week later it went to our Board and the Board threw it out, so my credability is zilch with the local authority . . . They've gone and committed themselves to a project they didn't have the money, they were only going because we said we'd give them the money . . . Basically the Board are mainly representatives from developers and they have a particular viewpoint on tenants and what tenants should have and they basically thought the scheme was too expensive and giving the tenants too much of a nice deal, didn't like it. . . . How can I keep my credibility with the community, I'm promising them one thing which can be overturned elsewhere by people who don't know the people involved, don't know the situation . . . It's the same as working in local authorities.

A journalist said he was regularly fed information from LDDC employees whom he characterized as "good professionals" frustrated by "the politicos" who "make the decisions" — leaks that were often laid at the Community Division's door. "It was largely party political", one of the Corporation's staff said:

> For example, if not enough money was being spent on, say, education or training or whatever and there was a budget bid going on, a leak to the press might just persuade somebody . . . All the leaks had a strong bias towards community matters, and amongst the Community Team you obviously have people who are politically biased . . . left bias . . . It's another way of putting pressure on, to get a bigger part of the pot.

It is interesting that those fighting to retain, or extend, a share of the regeneration budget for local people were regarded as "politically motivated" when others fighting to limit the Corporation's activities to the "bricks and mortar" were not viewed in a similar light. Regeneration, for the latter, as it was for many developers and ministers, was about "driving hard" and "if you even appear to wish to nod in the direction of other interests that was a sign of weakness (LDDC executive). "If it was a choice between infrastructure and community, infrastructure would always win", an LDDC Board member said, and the problems this posed for community programmes were self-evident:

> It's now generally accepted that refurbishment is part of our general regeneration remit if you like. Although having said that the message

coming down from the DOE . . . is major priority is transport, the infra-structure of the roads, the railways, and we have to achieve those first and then once those are achieved and if there's any money over we can spend it on community facilities. That makes it very difficult to plan ahead, . . . The DOE are very nervous about the way we're planning ahead because they are saying "Well, if there was a shortfall on the roads, sorry the money has to go on the roads. That's what we're here for."

The Department were "very nervous" indeed. "Ministers and civil servants became distinctly unhappy", an LDDC (1998:22) report suggested, and "felt it [the Corporation] should treat its partner and paymaster with respect in drawing up its budget." However, the Corporation stuck to its new housing and community remit and its draft corporate plan . . . pressed the case for further expenditure including a £60 million increase over three years in housing (LDDC 1998:22).

The Corporation became "hamstrung", an employee said, drawing out the contrasts between his experiences under different chief executives:

In the early days we used to do things and not ask, whereas [now] it's impossible to sneeze without getting D of E approval . . . A lot of things we did were risky and you know lots of things we did they thought we could get more money for this or that bit of land or we shouldn't have done this and we shouldn't have done that. The fact is we did it and then got rapped over the knuckles for it afterwards. But without that kind of risk-taking initiative nothing would have happened. Nothing's happening now. They've got us screwed down.

Despite the impression that the LDDC was largely autonomous in the early days, Sir Nigel Broackes said: "We were subject to far more scrutiny by Environment and by the Treasury — I emphasise that — than any normal mature public body and they were very much breathing down our necks at every step. . . . the differences were that we had a direct line to Michael Heseltine, the secretary of state, for as long as he was there." In the latter stages of Honey's regime, the situation was fraught. "Even with the benefit of hind sight, it is difficult to know how the LDDC managed to get so far out of line" was the LDDC's own analysis of this period. The situation got so bad that "the Minister of State, by then Michael Portillo, and the Department [of the Environment] rocked the LDDC by throwing out its 1990 Corporate plan and issuing their own precise indication of future priorities" (Hillman 1997:47).

Even those whose work Honey had supported became critical. "There is no doubt that whatever Honey's contribution to helping the Community Division under way . . . the perception that then became locked in at Board level and at DOE level was that he was just . . . too much the local authority man in the end", one executive said, and he "became a negative that we all suffered under."

The contrasts between Olympia and York's approach, which paid attention to the fine detail on a personal and small scale level, and that of the Corporation were marked. One of the Olympia and York executives said:

By developing Canary Wharf . . . our contribution is pretty massive to the regeneration of the area because it makes it a magnet . . . not just for the

Island but for the whole of East London . . . at that level these are the big regeneration issues. The transport that's going in is regenerating the whole area . . . but at the same time the effects on individual people's lives in the short term is . . . very much more mundane and its very much more the effect of money we give to a summer project . . . which means that some of those kids get to go to the Lake District on a camping holiday for the first time in their lives or that the Educational Trust that we fund is able to pay for people to stay on after school leaving age . . . all of that is part of regeneration.

The local authority model pursued by the LDDC was less personal and more bureaucratic: "There's something in the process", a consultant said, "which means that what they do, you appoint someone who then appoints someone who then appoints a deputy, who appoints somebody else who they then say we need a report . . . get a consultant" and that was going on "right throughout the organization." He said:

I think I could identify some households which have ten thousand quids' worth of benefit. They would be the people who have moved into new homes — a few. But I could also show you some places where £100 is going to change their lives for the better, where £1,000 is more than they've ever seen and I think that's the tragedy. That type of expenditure is much harder. It requires a lot more work. It's much easier to give someone else a million pounds . . . the money doesn't go to local people . . . what happens is the money goes to set something up and someone's appointed to run it and . . . they . . . appoint people to do a report, who then suggest you do this, who then suggest you do that and you advertise it, you do this and at end of the day, at the point of service there is sod all left going into Dot and Doris' eighth-floor flat.

The money, the same consultant continued, had been spent in the wrong ways "for the wrong reasons, normally out of ignorance and lack of understanding, or knowledge of people or the area". During Reg Ward's period, he said,

the ways of dealing with community issues, they may not have been brilliant, but they were much more effective . . . we didn't write reports . . . justifying . . . this organisation . . . he'd come to an unwritten agreement and see what we could do on a certain number of issues. It was fire fighting if you like but at least there was communication . . . now there's no communication just aggressive predetermined narrow instructions, it's ludicrous. We didn't spend any money on them and now it's money wasted, wasted, wasted.

The Corporation "spent an enormous amount of money making a lot of enemies" was the verdict of one critic, which was a commonly held view. A letter in an Island newspaper written by a local resident said it all: "Come back Reg Ward, all is forgiven." When I asked one prominent activist if Honey and his colleagues were an improvement on the previous regime his reply was emphatic: "No, oh no!

No way . . . they couldn't actually talk to people." "There is only two people who I can honestly say . . . within the LDDC that I've had that sort of relationship with and one was Reg Ward"; the other was Eddie Oliver, his deputy. "There are some very competent and caring people . . . within the Corporation," he continued, "but they are swamped."

The Corporation "liaised with the front people. . . . They've dealt only with what I call the Mafia without ever bothering to try and really get themselves genned up about the Island *per se*", a newcomer said; yet "what the grass roots feel . . . is nothing to do necessarily with their representatives." This view was shared by a community development worker:

> What typically happens in organizations like the LDDC is that they talk about community consultation but they actually consult community leaders who don't necessarily speak to the community. I think the best favour they could have done themselves was to really listen to local people and it wouldn't have taken that much effort. . . . People in general think that local people can be won over with money. The thing that's happened is local people will take the money but they hate the LDDC. Now there are lots of reasons why they shouldn't hate the LDDC because there are things about their lives that have improved because the LDDC is there and because they've given the money and what have you. The fault is not that they've not given enough money, it's the way that they've treated people.

Why could a man who was responsible for setting up a system from which local people might gain be seen as such a negative influence? In part because he raised expectations and found himself unable to deliver, which constituted — as a housing officer said — "a double blow". There were more broken promises, as this local journalist explained:

> The promises made to the community were just not fulfilled. It's as simple as that. I'm not saying they didn't get anything, they did. They've got more now than they've ever had, but perhaps not their fair share some might say, they do say. . . . Reg . . . gave more to the community while seemingly opposing their demands than Honey gave as a bottom line amount with all his grand promises.

Furthermore, in their attempt to direct resources to the most needy groups, the Community Division challenged the power of the established Island community and not surprisingly angered them. "The way the Corporation used grants budgets in the past", one of team said, "and the way it handed control of those budgets to key individuals in the community created the myth that there was some kind of moral responsibility on our part to perpetually sort these groups out"; and for many years the LDDC had became a primary source of funding for the Association of Island Communities in particular. "High-profile political disagreements between the local authority and [those] people who ran as independent" meant they had no possibility of getting local authority funds. "Given that those individuals are connected with just about every long-term project on this Island", a Community Division employee said, "you can appreciate that they didn't really have anywhere else to go but to the LDDC in a quite large number." This source of funding was drawn to a close by a consultants' report in which the Association of Island Communities was characterized as "a white elitist body" that was unable to respond to the changing climate on the Isle of Dogs and which had failed to embrace membership from the Bengali families moving into the area, or the poorer sections of the white working class to the north of the Island who came within the local authority Isle of Dogs' neighbourhood boundaries though it was not part of the "Island".

It was not the Division's prognosis which was problematic but, as a businessman pointed out, "the actual handling of the relationships" that went "down hill" and eventually generated yet another conflict in the already complex politics of the area.

Back to the margins

What began with promise ended in acrimony. The revised 1990/91 LDDC Corporate plan imposed by the Department of the Environment once more placed emphasis on

> the transport programme and other projects that will have the maximum impact on economic, physical and social regeneration. This includes building confidence in London Docklands, attracting inward investment, business support and housing refurbishment, as well as training and other schemes that will allow local people to take advantage of the opportunities provided by the regeneration of the London Docklands economy. (LDDC 1998:22)

Honey left the Corporation in 1990. Eric Sorenson, his successor arrived a few months later. A former senior civil servant in the Departments of the Environment and Transport (LDDC 1998:22) he was regarded by one developer as "an overdue response" from government: "The thing had gone off the rails and they'd have to put in a man from Head Office." "I wouldn't have thought that was a perfect solution," he continued, "but it was certainly the proper thing to do when they got into this mess".

"Bearing in mind that you're starting off with a task force attitude, long term you've got to finish up with the thing fitting nicely back into local government structures", a Board member said. The changes in chief executives reflected the life course of a short term organization like the LDDC:

> There comes a point where you want to consolidate and that's when Michael Honey came in . . . as we did when Christopher Benson took over from Nigel Broackes. . . . [But] then we . . . hit this period of recession . . . the majority of our money is having to go on infrastructure and therefore it means even tighter control, it needs an even closer relationship between ourselves and the people who gave us the money, who are Department of the Environment, and that is where Eric Sorenson fits in. So there's, as I see it, a total logicality of the various phases with the type of approach.

The Corporation had come full circle. Whatever the verdict on Honey's approach, Island residents had for a brief period been on the regeneration stage, but when the priorities changed they once more learned that they were the last to be considered and the first to be axed when public spending required it. "Who suffers?" one local asked. The suspicion, and for some the firm conviction, was that it would be the local community.

Sorenson was measured in his response to a question about how he felt the community fitted into the Corporation's remit: "One of the perennial difficulties with . . . any regeneration initiative is trying to establish what the boundaries are so that the organization . . . doesn't become responsible for all aspects of what goes on in that community", he explained. So the solution was partnership between different agencies, co-ordinating their activities to achieve the wider goal of regeneration. The Corporation's contribution would be primarily physical:

> We . . . had to rein in [the] . . . community effort because the budget on the transport side blew off course. We had to re-establish what the shape of our programme should be. We had some major transport infrastructure projects, which have to be completed to good time, to budget, and that didn't for one reason or another leave as much room for the community effort. I think it's quite difficult to leave an area in a satisfactory way if there are a wide range of major services and community inputs which are clearly not satisfactory in terms of what you want to achieve under a banner of regeneration. You can actually encourage building office blocks, you can put in a new road, but if people do not have the opportunity to take the new jobs because there's not enough vocational training or education standards are still very poor compared with the national average, then it's very difficult to convince people that the whole exercise has been a success. And that's why it's important that it's a joint effort between the regeneration agency on the one hand and the local authority on the other and to make sure that they all relate to joint goals. (Eric Sorenson)

By 1996, after 15 years of the LDDC, 32 per cent of the Corporation's expenditure went on "roads and transport", 8 per cent on social housing (much of that linked to the road), exactly the same percentage as administering the LDDC. Only 5 per cent of expenditure went on "community and industry support". One assumes that the community's share of £108 million of expenditure in this area was not significant enough to merit its own classification (LDDC, Key Facts and Figures 1996:19) and certainly Honey's hopes of £50 million a year turned out to be an impossible dream.

Although the community was the last to be considered and the first to be discarded when politicians decided that the cost of social regeneration could not be sustained, this time anger over broken promises and raised expectations was not vented on the Corporation or developers. That fight had been waged and lost. The agenda had changed. The marginalized and frustrated sought a different path and perceived an altogether more threatening force in a conflict over increasingly scarce resources. The development and the issues concerned with it were no longer "the enemy": there was a more immediate and powerless target, as the next chapter describes.

Chapter 7

It all turns very nasty:
"Obvious, visible and on our doorstep"

The preceding chapters have described the redevelopment of the Isle of Dogs, the motivations of those who steered it, those attracted by it or forced to accommodate it. This chapter focuses on two issues that came to dominate the lives of many working-class residents on the Island more immediately than the development itself: the battle over public housing and emerging racism. These separate but interlinked issues personified the conflict-ridden and contested nature of Island life in the 1990s and revealed the extent to which some were willing to go to retain what they regarded to be *theirs* and to make their frustration and anger felt.

Although it may seem difficult to imagine that development on the scale of that undertaken in the Enterprise Zone on the Isle of Dogs could be distant to people living alongside it, there were a significant number for whom the development was a peripheral issue. "It wasn't a huge topic of conversation," said one resident, "[people] weren't really thinking about it." While another articulated a feeling experienced by many: "The thing is just so big that you can't relate to it" (Figure 7.1). For some, especially the most marginalized, "the issues of development . . . should there be a development and how should it happen" were almost an irrelevance because, as a clergyman explained: "It's more a case of how do I survive this week . . . the development and new houses and all the rest of it, you might see all these grand ideas but they are just abstracts. I'm just the person that gets pushed around."

It became very clear, however, that there were limits to the extent to which some local people were willing to be "pushed around", especially when the crisis in the availability of local public housing threatened them directly, providing a focus for anger and frustration as this Islander explained:

> It took a while for it [the development] to sink in. There was a horrible situation where it was I'm all right jack, if it didn't affect me, it didn't matter. But then as people's children were leaving home and wanted to get married . . . and find a flat, then they started realizing how bad it was. Then there was anger.

As this account has already demonstrated, anger and frustration are powerful emotions and, alongside the increasing fragmentation and marginalization of the white working class on the Isle of Dogs, produced a volatile and threatening

Figure 7.1   "Development just so big you can't relate to it", Janet Foster

combination. However, this time anger was not directed at the LDDC or developers — that war had been waged and lost. It was directed at an altogether more vulnerable target: Bengali households, frequently forced to move to council housing on the Island against their will. They became scapegoats for "local" white people's frustrations although, as Murshid pointed out, the blame lay elsewhere:

> ... often [a] deep sense of grievance is not directed at the source of the problem, or at, for example government policy. It's directed at sections of the community that are deemed to be vulnerable, and deemed to offer some kind of a scapegoat, to be used to say that, yes, these people are perhaps responsible for our poverty and for our lack of opportunity ... that we've seen on the Island ... People feeling dispossessed, people feeling alienated from not having access to resources and so on, and requiring to pin that blame on someone. And not being able to pin it on the right people, or on the right institutions. (Murshid 1993:18)

More than a decade of development had fundamentally changed the Island and, in conjunction with both national and local trends, created "a community so diverse that it might almost be described as three communities". A recent report on the Island suggested: "middle class professional people ... buying or renting homes in the private sector ... incomers to social housing ... [are] increasingly likely to be among inner London's most deprived people", a significant proportion of whom are "from ethnic minorities", and "the indigenous population, mostly white and traditionally unskilled or semi-skilled, ... alienated by the changing environment and by rapid development of homes and jobs which seem designed for other people" (Lupton 1997:viii).

Given this background, the battle over public housing was a bitter one in which "local" people (now code for the white working class), but especially Islanders, would not, as one said, "accept ... that their access to council accommodation [was] effectively prevented by those who have no links with the area and in many cases did not want to be housed there". "Island" people felt that in a competition for scarce resources their needs should be met first, even though the greatest housing need lay elsewhere, as this community activist explained:

> There's two points of view. There's the point of view that property developed on the Island, this is the Islanders' perception of it, should be for Islanders, people who lived here and worked here all their lives. The attitude of the local authority is that it's gotta be on a borough-wide basis. So, for example, you could have a family living in Bethnal Green who are in desperate circumstances, would be top of the list in housing need [and] would be nominated to come to the Island. ... Although one can see the argument for giving to those with greatest need ... there's also a lot to be said for the Island being given special opportunities because of the way that we've had to put up with all the mess and the changes that are taking place around here. (Foster 1996:154)

The fact that this time the newcomers were not only "outsiders" but from an ethnic minority unleashed racism along with resentment and anger.

While their claims to special status were frequently articulated, residents who had lived on the Island all their lives were rarely supported and it was often difficult to establish where discontent and anger over housing ended and racism began. Islanders, however, strongly denied the charge of racism and consistently argued that this latest set of incomers were no different from their predecessors:

> The Isle of Dogs Neighbourhood Committee [predominantly elected Labour councillors said in 1986] "In future no house would be given to the local community, it would be given to the Borough, . . . and more than that it would be given to the homeless" . . . The Neighbourhood calls us racist . . . When the GLC plonked 4 big tower blocks on the other side of the Island and bought in people from the other boroughs . . . because they [weren't] West Indians or Bangladeshis or the Chinese even they didn't see it as racist but now we're complaining . . . we're . . . being told ". . . the only reason you're saying it is racist." Nonsense, we're saying it cos they're bleeding dumping people on our patch, irrespective what colour they are, it's totally immaterial. (ex docker and community activist) (part quoted in Foster 1996:152)

There were some similarities in these arguments, as Massey points out, with the defence of working class communities from "yuppies" and earlier examples of ethnic minority settlement in other areas of the East End of London. However, in both cases the defence of the area was based on static and unchanging notions of place:

> In the 1980s when certain East End communities in the Docklands of London resisted the encroachment of new developments and, quite specifically, of "yuppies" there was a tendency to make the case on the basis that this was "a working class area" (yuppies, in other words, had no place there). This was problematical on [at least] two counts. First it was a claim for timeless authenticity [as a working-class area — implication: it should not be changed]: yet a couple of centuries previously the Isle of Dogs was fields and farmland. Second, it was an essentialist claim, . . . best illustrated by going back some fifteen years previously. Then, similar communities in nearby areas had resisted another "invasion". This time it had been by ethnic minority groups; and this time the claim was that the place was a *white* working-class area. The political left, on the whole, supported "the local residents" against the yuppies but had resisted the racist version of their claims to exclusive ownership of/right to live in that place . . . The real issue was the politics and social content of the changes under way, including their spatial form, rather than a fight over "the true nature" of a part of East London. (Massey 1994:121–2)

But there was an additional layer, too, in which the words and actions of some Island residents were racist *per se*; attitudes exemplified in the comments of this Islander:

> We're not racialists not by any means. Well you can tell we're not racialists because everybody's got feelings haven't they and it's not their fault if

they're given things. They come to expect it without working for it. We know that what we want we've got to work for but they come over here and it's given to them . . . For all they know we're gettin' the same as them. But we're the ones that know we're not in'it. You can't blame them if it's given to 'em. (Foster 1996:155)

Racism was an unquestioned and deeply embedded feature of white working class culture on the Isle of Dogs, which many felt resulted, in part at least, from its insularity. "This is a very racist area", a local journalist said. "Dockers were racists . . . and they supported Enoch Powell . . . when he made his first speeches on racism." These speeches presented a vision of Britain torn apart by racial conflict as the numbers of ethnic minorities "from different cultural, racial and religious backgrounds" swelled, undermining "the whole social and cultural fabric of British society" (Solomos 1989:124) and resulted in perennial concerns "from the late 1960s onwards" about "the size of the black population" as well as "wider political common sense about immigration" (Solomos 1989:124). When Powell was dismissed from the Conservative shadow cabinet,

the first demonstration of support from the electorate came in the form of a march by London dockers to stake the claim of "Enoch for PM". While only a small proportion of the work-force took part, the dockers were one of the few groups with the self confidence and motivation to demonstrate a point of view that, according to Schoen (1970, p. 37) was shared by between two-thirds and three-quarters of the electorate. (Hobbs 1989:129)

"And that still exists", a journalist continued "You get the feeling it's a bit of a powder keg [here]." As if to demonstrate his point, Powell's infamous speech resonated in the words of this Islander who said: "These do-gooders have gone too far . . . they've created the problems. . . . They [Bengalis] will never integrate in this country and I think it will finish up . . . I mean Enoch Powell predicted this 20 years ago and I think his words will come true. I think this will finish up [with] race riots as we've never known race riots before." "People are so racist on this Island, it's incredible", a community worker said; "people say it's the same everywhere but I think being such a close knit community, it's so insular, it's got to be worse here . . . Everywhere else in London people have sort of mixed but they haven't here . . . It's so much more here than I ever hear anywhere else . . . It's sort of like delayed reaction."

While the perception may have been that racism was worse on the Isle of Dogs, the problems of course were not restricted to the Island. Even in other areas of the East End, Spitalfields for example, where Bangladeshis had been firmly established for more than two decades and, in the 1991 Census, comprised almost a quarter of Tower Hamlets' population (Eade 1997:138), racism and racist violence were ongoing problems (see Husbands 1982, Fielding 1981, Thompson 1995). However, the concentration of Bengali households on the Isle of Dogs was certainly much lower than in many other parts of East London and it was an area that Bangladeshis were reluctant to move into. Consequently, even after changes in housing allocation that forced Bangladeshis to move into areas like the Isle of Dogs, who came to comprise 11 per cent of the local authority neighbourhood by the

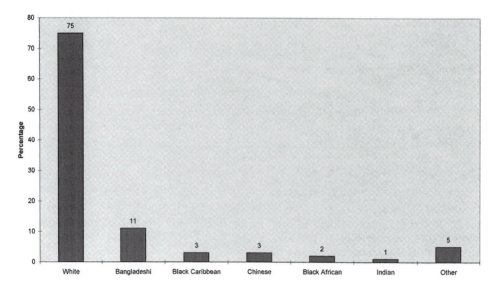

Figure 7.2 Ethnic composition of Isle of Dogs. *Source*: 1991 Census, small area statistics. Reproduced with the permission of the Controller of HMSO. © Crown Copyright

late 1980s (8 per cent Millwall ward) (OPCS 1993), over 70 per cent of the Isle of Dogs population remained white (1991 Census), see Figure 7.2). In fact Millwall ward, the area most commonly defined as the "Island", "has the fourth highest proportion of white people and the fourth smallest proportion of Bangladeshis" (Lupton 1997:10) of the Tower Hamlets electoral wards (OPCS 1993). By contrast, in Spitalfields, adjacent to the City of London, Bengalis formed the majority population (61 per cent) (OPCS 1993 in Eade 1997:138). Given the relatively small but rapidly growing concentrations of ethnic minority households, it is unsurprising that the *perceptions* of the type of problems caused by "new" minority settlement were viewed differently.

A newcomer who had close contacts with the established Island community said: "The Islanders absolutely loathe and detest what they call 'pakis', Asians. They won't have anythink to do with 'em, hate beyond venom, . . . there's real venom with that" (Foster 1996:154). This view is reinforced in the comments of an Islander who said: "People who have strong roots don't want to mix. They want to be with their own kind."

But what was their *own kind*? The area had historically evolved as a melting pot of different people from different places. The Huguenots, Irish, Jews (Hobbs 1989:93–101), Chinese and Somalis, for example, had all settled in the East End and the area personified by "cockneys" and pearly kings and queens concealed considerable ethnic diversity. In fact some "local whites" with racist views towards the Bangladeshis were themselves of Irish descent, and whilst defining the Island as a "white" place, even "old" established Islanders recalled seeing people of different ethnic origins in adjacent areas (the Chinese in Limehouse, for example) and in and around the docks themselves. Furthermore, many of the first generation Bangladeshis who settled in East London were seamen ("lascars") on merchant vessels (Adams 1987), as this Islander recalled:

> The only Indians you saw was the lascars that came off the boats . . . you always had a fair sprinkling of different nationalities from the boats that came into the dock because the Swedish boats, the passenger boats, came in every week and they were here for four days so you always had a fair amount of Swedish people and Norwegians and Finns, and you'd get the lascars that came over on the boats . . . Apart from that there was hardly any mixed races in the school.

Another said:

> The only place I would have been frightened of would have been Penny-fields around that area where the Chinese were, that was a little bit dicey round there but round this way there was nothing and we used to have boats in regular laying in the dry docks here and we used to have Peruvians and all that sort of thing and they walked up and down here, no bother at all and we was never frightened of them. . . . when I used to walk through the docks and we used to see the lascars and that on board ships and they never never, never bothered us at all.

A third Islander, several years their junior, said:

> We had the first sort of black and white couple move in along my landing in about 1961 . . . the fact that he was a black man living with a white woman for a dock area it was really, really unusual. I don't say that the Island didn't have its fair share of prostitutes as you must have in all dock areas but on the Island it wasn't very [noticeable], you didn't see that. Up at Limehouse, yeah, and Cable Street at Stepney. I was never allowed to go to Limehouse on my own. My dad used to say I'd get in the white slave trade traffic. . . . He used to say they'll stick a needle in yer leg, you'll collapse and you'll find yourself on a ship somewhere. People believed this years ago. But you see the only black people you ever saw were the lascars off the ships and you only saw them at Limehouse. Chinese people melted in the community. I'd always seen Chinese people cos they was always there at Limehouse. When I went to secondary school my best friend was Chinese so you always accepted them being there. But when I was at school there was one black boy in the whole of the school and he was an adopted child to a white family and then when I went to secondary school there was one black girl in the whole, we're talking about a school of 900 girls. So it was really [white]. That's what I keep sayin' to 'em now when they're moving these Asians in. . . . these people are so obviously foreign. They're gonna get much more resentment and then of course they're building brand new homes for them . . . and it rubs people so much up the wrong way and it doesn't make for happy relationships. . . . I

mean people would go up to Stepney and say "Oh it's like India up there" but now it's on their doorstep they don't like it.

The imagery evoked in the extracts above links those of different ethnic origin, among other stereotypes, with fear, sex and subservience and a powerful image that, as long as *they* were somewhere else, on the ships in the docks, in localities other than the Island, contained in a physical space, their "foreigness" and exist-ence was not a problem. However, when it came to occupying the *same* territory the problems began and the Bengalis found themselves blamed for fundament-ally changing the nature of *their* place, as this Islander clearly demonstrated: "We didn't have hardly any Asians living round 'ere at one time and that is the truth cos they used to say 'One thing about the Island you haven't got that, over run', yer know, but now it's getting outta hand." Her friend pointed out that "there were very few [Asians] compared with other parts of the East End like Bethnal Green". "Oh it's terrible up there innit", she continued: "Terrible . . . but this is what I think is the root of why it's all changed, it's nothing to do with the develop-ment, nothink whatsoever, it's the Asians that's coming in."

Rex (1986:105), in an analysis of racism and it roots, suggests that minority groups have always served important functions for their oppressors. "The Jews were the chosen scapegoats of a Nazi Government facing economic and political crises. 'Racism' in South Africa was the inevitable accompaniment of a situation in which White employers exploited Black labour". While Britain, steeped in a "colo-nial" past in which "colonized people were regarded as inferior", saw black people as the solution to labour shortages in the 1950s and 1960s in occupations that whites largely eschewed. The problems in any of these situations, as Rex (1986:105) points out, "were not those which arose from racism either on the psychological or theoretical level, but questions of social structure and inequality".

For many years this inequality was expressed in residential segregation because incoming minority groups were discriminated against in the housing market, denied access to local authority housing and forced into privately rented housing (see Dock-lands Forum 1993:3; Ratcliffe 1997:87; Glass 1964). Smith (1989:170) suggests that

> the racial dimension of residential segregation is a product both of mater-ial struggles and ideological conflict. As a residential pattern . . . specifically *racial* segregation reflects and structures enduring inequalities in access to employment opportunities, wealth, services and amenities, and to the pack-age of civil and political rights associated with citizenship. As ideology, on the other hand, segregationism builds from the objective deprivation of black people to a subjective acceptance that racial differentiation has a logic of its own; it provides a reservoir of common sense justifications for discriminatory policy and separatist practice. The study of racial segregation is therefore about much more than housing and urban research; it is an issue which penetrates far beyond the analysis of neighbourhood structure, organization or change. (Smith 1989:170, emphasis in the original)

Arguments about neighbourhood defence were not new, either, as Cohen (1993:5) illustrates with an example of opposition to "Jewish immigration into the East End of London" during the 1880s that has some direct parallels with the Isle of Dogs more than a century later. "This great influx is driving out the native from the hearth and home", it was said.

> Some of us have been born here. Others have come here when quite young children, have been brought up and educated here. Some of us have old associations here of such a nature that we feel it hard to be parted from. And all this is being jeopardised by those who come amongst us with their alien ways, and who dare to claim our native heath as their own. (Quoted in Cohen 1993:5)

The complex structural and historical processes that shaped the experience of any minority ethnic group were largely hidden, unacknowledged or not recognized at a local level where the focus was on the more immediate and readily tangible (though distorted) picture. However, "the complex totality of economic, social and political processes which can help explain the institutionalisation of racism" (Solomos 1989:234) provides a very important framework for considering the following account in which those who were already structurally disadvantaged found themselves further victimized by processes well beyond their control.

I am conscious that some commentators (for example, Lawrence 1982) have suggested that accounts like the one that follows focus attention on ethnic minorities as individuals or groups rather than on the institutional racism that has caused the difficulties they face. Furthermore, Lawrence suggests (1982:100–106) that community accounts often present minority groups in a "passive" way "with little or no account taken of their capacities to respond positively and defensively to their historical experience of racism, either as individuals or collectively" (Solomos 1989:12). My only defence to this would be that I began this book wanting to understand the Docklands story from a variety of different perspectives and that account would not be complete without the Bengali families who were forced to move to the Isle of Dogs. Even though their accounts of victimization and harassment do, by definition, demonstrate their relative powerlessness and vulnerabilities, these experiences deserve attention and their stories, and the events that unfolded on the Isle of Dogs aptly demonstrated the powerful structural and external forces that had shaped their histories and continued to shape their present.

Not enough houses to go around

In order to put the debate about public housing on the Isle of Dogs into context, it is important to briefly discuss the underlying factors that contributed to the tensions. The LDDC's approach to housing was, as I discussed in Chapter 3 and 4, based on the premise that there was an "unhealthy" concentration of public sector housing and that the area would be better served by encouraging private sector investment, widening tenure, expanding the resident profile and encouraging mobility, which had a profound impact on changing the profile of the area. Yet even after a decade of development, census data indicated that the percentage of home ownership on the Isle of Dogs was just 30 per cent, less than half the national average (Census 1991, in Lupton 1997:20).

The LDDC's approach to housing was the antithesis of what local Labour councillors wanted. In their solutions for Docklands' difficulties new social housing played a central role (Brownill 1990) despite the fact that when the LDDC began developing the area in the early 1980s the Isle of Dogs was one of the least popular and most difficult to let housing areas where "the legacy of landlord quality and management by the local authority in some parts of the Island was one [with] which the tenants were deeply dissatisfied" (Senior LDDC figure).

"When you're driven by the tradition of providing public housing and . . . by the political significance of providing new housing and when you've got apparently an exhaustive list of people in housing need, then it's much much easier . . . to provide new housing rather than to tackle housing", Reg Ward said. "Who's going to stand up politically and actually say . . . in fact we don't need any more public housing, when you've got all these big lists and an obvious need?" However, getting a "balance", he continued, was essential to the long term success of the area:

> You couldn't leave this area as a total commitment . . . to local social housing with a far too high percentage of dependent people. You just couldn't. The path would have been forever downwards . . . If we want this area . . . to be successful . . . it has got to have a broad thrust [with mixed ownership] . . . There's a point at which you have to accept that there are some limits on the capabilities of housing.

Even if the basic philosophy of mixed tenure in terms of housing and land use was right, the Docklands boom and developer's frenzy, as I described in Chapter 4, put an end to any notion of "balanced development", at least in the short term, as this former councillor explained:

> The housing policy that the Corporation have done especially on the Isle of Dogs is wrong. They've resulted in, as it were, statistically balanced communities — you can say yes there are more private sector housing, where there's a balanced public/private sector community there's a balance at a gross numerical level but, in terms of the environment, they've created . . . two communities there . . . There are so many small flats . . . [alongside] a relatively deprived community . . . [It] is just not very healthy.

In almost a decade, of development not a single home for rent had been built on the Isle of the Dogs (Foster 1996:153). Although this was a feature of other local authority areas across the country because of central Conservative government policy, the lack of new-build rented housing stood as a startling contrast to the activity in the private housing market on the Isle of Dogs. Nevertheless, the council remained a major housing provider, with almost 50 per cent of residents still in local authority accommodation (Lupton 1997:20), despite its dwindling percentage of the total housing stock during the LDDC's reign. The Conservatives' right to buy policy for example had reduced the housing stock in Tower Hamlets between 1981 and 1993 by 28 per cent (London Research Centre Housing Statistics, 1993 in Lupton 1997:21) which, combined with long housing waiting lists, created competition for and pressure on public housing that had not previously existed since the end of the war. "Since 1981, 1,150 council homes have been sold under the Right to Buy programme [on the Isle of Dogs]. 3617 homes have been built, but only 690 [all housing association] were for rent in the social housing sector" which "resulted in a net loss of social housing" (Lupton 1997:iv).

The development also added to the strain by increasing the numbers of those wanting to stay. "There is pressure on local authority stock", a former councillor explained:

> There's a long standing community which wants to stay. I may say that's new. . . . That's to some extent to do with the Corporation's success or may be general transformation because previously people were leaving the Island with a GLC housing policy of giving away empty flats because there were large numbers. Nowadays people want to stay. People are buying their flats and there's pressure on housing. That's a very significant change over the last ten years.

Consequently, some people who wanted to stay found themselves unable to get public housing, or once they were in it were unable to move, like this couple: "It was desperate. We was on the ninth floor. It was only a one bed-roomed and we had the first baby and then the second one come along — me and Jo and two kids in one bedroom. We was at the council left, right and centre to get us out and they was putting us back and back."

Despite the LDDC's dream, or as an illustration of the problems some would ever have in realizing them, less than ten per cent of households in the rented sector could afford to take out a mortgage or buy "even the cheapest house" (Lupton 1997:24). Not surprisingly, therefore, an acute interest was taken in the public housing sector and the picture for many seemed bleak. A MORI survey (1996 in Lupton 1997:25), for example, indicated that over a quarter (28 per cent) of Isle of Dogs residents believed that both their own and their children's housing futures were worse than 12 years before.

The crisis in housing came about at its simplest level because there were more people seeking housing than properties available. At the end of 1996 there were almost 7,500 people on the Tower Hamlets waiting list, of whom just over a thousand (15 per cent) were from the Isle of Dogs (Lupton 1997:26). Approximately 1700 others were on the transfer list. Consequently 13 per cent of Isle of Dogs residents were on the housing register (Lupton 1997:26). In 1996 only 275 lettings were made on the Isle of Dogs where ten people were seeking to be re-housed for every one letting, a ratio higher than "the Borough as a whole (7:1)" (Lupton 1997:26).

In times of scarce resources, the Labour-controlled Isle of Dogs Neighbourhood implemented a needs-based housing policy (Foster 1996:153) with priority going to homeless families, the vast majority of whom were Bangladeshi. These families had no special connections with the Isle of Dogs. In fact most, having heard of its racist reputation, were far from happy about being housed there. But they had no choice because the central Liberal administration implemented a one-offer-only policy. "Most of the Bengali family living here, most of us were homeless", a Bengali resident explained:

Council policy has [it] that homeless only get one offer. This is why we come to this area. None of us likes to come here because . . . our facilities in Spitalfields . . . At least once a day I must go to Spitalfields either for shopping or to see my relatives, or to look for work. (In Foster 1996:156)

A Bengali women said:

I did hear rumours about the Isle of Dogs being very racist and Bengali[s] wouldn't live on the Island 'cos of the racial attacks. So . . . I was having second thoughts, but you know, there's nothing you could do you had to take the accommodation that they offered you. Or you was out on the streets.

Irrespective of their wishes or fears, Bengali families were forced, as a clergy-man explained, onto the Isle of Dogs "to essentially a community that didn't want them and secondly that they didn't even want to come to" (Foster 1992:178) with devastating consequences for some individual households.

Local politics and their impact

The tensions that existed between elected representatives and some local people over their refusal to negotiate with the London Docklands Development Corpora-tion (see Chapters 3 and 6) left many feeling neglected by the one group who could represent their views even before the housing crisis emerged. "The Labour Group was very political", a committed Labour party supporter said, "and not community minded."

One Corporation official said that local politicians "had a policy of putting trouble on our doorstep and walking away". "I was fearful", he said, "that they'd start moving difficult tenants down to their blocks in the LDDC territory and they did. I'm talking not just of coloured people but difficult people." The association between people from different minority ethnic groups and trouble in this context was itself racist and inaccurate. It was not the Bengalis but some of the local white population who caused "trouble". Furthermore, it was the intensity and type of development generated by the LDDC and government housing policies that con-tributed to such marked social divisions and competition. As one of the Labour councillors, Julia Mainwairing, elected in 1994, said:

We've had another layer here which has led to even more lack of account-ability and that's the LDDC, which has poured millions and millions of pounds into the Island, and local people have had virtually no say in how that money has been spent and what you've got on the Island are glar-ing inequalities — you know there are little fortresses where people live in million-pound houses on the Island and at the other side you've got large old GLC council blocks where they have to padlock themselves in because they are frightened and scared people. (London Weekend Televi-sion, 17/7/94)

Despite the obvious relationship between central government and Corporation policies on the local situation, it was the political conflicts between the Labour-controlled neighbourhood councillors and the Liberal centre, and their differing housing policies, that provided a tangible focus for residents' anger and, in so doing, deflected attention away from the Corporation and broader social and economic changes that were having an impact (see Chapter 9).

When their own elected officials seemed to have turned their backs, the established Island community perceived themselves to be caught in a hostile climate between two enemies: the non-listening Corporation and uncaring neighbourhood councillors preoccupied with political point scoring. "You've got a housing policy that actually works against indigenous, no doubt about that", a local explained; "the highest points take all throughout the borough . . . You can't compete with the squalor of Spitalfields, course you can't, and no more should they be living in crap like that. [But there's] a minute percentage . . . of rehousing especially for our youngsters . . . so there's that unrest." Another said:

> The Neighbourhood are having a go at us because we're saying [we want housing but] . . . we're in a . . . terrible situation, like morally, because you're being hit for somethink that you know is not your fault and we've always spoke for our people who've lived on the Island . . . So we're in an embittered . . . position. . . . We can see what's building up . . . We feel it and see it. We know there's a resentment building, . . . quite a lot of anger's building up . . . People feel embittered or endangered . . . [and] they're likely to sort of fling out. People react . . . when they feel they can't win.

The political climate was further exacerbated by the contrasting policies of the Labour and Liberal neighbourhoods about housing homeless families. "There's a sharp political difference between us and the Labour party", a former Liberal councillor explained:

> Our view is that our first line of responsibility is towards the existing community not towards homeless families. We think it's an immigration problem . . . especially in Tower Hamlets . . . which has been foisted upon us to deal with . . . without any resources and that puts us into an impossible position. All that happens is for every homeless family we house, more turn up. So it's an absurd situation where national action is required. Our view is that the local community . . . have been there for a long time. . . . People regularly say "I've been here 25 years." They've never been one penny behind on the rent that whole time, ever. It's impossible to get a move. They feel, rightly I think, that if you've been in one council flat and you've paid your dues you're entitled to a shot at the next one. . . . and it's deeply irritating, really deeply irritating to see young people being housed before them and to cap it all to see homeless families under Isle of Dogs policy moving into new houses . . . That's a very serious political problem.

Derek Beackon, the British National Party councillor elected in September 1993 to represent one of the Isle of Dogs wards, called the Liberals "sneaky racists". "We are openly racist", he said, "unlike the Liberals who are sneaky racists.

They're racists on the door to get the white vote, then when they get in they put all the Asians first" (London Weekend Television, 1994).

However, there were differences at neighbourhood level. The Isle of Dogs "needs-based" policy inevitably meant that large numbers of homeless families were being housed on the Island in homes that "local" people also wanted. "Everyone's frustrated that they can't get better housing . . . [that] their children can't get housing", a former Labour councillor explained:

> I think some people scapegoat, [and] neatly go for what seems to be the softest target, which is the Bangladeshi community. . . . There is this sort of resentment . . . [that] they seem to be able to get housing where other people can't. . . . They tend to forget that there's a lot of Bangladeshi families in bed and breakfast . . . I think there is . . . and always has been . . . a small percentage of people who are extremely racist . . . [but] I think the majority of the community, once you actually start explaining the issue, start putting things into perspective quickly and see that really the housing crisis is nothing to do with Bangladeshi people.

Although this analysis was right, unfortunately — as the following account demonstrates — many people did not put the battle over public housing into context. In fact, the agenda that some white Islanders read into the needs-based housing policy was astounding, as the following remarks demonstrated:

> This Asian thing. . . . The local councillors we had at that time were Southern Irish. Now how [come] they're councillors? They're from Southern Ireland and they in their wisdom decided that we should have at least 5,000 Asians coming down because it would even out the rest of the borough. I mean is that their way of getting at us?

The battle over public housing in the area was also complicated by a "sons and daughters" policy pursued by Liberal-controlled neighbourhoods that did offer some element of housing to those with local connections, a policy which was condemned as racist by the Commission for Racial Equality. Not only was it discriminatory but it also meant, as this housing officer argued, that these areas "refus[ed] to take their quota of homeless families". These contrasting political policies in neighbourhoods under different political control heightened white Islanders' sense of injustice about access to local housing. "They are sending a lot of Asian families onto the Island", an Islander complained,

> but there's kids of our own who can't get a flat. My daughter's brother in law, they're expecting a baby next month and they can't get a place. They're living with his mother . . . They've been told they don't stand a chance of getting a place. Well this is not fair. I think they should cater for our own. . . . we do feel the Island's a special place . . . [and] I think there's gonna be a lot of aggravation before very long over this.

Those who supported Islanders' claims to public housing emphasized the need for a "balance" between local needs and those on council waiting lists. "The Isle of Dogs is like a village", a prominent community worker said:

and in the village there are people who've lived there for generations. . . .
If there are three . . . families who have housing need and there are five
plots, there's no problem. Bangladeshis come in and so we've got three
there . . . and you have the other two. What happens in fact when three
Bangladeshis come in? Now that is to me a simple decision you say look
we're very sorry we've got three here we can take two of you and the other
must go somewhere else. Now the Labour Militant group doesn't do that.

However, the problem was far more complex than this. In a 14-year period
(1981/82 to 1994/95) Tower Hamlets had experienced more than a 200 per cent
increase in the number of families accepted as homeless (DCC/Docklands Forum,
1995). Even in the face of such pressing need elsewhere Liberal councillors had
pursued a sons and daughters policy. Yet the dilemma remained, as this local
authority officer pointed out: "if the Liberals were in down here and they stopped
re-housing Bangladeshi families, where are they supposed to go?"

The question of equity did not concern many local whites, and not surprisingly the needs-based housing approach that denied them access to public housing in the area inevitably led to criticisms of the councillors pursuing these policies.
"People locally here vote Labour," a resident explained, "but if you ask them about
housing policy [and] which they prefer they will actually say we like a Liberal
Democratic policy". The housing crisis therefore led to another shift of allegiances,
and this time the local authority, not the development, the newcomers or even the
Corporation, came to symbolize the neglect of locals' interests. "Most people's
anger actually it changed away from the yuppies towards the council", an Islander
explained, "because the council were just so extreme for the minorities it was just
untrue. . . . The housing shortage is so bad that when they start givin' 'em [to
Bangladeshis], I mean it's not the odd one or two you can handle that, but completely givin' it to the ethnics then that causes anger" (Foster 1992:179).

In the short term, Islanders retained their political allegiances in spite of
Labour's housing policies, but they were none the less conscious of the contradictions: "The problem can be solved . . . by voting 'em out. It's the council that we
elect", an Islander explained. "We know they're gonna do it and we still go and
vote for them because they help us out with other bits." "But", he predicted, "there
will be racist problems"; and sadly there were. In fact, racism and the housing crisis
as combined forces led to fundamental political changes not from Labour to
Liberal but from Labour to the extreme right.

Myths and realities

Although Bengalis were popularly viewed as getting preferential treatment, (Foster
1992:179) the majority had the poorest access to housing, and found themselves
on the worst estates after the longest wait (see CRE 1988:49, Ratcliffe 1997–8). A
Bengali teenager summed up the situation well: "The working class people get it
drummed into their head that the Asians are destroying their lives, they're gettin'
all the houses, all the jobs. They don't know the facts right" (Foster 1992:179). A

tenant support worker vividly described the problems of two families whose plight was not atypical:

> If you go round these estates, I mean, on this estate alone there's an Asian woman, she's got eight children (in a three-bed maisonette), she's diabetic and . . . her husband suffers with his chest and asthma. The youngest baby's . . . Downs syndrome. She's got a married daughter and a new-born baby living there . . . I went to [housing] last week and said "Can't you help me get a transfer?" . . . They can't get out, the wife's expecting again, they can't get another cot in the bedroom. . . . [In another flat] the place is absolutely dripping with damp, water's pouring in, she's infested with cockroaches, she's got TB, she's four floors up . . . She's got two children, [and] she's already sharing with another family, and . . . they can't even re-house her.

According to the criteria set out by the Census, deemed as more than one person per room Tower Hamlets has the highest level of overcrowding in England and Wales (11 per cent) (1991 Census in London Research Centre 1996:89), almost twice as many households as comparable inner London boroughs (5.6 per cent) with 50 per cent of its child population living in overcrowded housing conditions (London Borough of Tower Hamlets in Lupton 1997:23). The two electoral wards in the Isle of Dogs neighbourhood area are "among the most overcrowded in the country" (Lupton 1997:23).

Despite the statistics, once many of the white working class had found such an immediate and identifiable enemy they were not willing to let go and seemed oblivious to the difficulties that Bengali households experienced. "When you talk to the Islanders now . . . you never 'ear 'em say anythink of how bad this is [the development]", an Islander told me, "they're more against there's so many Asians coming on to the Island." And as the number of Bengali households rapidly grew so did the perceived threat. "There's good and bad in every race", she continued:

> I've got nothink against them, I mean live and let live, but down my street . . . where there's one moving out now they're . . . going in all over the place . . . and the Islanders are saying "Well we're not going to put up with this" . . . You don't get no trouble but the people, the old Islanders, don't like it one bit . . . What can you do?

Her employer replied: "You can't do anythink about it, the law . . . protects everybody except the indigenous white population, I mean you're always on the wrong end of the stick."

The majority of white tenants simply did not understand why Bengali tenants should be housed in front of them as this woman explained. "It's Pakis got that place".

> They don't look at the priority of it. That family could have five children living in a one bed-roomed place . . . and they're standin' there sayin' "my daughter could have had that." Well course yeah, give a young couple a four bed-roomed house that people are cryin' out for . . . come on let's talk sensible. But they don't look at it like priorities. They go, "Pakis got it".

With, as one council officer put it, "more than the average amount of things happening to people ... which aggravate and make them feel [angry]" the housing issue tipped them over the edge and came to embody the conflicts between established Island residents and "outsiders" about their "rights" to *their* Island, which they were not willing to give up without a fight.

Fighting back

> I have no objections as such to Bengali families coming down here but I don't think that the people down here should be pushed out for the other people.... People that have lived here all their lives ... are suddenly in a situation where they cannot be re-housed and there's somethink wrong somewhere when that happens.... I mean it was always a tradition that the children could always get somewhere to live so the family stayed together. (White Islander)

> The people of this Island ... they don't like to see any strangers, they don't [want] to mix.... When we came first ... people's had a terrible look, ... terrible look and say "why these bloody strangers come from ..." ... and some peoples are quite unfriendly here and they don't like to see any strangers. We tried ... to introduce ourselves with the neighbours but some peoples will not answer me even if I said "hello, good morning" ... First of all it was very hard for us ... there was lots of racial harassment, racial abusement and racial attacks. (Bengali resident, Isle of Dogs, in Foster 1992:178)

The battle over public housing was most intense in the parts of the Island where council accommodation was considered most desirable. These areas were predominantly white. Until the homeless families policy began, local whites did not see themselves in direct competition for housing because, as this Islander openly admitted: "Ten years ago when the Asians and that ... moved in ... they were put in dumps. No one ever cared." The difference, under the needs-based policy pursued by councillors, was that Bengali households were occupying most vacated council properties. "Now ... they're having to put them in more decent places", she continued, and these were homes that white households wanted, as a Labour councillor explained: "The feeling on the Isle of Dogs is that you can move as many 'Pakis' as you like on to —— [estate] cos they don't want [that] for their sons and daughters anyway. They don't want to live there ... but if you move a 'Paki' family onto —— [estate] that's where they want to live" (inverted commas in original).

Bengalis themselves realized that these ideas underlined white hostility, as this young man explained "[The] number of Asian families [that] live in Tower Hamlets are making these perpetrators think that this is a problem for them. As they see that most of the Bengali families in council flats, they're getting the same facilities as they're having and ... it's something they can't tolerate" (in Foster 1996:159).

Established white residents, determined to secure their place in the locality, felt that it was the local authority's responsibility to support their needs above those of the homeless, as this councillor's example aptly illustrated:

> A nice three-bedroomed maisonette had become available . . . and there was a [white] family living in [a tower block] and their suggestion was that they should be given [the maisonette] because they were Islanders and . . . you could give [their current flat] to 'Pakis' off the homeless list . . . Because they were white they had the right to the better housing and they didn't give a shit about what happened to the tower block flats . . . They're people who wouldn't actually consider themselves to be racist.

Consequently, the greatest conflict occurred in the most "desirable" and predominantly white enclaves. "It's a completely different area that side," one tenant remarked, "a lot of them are more affluent on that side of the Island, a lot of them are old Islanders who have been here for generation after generation." One of the first Bengali families moved to a property in this area was greeted with a pig's head outside their front door; another had a live firework put through their letter box and found their telephone line cut off. In the same neighbourhood, an estate officer and a prospective Bengali tenant were confronted by an angry white resident and told: "Don't give any flat to any Bengali families because we won't let them stay here" (Foster 1996:156). "The tenants . . . [are] very racist. They don't want Bangladeshi families to be there", a Bengali worker said, "because they think the Bangladeshi families are taking their housing."

Although frequently vocal about the housing situation, all of those I spoke to denied that they were in any way involved in harassment. Extreme right groups did infiltrate the area and did generate trouble, but some of the perpetrators of racist violence were young local whites: "Some of 'em actually know that it's their sons that are going out and doing it [harassing]", a tenant said, "and try to defend what their sons are doing, that's the horrific part of it."

Whoever was responsible for the harassment, the impact on those victimized was the same. Some Bengali households received both verbal and physical abuse and many, not surprisingly, were concerned about their safety, like this man, speaking through an interpreter, who said:

> Not a single time [has] he found it safe to go out with his kids and living inside all the time his kids became sick and the doctor was telling him "Why don't you take your kids out?" and he said . . . "I can't go out cos it's not safe to go out" . . . They can't use the stairs [on their block] safely as . . . kids about 12, 13, 14 years' old gather round in the stairs. They're kicked [by] the kids and also sometimes [they] throw stones and things like that.

This was not an isolated case. Young, school-age children were often victimized and found themselves excluded from public facilities. In an account about her experiences of living on the Isle of Dogs, a 15-year-old Bengali girl wrote: "There is a Park . . . where most of the white kids go to play (Foster 1996:160). But if an Asian kid go by themself they get called names sometime beat up." Another resident said:

— Park, . . . was a no-go area for Bengalis, one of the women that comes to our group, she . . . took her two children down there one day and they had stones thrown at them, called "Paki". She said . . . "Oh they're just ignorant people" but she didn't want her children to be subjected to this kind of thing so she didn't go there anymore.

Adults and children were verbally abused walking to and from local shops and food, drink and other objects were thrown at them. One of the women said:

When we first moved in . . . we had no trouble . . . and then it all started . . . you know the racial attacks and them knocking on your door and calling you names as you walk down the street, . . . my kids are not older enough to go shop[s] but the ones that are a bit older, who goes to shops they get attacked if they're coming back for shops or they're going to shops.

Even in their own homes some were far from safe, as this Bengali woman explained:

A family in — . . . they'd shoved I think it was a scaffolding pole through the glass door and the fellow was really that scared that he just left his property and went with his family. . . . I thought that was wrong . . . 'cos if they see that they've done that to one family and they've made them move then they're gonna try to [do it to] all the rest of us, you know, they try and make us leave as well. . . . I understand that he was scared . . . he was really scared that this happened to him but I don't think he should have left really. (Foster 1996: 160–61)

There were also serious physical assaults: "We moved a Bengali family into a block of flats," a councillor recalled,

the son was coming home at about 8 o'clock in the evening. It wasn't dark. It was in the summer and he has the shit kicked out of him and he ended up in hospital for a week. . . . The following Saturday morning his elder brother got up and somebody had broken into his car . . . and with a tin of red paint had scrawled "paki" across the front windscreen, [and] sprayed all the seats.

In another case a youth was set upon on his way home from school:

Three or four youths . . . attacked him, took his wallet and even broken his jaw. He managed to get to the front of [a block of flats] and then he must have just fainted because . . . he was unconscious for about two minutes then he . . . came round and got home, . . . His father literally phoned the police four times that night and no-one came up . . . They even turned round and told him to take him down to — Police Station which he did and then they saw that his jaw was broken, he was black eyed and sores all over, where they'd punched him and kicked him, then they got an ambulance to take him from — Police Station to hospital and they called the police four times that day and nobody came, that's . . . how the police were. I think the police are racist as well, aren't they. (Bengali woman)

This incident was seen by many to exemplify a commonly held belief among Bengalis that the police were unsympathetic and many said that they did not report racial harassment because of this — a problem which was intensified by difficulties in communication. "It's most probably because the majority of them [police officers] are white", a young Bengali man said: "No offence to no white person . . . [but] if you are a victim of something you understand the anxiety." Another Bengali man said it was not simply they who were experiencing difficulties but the Vietnamese too [a group I was unable to access]. "They are suffering like us", he said, "but they don't report it." A survey conducted for the Metropolitan Police by MORI (1997) revealed a marked reluctance on the part of minority ethnic groups to report incidents of racial harassment to the police. Only 58 per cent of Asian respondents in the survey said that "they would be certain to contact the police if they were a victim of racial harassment" (MORI 1997:8), which suggests fundamental structural problems in, and perceptions about, the relationships between the police and ethnic minorities that are far broader than simply delivery of local police services (although the quality of service at a local level may also have an impact on willingness to report).

Even those who did report had little confidence in the process. One Bengali man, for example, described his own experiences of harassment, including an incident in which his car had been vandalized, where he was able to identify the perpetrators. His verdict was that the police were not doing enough. In this particular case he discovered that his white neighbour provided the boys under suspicion with an alibi. Although racism pervaded all parts of the class system, he believed that "it is the poor whites that cause problems of racism, not the rich or middle class, who even if they are racist, keep it to themselves. It is only the poor whites who are outwardly racist" (Research notes).

The police powers to deal with some areas of harassment were often overestimated by victims, and some police actions, for example, response times to an incident, were frequently interpreted as racist *per se* when in some circumstances these may have been influenced by operational constraints that affected all people whatever their ethnicity. Despite this, it is easy to see why the Bengalis felt vulnerable and unsupported and interviews with local police officers revealed that the police did not fully appreciate the difficulties some Bengalis were experiencing. At the time of the fieldwork, an officer said that "most incidents were minor" and that "they were blown out of all proportion by one particular community worker". Consequently, "The LDDC Community Division thinks racial harassment is a problem" (Research notes). Like many white residents, this officer and others I spoke to felt that community activists (some of whom were openly anti-police) and local politicians were exaggerating the Bengalis plight for their own ends. Furthermore, they argued, most of the harassment with which police officers dealt was perceived to be "low level", largely the responsibility of children and therefore a low priority.

However, minimizing the problem, not being seen to be taking the harassment seriously, or explaining it away as a part of a broader political agenda was problematic on a number of fronts, not least because it immediately reinforced the established Islanders' perspective. "[Racial harassment] has got out of order now", an Islander explained, "because it doesn't matter what happens, if you're black it's a racial incident. . . . I mean surely two people can have an argument without it being a racial incident." The woman went on to describe an incident in which two children, known to her on the estate, had been putting bread rolls through tenants' letters boxes, including her own. "They'd done it in everybody's door but because they done it in the Bengalis' that was a racial incident and that's where it's blown out of proportion." Whether this particular incident was racially motivated is not the issue. It is the victim's *perception* of the events which is important. It is entirely understandable that for a family who were fearful, who may have been repeatedly victimized, and live in fear of further attack, such an incident, however trivial it might have appeared and whatever the motivation of the offenders involved, would have a very different meaning for the victims and provoke further fear, anxiety and stress. Consequently "low level" harassment might not have seemed important in itself, but its cumulative effect on the victims certainly was. Indeed Bowling (1998:158) highlights that part of the difficulty with the police response to racial violence is that each occurrence is treated as an "incident" rather than being seen as a process. "Conceiving of violent racism . . . as process" he writes allows a more "dynamic" understanding of

> The social relationship between all the actors involved . . . can capture the continutiy between physical violence, threat, and intimidation; can capture the dynamic of repeated or systematic victimization; incorporates historical context; and takes account of the social relationships which inform definitions of appropriate and inappropriate behaviour. (Bowling 1998:158)

At the time of the fieldwork, one white Islander's verdict on the police was: "We've always been lucky on the Island with the police force, whether it's because we're cut off from the rest or not I don't know, but they call it a 'country posting' down here, the policemen, because you're dealing with the same people all the time and they know whether the crime's being committed locally or whether it's outsiders come in and done it." The difficulty was that the police did not have the same degree of insight into the victimization of the Bengali community. Even if the harassment was predominantly low level, its impact on people's lives could be considerable. As one man simply said: "We are suffering."

The reactions of some local people at one of the schools to a serious incident of harassment in which a young Bengali man was beaten up, described by a white parent below, is perhaps indicative of the complex attitudes that existed regarding racial harassment:

> A couple of boys beat up this Indian boy, like nearly killed 'im. He was in hospi'al for weeks. [The mums in the school said] "Oh well they're Pakis, what does that matter?" . . . Some of 'em have said "Oh that was bad because they really went too far." He should have had a whacking, for nothin', it's alright givin' him a whack, but not to put 'im in hospi'al for like four weeks and fracture his skull and nearly kill 'im. Don't nearly kill 'im! (In Foster 1996:157)

Incidents of harassment (see Table 7.1) involving extreme physical violence were relatively rare, but alongside other forms of harassment their impact was very significant, as a community worker explained:

Table 7.1 Racial incidents recorded for the Isle of Dogs, 1982–1997, by incident type

	1982	1983	1984	1985	1986	1987	1988	1989	1990	1991	1992	1993	1994	1995	1996	1997
Serious assault	3	2	7	1	2	4	8	19	16	26	33	56	6	5	4	7
Actual bodily harm	0	0	2	1	0	7	2	3	8	12	7	12	41	23	7	12
Common assault	N/A[a]	N/A	N/A	N/A	0	1	0	6	1	0	0	2	38	16	6	13
Arson	1	3	12	8	10	7	2	11	5	13	8	32	5	4	2	2
Criminal damage	1	0	1	0	0	4	0	4	2	1	3	5	39	15	10	8
Graffiti and daubing	1	0	5	2	2	5	19	21	28	20	37	33	4	3	6	0
Threats/abuse	0	0	2	1	0	1	2	0	3	2	1	1	41	34	8	17
Disputes etc.	0	0	0	0	0	0	0	0	0	0	0	0	6	5	2	4
Total	6	5	29	13	14	29	33	64	63	74	89	141	180	105	45	63

[a] N/A = not available.

Source: Metropolitan Police Statistics 1982–1997.

They're afraid of racist attacks and that's because they hear about some of the things that go on . . . In a way whether it does or doesn't [happen] is neither here nor there, it's the fear of it because it stops people going out and it stops people being as friendly as they might be with white people. . . . In Spitalfields it's OK, there are plenty of Bangladeshi people about, you feel safe. Down here you're that much more isolated. (Foster 1996:156)

Despite the unifying factor of racism, the Bengalis on the Island were divided. This was illustrated when attempts to form an Island-wide Bengali Association with representation from each of the estates in the area went ahead without the participation of one group who perceived the forum as usurping their role. Political and religious divisions also existed. For example, when Bengalis were moved to the Island as a result of the Docklands Highway, a tenant support worker explained that "established Bengali groups on the Island" did not "necessarily offer them any support . . . [because] these are new Bangladeshi people who've got to break into those [groups]". One of the Bengali women I interviewed said:

I wouldn't like to go somewhere not like Brick Lane, or somewhere like that, 'cos I think that there is too many Bengalis there. I mean like with this [the Island], where there's an area where there's too many Bengalis I mean then there's problems starts up 'cos there's Bengali and Bengali, you know, start over little things . . . I'd prefer it not where there is too many Bengalis, just a few. But I rather like this area, you know, it's quite nice . . . with these other developments going on.

The isolation experienced by some of these families was acute, especially for women who could not speak English. "So lonely", a young Bengali woman explained: "When it's a sunny day I just come out in the garden. There's my neighbour just talking . . . but these ladies because they can't talk they just don't come out. [They say] 'Oh no, if I go out and my neighbour talks to me, what am I going to say? Just smile at her and that's it. I can't talk to her'". Many white people did not appreciate the barriers that language and the feeling of discomfort generated by their unwelcome reception caused. Instead they often accused the Bengalis of not wanting to mix. "I don't think they want to integrate. I think they want a separate identity", an Islander said:

I think their cultures and their lives are poles apart. You see the West Indians, although they have . . . up to a point a different culture, it's much closer to our own isn't it . . . but when you get the Bengalis they insist on wearing their dress, all right if that's what they want to do . . . I think really . . . they would be much happier if they did all live closer together . . . Don't you think when in Rome do as the Romans do?

"It's not a colour problem, it's a cultural problem", another Islander said, "because their culture is completely different and they don't want to integrate." But this was not the way Bengalis saw it: "I don't know why white families think that Bengali families don't want to integrate", a Bengali community activist said: "It's mainly the communication problem." Of course, integration and retaining a separate identity are not mutually exclusive and the denial of race and assertion of culture as the "problem" is according to some commentators part of a "new racism" (see, for example, Solomos 1989 and Gilroy 1987:55–6).

Far from being shamed by the events that occurred on the Island, most of which those interviewed said they knew nothing of or that politicians and others were exaggerating for their own ends, the anger and frustration about the housing situation intensified. "They're [the council] treating us like pieces of dirt really", an Islander complained. Another said at a community meeting: "Everybody's under attack. Racism isn't on the Isle of Dogs. It's outside in the media." Even when the plight of the Bengalis was graphically described, many remained doubtful about the validity of these accounts, as this white tenant worker explained:

> We put forward this report on the racial attacks to our TA (Tenants Association). . . . A lot of them were horrified and said "Does it really happen?" I've spoken to what I class as decent people, you know, down here and they 'aven't got a clue . . . about the extent of these attacks . . . But . . . if I'm honest, the majority of people, if you said to them about the state of the Bengalis, you know, the majority of people on the Island . . . [and] in the East End as well, . . . would say "So what? It's their own fault, shouldn't 'ave so many kids and we didn't ask them to come 'ere in the first place, they come over 'ere and they're just taking resources what we need" and that would be the attitude of the majority. . . . That's what . . . people think. (In Foster 1996:162)

Another white woman said:

> You don't think about these things you see . . . I didn't even realize that . . . it was happening, naive probably . . . You can't blame other people for not realizing but you don't, not unless you think about it. Never having been in another country where I couldn't speak the language I didn't understand. (In Foster 1996:162)

"Understanding" for this woman finally came when she helped her Vietnamese neighbour to improve his English. It was then she began to reflect on the problems other ethnic minorities might be experiencing.

> He comes and see me every Saturday . . . and we work on his English . . . I said to him one day, do some writing . . . and he wrote me some stuff . . . about what it was like to be a foreigner and I read that and I thought what must it feel like to come into a country, you don't understand the language you don't even have a clue because even your writing is different . . . what must it be like to come into a society like that. And that's when I started to realize. (In Foster 1996:163)

Although it sometimes appeared that racism was so prevalent that all Island residents were tainted with it, there were individuals from all walks of life who detested it and, although the East End is infamous for racism (see Husbands 1982, Tompson 1988, Jacobs 1988, Thompson 1995, Fielding 1981), the area also had a history of anti-racist campaigns including the fight against Oswald Mosley and the Fascists in the 1930s. "I feel ashamed sometimes, I really do feel ashamed", a newcomer said about the racism. "There's a lot of racism . . . and that really upsets me", said another, "cos I think Bengalis should be allowed to sit in their house, they shouldn't have their doors kicked, they shouldn't have lighted paper put through their letterbox, they should be allowed to walk down the street without getting beaten up, it's abysmal it really is." However, these views did not form part of a strong collective voice against racism on the Island and did not attain any public profile except from the very local labour councillors, whose credibility was undermined by their stance on housing allocation, and the Community Division of the LDDC, whose views were also resented by many of the established Island community.

It may also have been that some of those with anti-racist views were themselves ambivalent about the local housing situation. An Islander in his thirties, reflecting on his experiences of growing up in the area, said that he "could not understand why people were beaten up because of the colour of their skin" and recalled Asians getting beat up when he was at school just because of their colour and described how he often used to have rows with his mates who after a drink in the pub started talking about "shooting the 'Pakis' and sending 'em home" (Research notes). Yet even he felt the Bengalis got preferential treatment, which "made them all resentful".

Although often simplistically presented as white working class on Asian, racism in the area was more complex than this, as a community worker explained:

> It quite upsets me, you know . . . it's so intense the feelings towards Asian people . . . my own kids a few times I've pulled 'em up and said "'Ere, you know, it's really racist what you've said there." It's when they hear it all the time and they start not to think. . . . I was saying to my brother about how upset I get about it and he said he can remember . . . as a kid, you know, it was Jews and he said he grew up only connecting Jews with "dirty Jew bastard" 'cos everyone who spoke about Jewish people that's what it was, you know, . . . I think it's going to take about two or three generations for [them to be accepted] . . . much the same as the West Indians, the Africans, they're accepted much more now, but then . . . again . . . they've got someone else to fire at, the Asians. It's quite frightening, Afro-Caribbeans, how prejudiced they are, how racist they are towards Asians as well.

The particular set of circumstances that pertained on the Island at the time of the Bengalis arrival certainly contributed to the racism they experienced, but racism occurs without the powerful underlying factors that existed on the Isle of Dogs. Racism and its causes are very complex indeed and, as one local depressingly observed:

> No-one is ever, ever gonna solve the problem of racism; racism has been with us ever since time began. Not only in our society, it is in every society. While you've got human beings living on the earth you're gonna have it. And the Bengalis will be harassed until someone less popular moves in and then everybody will harass them. But the thing that makes me really quite sad is if you look back twenty years to when West Indians and Africans were being harassed . . . and look at what they had to put up

with . . . now look at them . . . they harass Bengalis. The Jewish people who have been harassed since time immemorial . . . hear them speaking in a derogatory way about people of another race. How can people do that? I don't understand. Having been a victim, how can you all of a sudden be an aggressor, it's crazy.

Despite the general pervasiveness of racism throughout society, its intensity on the Isle of Dogs was almost certainly fuelled by the rapidity of change in the area and the conflicts it highlighted or in some cases generated.

The Bengalis themselves often differentiated between white racists and sympathetic white residents. "There not all racist. I [k]now people who are very nice friendly", a Bengali pupil wrote: "Most of them are white. Even though I am Bengali when people say everyone is racist round this are[a] I don't agree because there are some very nice people who are helpful, kind and understanding" (in Foster 1996:158). "I actually don't blame the people here . . . only numbers [individuals] of people", another Bengali resident said. "Most of them are good people." Nevertheless, the suspicion remained that "some . . . white people" were unwilling to challenge the racism of others. "Why don't they speak up, why are they so quiet?", a Bengali man asked (Foster 1996:158).

Not surprisingly, given more pressing and immediate concerns, the development played little or no part in Bengalis' perceptions of the area: "Most of the Bengali peoples does not like it because they consider themselves 'Oh we are very poor and helpless with this'", one man explained, ". . . so they are not thinking how to utilize them . . . they are not thinking about that because . . . they are not really understanding what is going on . . . Here . . . [is] not a safe and happy place because we have not had our shopping places here, we have not enough religious places here, we have not got enough education for our childrens."

"Island homes for Island people"

Islanders' concerns to protect their territory and deny the Bengalis access to local housing that they wanted for themselves brought them into head-on conflict with the local authority over the allocation of a prized 1926 council house. "Everyone but everyone who's ever lived on the Isle of Dogs would give their right arm to live in one of those houses", an Islander explained. So, not surprisingly, a white couple trapped on the ninth floor with two children approached the council when they discovered the house was available:

[We asked could we have that?] "Oh no, it's already been allocated" and then you find Bangladeshis have got it . . . Now I'm not racist but that tends to make you racist when you've got on the one side the LDDC who are only pulling for the rich and then you've got the council on the other side who are only pulling for the ethnics and we're the poor sods in the middle and we could do nothing. We were still banging our heads up against the wall.

The anger over this allocation was palpable. Residents in the street, "most of them . . . born and bred Islanders", signed a petition requesting that the house be allocated to "locals". "It weren't very well worded but then we're not really good at things like that", one explained. "All it said was 'We, the local people, believe that the 'ouse there should go to local people not Bangladeshi people' . . . It was blunt and everyone of them locals signed it. Give it to the council."

The council condemned the petition as racist and insisted that the house would be allocated to a Bengali family.

> The Neighbourhood said "This contravenes the Race Relations Act" . . . and asked people to withdraw their names from the petition . . . People on the whole didn't remove their names from the petition so they were visited by officers of the council. Those that hadn't then removed their names were visited by members of CRE and then the CRE threatened prosecution. . . . Some people who signed were council tenants and if they didn't remove their names their tenancies were put in jeopardy. [Parish priest]

In order to protect the family the council erected fencing around the property and security cameras (the cost of which was vastly exaggerated in many of the accounts), a move that was felt to be a gross overreaction. One of the Islanders who lived on an adjacent street said: "Whereas the neighbours would have got to know each other, they are cut off now [because of the fencing] . . . no way would anyone have done anything to that family . . . it's not Islanders who are doing it [harassing]." Another resident said: "No one had harassed this bloke. [But] the Neighbourhood installed seven thousand pounds worth of camera equipment on that man's home . . . He had never been physically abused or harassed but just in case they put it all up."

It is understandable that the council wanted to protect the family. However, as the local vicar said: "It did seem . . . that it inflamed the situation rather than calmed it down." It also seemed to reinforce the misconception that where the Bengalis were concerned the council was willing to invest money and provide housing while the indigenous population could be ignored. This view was aptly illustrated by an Islander who described the problems of an elderly white tenant in the same street of prized council housing:

> There was an old lady . . . on her own and she was robbed three times and she asked the council for a decent gate and they wouldn't give her one. They moved the Bangladeshis in and they put two, what were they £30,000 them cameras on the door so that no one could touch 'em. And an old lady who'd lived here all her life, bought up her family on this Island, robbed three times asked for a decent gate. I mean how much does a gate cost? No they can't do that but they can put in £30,000 cameras up for Bangladeshis so they don't get attacked.

In trying to unravel the complexities of the housing issue, which at face value looked like straightforward racism, the local vicar raised a number of important issues: "To start with all my sympathies were with the council," he said, but as I talked to the people in the congregation a number of things emerged. One is the kind of houses in — road were the sorts of houses that people on the Island prized above any others because they're houses with gardens and they're quite large houses — they're the most sought after property. So people were saying the housing waiting list in this borough is so appalling that we all have difficulties with housing and then when a "paki" family gets moved in you just feel that really is the last straw. My reading of that is not so much the racism but the frustration of the long housing waiting list and the fact that your sons and daughters, your children can't get housing . . . Secondly people were saying you know the change going on around here which is out of our control is enormous and it's not surprising that when we feel as though there's something that's obvious, visible and on our doorsteps we react badly against it. Again it was one more bit of change that people didn't like but which they felt they could kick against, whereas you can't kick against the LDDC, not successfully . . . in the end it was said that only one family kept their name on the petition and they·were the immediate neighbours and when one of the children in the new household had an accident and had to go to hospital . . . it was the bloke that kept his name on the petition that took the child to casualty.

Cohen provides an example from a different historical period during Mosley's antisemitic campaigns in the East End in the mid 1930s that has parallels with the account above but offers a different interpretation:

In 1936 Mosley's British Union of Fascists proposed to march through the heart of the Jewish community in Whitechapel . . . The Mosleyite action provoked a counter mobilisation from left and anti-fascist organisations, which was also supported by sections of the local labour movement. The organising slogan was "They Shall Not Pass". . . . things came to a head in the famous battle of Cable Street, where the BUF were finally stopped.

Now at that time there lived near Cable Street a working class family, who were well known for their antisemitism. Some members of the family were active trade unionists and had supported restrictive practices to prevent Jews being taken on in East End trades. The younger members had been involved in street brawls with the local Jewish gang. It might have been anticipated that they would welcome Moseley's intervention. But not a bit of it. Like many east enders, then and since, they were fiercely patriotic in both a local and national sense. They regarded facism as a foreign ideology and Moseley's march as an invasion of their territory. On the day of the march, the entire family stationed themselves on the top of one of the Jewish houses overlooking Cable Street. They spent their time ripping the slates off the roof and hurling them down onto the heads of the Blackshirts, yelling: "They May be Yids, but they're our bloody Yids". In this way they managed to simultaneously attack Jewish property and protect Jewish lives, making their area an exclusion zone for fascism, whilst at the same time asserting their claims as an imaginery ruling class to exercise a form of quasi-colonial jurisdiction over its "native" inhabitants. (Cohen 1993:46)

Although, as Cohen (1993:46) also argues, it is relatively straightforward to identify "people who commit acts of harassment against ethnic minorities . . . tackling the culture of racism which supports these activities is quite another matter" and on the Isle of Dogs it was this latter culture of racism that was so prevalent and went largely unchallenged.

"I understand how it must feel to be a long-standing member of a community, have your parents and your parents' parents, and you know you have that whole history of belonging to a place and seeing it degenerate and then seeing it became flavour of the month with the government and the amount of money that's being pumped in here and yet you're not getting a slice of the cake", a women who worked closely with established Islanders said. "I really can see the frustrations that are being felt by local people like that, but on the same hand being black myself and being part of the community I still can't condone their behaviour at all."

Others, like this local councillor, were not as forgiving. "They think because they have lived on the Island for a long time it gives them the God-given right to live here and have their sons and daughters living here. We can't work with them. That is the only explanation they have. They are confrontational and there is no compromise." Yet the councillors' inability to work with them was to have devastating consequences and, as Cohen (1993:46) argues, "we always have to deal with impure realities which do not fit into neat categories of what (or who) is racist or antiracist" and "containment" may be as much as we can realistically hope for (Cohen 1993:47). However, he continues that there are "difficulties in devising effective strategies of containment" including

> that they are not supported by the Left. Here the done thing is to talk about challenging, confronting or smashing racism; to try merely to contain it, or limit its effect is seen to be wimpishly reformist, even to be colluding in its continued existence. The idea underlying this response is the fantasy that the container is contaminated by the thing contained; there is however very little evidence that youth workers or educators who engage with the problem, become apologists for racism. They are however likely to reject anti-racist demonologies in favour of a more complex understandings of what is going on. (Cohen 1993:46–7)

Ted Johns, a veteran community campaigner, suggested that an accumulation of events, including the housing crisis, made local people feel that they had lost control of their lives and that these feelings encouraged extremism:

> There is a feeling that you can't do anything about it . . . and that's quite worrying because that does lead to Fascism, I mean, certainly before the war in Shoreditch and places like that, Bethnal Green . . . was the 'otbed of the Fascists . . . [People] thought well we can't do anything about it so we will go for a strong lead . . . I think that's what could be happening now.

This historical parallel proved all too apt, and in September 1993, after the fieldwork was completed, over a thousand Isle of Dogs residents expressed their extreme dissatisfaction at the ballot box and voted for the first British National Party councillor in Britain, which visibly demonstrated the extent to which some were willing to go to make their feelings felt.

Meanwhile, on the estate of prized council houses, the new Bengali tenant was ostracized. "Still people don't talk to him, you know," a Bengali support worker said, "he's isolated in the middle of [the estate] . . . if he goes out in the street the children, . . . sometime they throw a rotten egg over him. This thing . . . hasn't been solved." Yet Bengali families continued to be housed on the Island, with little or no support, in an often hostile climate where they were clearly not welcome. But they had no choice. As this young Bengali activist explained:

> People feel that this is the place they have to live and they try hard to adjust in this community because they can see that there is very few chances that they could move outside [the area]. So they try to adjust in the community. But once they get involved in any incident the distress of that makes them feel that they are not welcome here . . . even I feel that I would be better off if I moved . . . It's just terrible for people doing nothing being harassed just because of their colour. It's intolerable. (Foster 1996:159–60)

One of the council officers assessing the Bengali's situation said: "They're subjected to . . . the most appalling level of racial harassment. There's no school places for their kids. There's no places of worship. Aren't we just propagating the whole thing of alienating and isolating people by doing this?"

Pawns in many different games, the Bengalis themselves were not unsurprisingly far from satisfied, especially with the local authority. One man said he felt that councillors were "stirring up racism on the Island" and that it was "the council who are causing the problem, they are housing us here", a view with which white Islanders would have agreed. Others said: "Tell the council clearly don't send any Bengali family here because the white people they don't want any black people to live here . . . I tell some of the [white] people it's not our fault. The council leave us here. We had no choice at all. We had only one offer . . . It's not our fault. Go to the council."

But the council quite rightly did not want to give in to racism and generally viewed the sentiments of local whites as racist *per se* (yet offered little in the way of practical support to Bengali families themselves). The problem though, as one former councillor pointed out, was the message that this sent out to many white people in the area:

> The confrontational approach that the Labour party has taken is very counterproductive. It makes things worse to say to people: "Get stuffed, we believe that homeless families should have priority. We don't care whether you get housing or not." It's gonna cause terrible problems . . . It's causing very significant problems as far as I can see to nobody's benefit. . . . There's a lot of attacks on the Island . . . and more to the point there's resentment and . . . that resentment ought to be the target of race relations politics rather than the housing aspects because that resentment is what racism is about.

A glimmer of light?

The chronic shortage of public housing and lack of mobility gave the small number of new-build homes for rent on the Isle of Dogs great symbolic importance. For the Islanders they represented the opportunity for a few local people finally to gain good housing and when a low-cost housing project on a prime riverside site, developed by the LDDC in conjunction with the local authority and the Housing Corporation, was built, Islanders once more found their wishes in conflict with those responsible for allocating the properties. No special dispensation was to be given to Islanders or "local" people (see Docklands Forum 1993:10/11).

Rumours and misinformation about this particular housing scheme were rife and once more focused on the Bengali households who would benefit from some of the large six to eight-bedroom properties on the scheme, though the number of these was grossly exaggerated. On this occasion, disquiet about the allocation of dwellings led to the creation of a pressure group: the Isle of Dogs Action Group for Equality designed not to get equality of housing for all but to "ensure that new homes go to 'Island' people" (*The Islander* July 1992). Unlike the passive and unquestioned racism of many of the Islanders previously engaged in debates about housing, the Action Group had an overt and wider racist agenda, providing a platform for extremist views and discussion of issues like the abolition of immigration (even though there has been virtually no immigration, even secondary immigration since 1981 Nationality Act) and the withdrawal of welfare for the ethnic minorities (*The Islander* May 1993:1).

Another housing issue that caused great consternation surrounded those tenants decanted as a result of the Docklands Highway (see Chapter 6). As news filtered down to the Island about the deal that had been struck, Islanders once more felt the "goodies" were being handed out elsewhere. "I was fuming" one explained: "They're lovely little houses I'd love a little house like that." Another said: "There's been a lot of uproar over that . . . People resent the fact that they've always lived here and now they just can't find anywhere to live." But in this instance it was not simply a place to live, it was excellent housing to which Islanders felt they had a right. Once more the scapegoats were the Bengalis. "The vast majority of them on that estate are Bengali", an exasperated Islander explained:

> We have spoken to the LDDC and we've told them they will create racial problems down here if they're not careful but nobody seems to listen . . . When people have worked for what they've got and they can see it all being taken away from them and being given over to ethnic groups. It's not because people are racist it's because they're having what they worked for taken away from them. (Foster 1996:153)

Racism, though, did play its part. Bengali families were not seen to "deserve" such properties, as the views of this Islander made plain:

> Let's face it they [Bengalis] don't care, they don't care how they live. They wreck their houses don't they . . . So why don't they move into places like this and move the people that lived here like [our neighbour] . . . she'd

279

love a little house with a garden . . . or people like us with families who
lived here all our lives? We deserve a decent place you know.

"It'll be wrecked," a woman predicted, "it'll be a ghetto in five years' time if not
before 'cos they really don't care": yet more stereotypes without substance.

Those who moved onto the new estate were acutely aware that the Island, only
a short distance from where they formerly lived, was a different place. "People on
the Island are not particularly sympathetic to them", a community worker said. "I
know it's only down the road but . . . the Island is psychologically a different place
and it is a big thing for them to have moved down here."

The anger over this development did not subside when the decanted tenants
moved in but intensified when it was decided that the remaining properties would
go to homeless families, a decision that fuelled an already charged atmosphere,
described by many as a "powder keg". One local said: "The people will look for
someone to blame and the ethnics will cop it and the reason the ethnics will cop
it even more so than now is because the Neighbourhood has clearly demonstrated
. . . that they're siding with the ethnic minorities over the whites."

The sensitivities and anger surrounding housing on the Island eventually led
to a reversal of this particular allocations policy. Subsequently it was decided the
homes would to be let to existing council tenants on the transfer list and that their
vacated properties would be taken by homeless households. There was some sug-
gestion that this move, which clearly denied those with the poorest housing choices
an opportunity for good housing, was a response to pressure from the white community (Docklands Forum 1993:11). Yet despite the perceptions, both housing
developments mentioned above had majority white tenants (58 per cent), with just
over a fifth (22 per cent) Asian (DCC/Docklands Forum 1995), "slightly higher
proportions of ethnic minorities than on the Island as a whole" but representative
of "the population waiting for housing at the time (66 per cent white, 13 per cent
black and 21 per cent Asian)" (Lupton 1997:27).

Whatever caused the reversal of policy, it constituted too little too late. "Race
is now the biggest issue on the Island", a resident explained, and one that was
exploited by politicians of all political hues. "Island homes for Island people"
became a campaigning slogan for the Liberals in the 1992 local council elections,
generating considerable concern among the already fearful Bengali community.
"The Liberal don't want us here", one told me: "there's a leaflet it affect us. Island
for the people of Island, why does it say that?" On that occasion the Liberals'
manipulation of local feeling was not successful and Liberal candidates were condemned for their racist agenda. But the fury over housing and the neglect of
"local" interests did not subside and more extreme forces were at work behind the
scenes, fuelling racism and harnessing discontent. "You notice the old Bulldog
leaflet comes through the door a lot now", an Islander explained. "Maybe they'll
move in and it won't take much to get people together because it's brewing beneath
the surface." "They're sitting on a powder keg", said another white resident and
sadly this proved to be the case.

Although housing was the most hotly contested issue, competition for scarce
resources was not limited to housing. It also applied to other areas of local authority and latterly LDDC funding (see Chapter 6) where some of the established white

community groups lost out to more needy groups. This was not how they saw it, as a white Islander heavily involved in community issues explained:

> The Bengali people are getting thousands of pounds in grants and this is what creates the tension when people have worked for what they've got and they can see it all being taken away from them and being given over to ethnic groups. It's not because people are racist, it's because they're having what they worked for taken away from them.

Here, too, then it was perceived that "the race card" was being played, and at white people's expense, as this woman explained:

> I am a genuine nonracist. I don't see it in terms of colour, what I do see is in terms of people's interests and they use it. . . . Tower Hamlets had given £12,000 for the service that was taking Bengali pregnant women to hospital . . . Now I'm sorry but that is out of order, you get your own fucking self to hospital . . . That makes me racist over night not because I am but because I'm enraged that people are manipulating this community for political ends and it sucks.

This example was mentioned by an activist as well:

> Tower Hamlets maternity liaison . . . that is for ethnic minorities who feel threatened walking . . . to the prenatal clinic and it's a taxi service. . . . [A woman at the meeting where funding was discussed] went potty "I walked to fuckin' Mile End hospital, one kid in that hand, shopping bag over me shoulder and [pregnant], what on earth is going on?" . . . It's engineered to hit the people who can be the most vocal and that is not the ethnics it's your indigenous white.

This man believed that denying whites was a deliberate ploy by local politicians to generate anger that they might then use to their advantage

> [Councillors] know that if you get a bee under your bonnet and you strike at the very raw roots you'll get a bloody response because of the kind of animal that the Island is, it's still retaining something of a community and they know it and they're [councillors] working the system like you wouldn't believe to support their political attack on the centre.

The "politicians are lethal, at least the politicians that operate around here", another resident said, and the cynicism directed at politicians unfortunately automatically extended itself to the Bengalis who were mistakenly thought to be benefiting. In fact, it was a sad indictment on the gravity of the situation on the Island that it was necessary to provide a transport service because, as this community worker said, the reality was that Bengali women and children feared for their safety. "This is what we have to do . . . you need to pick them up because it's not safe for them in the streets." But the message that chimed with the voters was not one of reason.

The BNP "success"

In the local election of September 1993 the Isle of Dogs became infamous when Derek Beackon, the British Nation Party (BNP) candidate beat the Labour candidate, James Hunt, by just seven votes. "I didn't sleep very well after the last election", the defeated Labour candidate said: "The Labour party in two previous elections, their slogan was 'Housing for All, Vote Labour'. I suppose people came to the end of their tether and just decided housing for who, and then decided that they wouldn't trust Labour again" (London Weekend Television 1994). "The majority of people who voted for Derek Beackon, the 1,480 votes, were not racists", he continued. "They were intolerant. There was a protest vote . . . Frankly the Labour party should have taken its share of the responsibility for the BNP victory" (London Weekend Television 1994).

"There was a section of our community that were out and out racists," Peter Wade wrote, "but the vast majority were reacting to local authority policies that as they saw it treated them very unfairly, particularly on housing allocation. . . . the election of Beackon (BNP) was a one-off warning shot about these policies" (Peter Wade, written communication 27 January 1998). In a television documentary leading up to the 1994 election, some of those interviewed who had voted BNP reinforced this belief, suggesting that their motivations were to do with lack of access to public housing, not racism. "This is not a racial issue down here", one of the residents said, while another BNP voter said his was "a protest vote, but you can't protest twice". "I like to vote for a party I think's gonna help the people in the area, you know what I mean, but if it means racism and things like that", he said, "then I don't think I can go along with that" (London Weekend Television 1994).

It is naive in the extreme to vote for a party like the BNP with an overtly racist agenda, which includes repatriation, and to claim that racism is not part of this. As Bowling (personal communication) wrote: "The Islander's desire to expel the Bengalis from '*their*' Island resonates with the BNP's desire to expel all dark-skinned people from '*their*' country." This sentiment was aptly expressed by Richard Edmunds, "a senior BNP organizer", who told a crowd of supporters brought into the area to canvass for the May 1994 election: "We are doing this today for Britain and for the British people just down the road" (London Weekend Television, 1994). These essentialist claims lie at the heart of racist discourse. Yet, as the comments of young Bangladeshis growing up in the East End of London attest, they are, and see themselves as, "British" as one youth explained: "I am a British citizen. No matter what the whites say. I am British. Not English but British" (quoted in Eade 1997:157).

Even if voting BNP was simply a protest vote for some, in the months that followed Beackon's election it was apparent there was "a hard core vote", the former Labour candidate said. "It hasn't evaporated", he continued: "Within the last two years the BNP's percentage of the vote has gone from 3 per cent to 20 per cent to 34 per cent" (London Weekend Television 1994). However, in the May 1994 local council elections, Beackon was defeated and Labour regained control of the neighbourhood. But in the process the BNP gained 2,041 votes (an increase of 561) — a further protest or evidence of hard-core racism?

Although, as one of the local clergy at the time of the 1994 election said, "The BNP machine doesn't belong to the Island" and there was "a ceiling to their support in this community", which he rightly estimated prior to the election would be approximately 2,000 votes. This still represented 9 per cent of Isle of Dogs inhabitants who were receptive to the BNP's propaganda and fertile territory for them to gain further recruits.

On this occasion the BNP threat was beaten, with a massive turnout (75 per cent) and a vigorous Labour campaign supported to an unprecedented extent by the national party and also the anti-Nazi league. But Labour candidates this time promised not only new homes for the Island but "homes for local people" (London Weekend Television 1994).

From victims to aggressors?

Bengali youths growing up amidst the pernicious racism on the Island and the wider prejudices and discrimination suffered by the ethnic minorities in society generally were, unlike their parents, less willing to tolerate this treatment and envisaged a time in the not too distant future when they would fight back. "The environment they [the older Bengali generation] had and the environment we had is different" a young Bengali man explained:

> When they were living here and working in the factories and things like that, at that time the problem was that they weren't able to communicate with anybody and not just that they didn't face any problem because they were working most of the time and . . . very few Asian people was over here. Wasn't much concern of racism or something like that. The tension wasn't so high at the time. As families moved in to join them, now the people who are mainly racist find that it's a problem for them and they're trying to do something about it. Nowadays, even still, old people don't understand what the perpetrators are saying. It's the kids who goes to school [who] can understand it. (Foster 1996:159)

In areas like Spitalfields and Whitechapel where there were larger numbers of Asian teenagers, violence among young people was becoming a serious issue. Thompson (1995:111) in an account of Asian gangs in the East End, suggested that although "for the most part, the East End Asian gangs are fair-weather concerns — *ad hoc* collectives of schoolkids and unemployed teenagers who come together on cool summer nights" there were some highly organized and violent gangs including "the largest and most notorious . . . the BLM or Brick Lane Massif . . . who are believed to have been responsible . . . [for] a number of unprovoked attacks on white and black youths in the area" and, the author suggested, terrorised Asians too during violent robberies and in the pursuit of protection money. "It's gone too far", a young Bengali man explained:

I don't know how it started but maybe I think they faced same problems
as kids now face on the Island . . . and now they just can't tolerate it so
. . . the victims get gradually tougher . . . and I regret to say that some
Bengali kids in Spitalfields are committing crime now . . . Sometime they
don't fight whites, they fight amongst themselves . . . People always do worry
about what's happening in Spitalfields could happen in Island. (In Foster
1996:161)

And this is what happened. As tensions intensified on the Island, Asian youths
began to victimize white youths and white residents began to accumulate their own
set of horror stories about vicious attacks by Asian youths. In one case (January
1998) a white 14-year-old boy

was walking home with two young girls. As they passed the Bengali Cultural Centre a gang of Asians burst out of the door with an assortment of
weapons. [The young man] was hit with a hatchet, almost severing his
finger and caused a cut to the head that required 18 stitches. A police
officer commented that if he had not put his hand up to protect his head
we could have been looking at a far more serious incident. (Peter Wade,
written communication)

"We have a major problem in our community", he continued, "and if we do not
talk about both sides of the issue in a balanced manner we have no chance."

So yet another schism developed in the complex social relations on the Isle
of Dogs in which profound divisions existed within and between different sections
of the poorer population in public housing, who often perceived one another,
as Harloe and Fainstein (1992:263) remarked in a different context, "more as a
threat than as potential allies" and where "gender, and more significantly, 'race'
are particularly important bases for fragmentation". In the 1990s the Isle of Dogs
was contested territory and ironically the very people that some established residents were fighting so hard against were not only in many cases British but would
in another generation be Islanders born and raised there too (Foster 1996:165).

Conclusion

This chapter has outlined the conflict and competition that emerged as a result
of a combination of factors linked to the development, local political policies and
scarce housing resources in which the white working class, and Islanders in particular, perceived *their right* to *their place* to be challenged: first by the powerful
(developers, the Corporation, and later local politicians, the first two of whom they
felt they could do nothing about); and secondly by Bengali homeless households,
who were more powerless than themselves but whom — no matter how unjustifiably
— they felt they could fight against.

The social divide and racism that developed on the Island and the inability
of local people to move within the public and private housing markets might have

been avoided if the development had addressed local needs from the outset, as this newcomer argued:

> Improving the current housing stock should have been very well entrenched in the proposals . . . There is a problem because there's far too intense development here . . . for the people who have to live here, and that must be addressed . . . People . . . think of it as being forced out of their home area. . . . It really is important to generate an awful lot more, be it council housing, be it housing association.

As it was, a combination of government legislation with Corporation and local authority housing policies served to exacerbate and inflame the situation and squeezed the white working class out. Those who valued the "sense of community" wanted a more sensitive approach that balanced the needs of the homeless along-side the needs of the pre-existing local community so that what was special about the Island was retained. At one level there was nothing wrong with such an approach and, if demand for public housing had been less pressured, this could easily have been achieved. Once conflicting and competing demands on housing emerged and the absolute needs of one group far outweighed those of another, the problem became altogether more complex. But this was not how local whites denied access to housing perceived it. Just as they had once promised that they would not give up their Island without a fight when the development began, so they applied the same criteria in the newly emerging situation on the Island in the 1990s. However, the focus of the fight changed and acquired new forms and objectives that were exploited by some very unsavoury forces indeed.

No matter how local the agenda of some Isle of Dogs residents, the BNP's ambitions were anything other than local, and the extreme racist paths that some young white working class youths in the East End of London had embarked upon were mirrored in other parts of Europe (see Skellington 1992). In France, for example, Le Penn's racist agenda has gained electoral success; in Germany there have been attacks on the Turks and other minority groups, and even denials that the Holocaust ever happened. For the average Isle of Dogs resident, these places and issues might be regarded as in no way related to events in "their" place. But they are. "There is", as Bowling argued, "a 'macro level' discourse and practice concerning who 'we' (or in this case the 'authentic' Islanders) are which relates to white-ness and which is shared by many white communities at pan-European, national, regional, local and micro-levels" (Benjamin Bowling, personal communication).

It took the election of Britain's first BNP councillor on the Isle of Dogs in September 1993 for some to finally realize just how easily the vulnerabilities, prejudices and grievances of a relatively small minority could be manipulated for political advantage. In the context of Island politics, not addressing the discontent, frustration and powerlessness of the white working class had devastating consequences. From their perspective, their hopes and aspirations were relatively modest: access to good housing, the opportunity to stay near family and friends if they so desired, and appropriate services and facilities in the area. But these dreams were overshadowed in a development process that ignored them and by a local authority who housed the homeless ahead of them. They perceived themselves to be in a no-win situation.

Unlike the Corporation and the local authority, the BNP appeared, for those who wanted someone to blame and simplistic solutions, was willing to listen. A salutary lesson indeed about the need to work with different and sometimes diverse and conflicting demands in a locality rather than simply ignoring or imposing policies upon them, no matter how difficult this might have been. A small minority had once more demonstrated that they were willing to put up with only so much before fighting back.

Chapter Eight

"A different place altogether"

I . . . very much doubt whether there's ever been a community which within a matter of five years has gone from our present size . . . going up to 150,000 coming in daily. . . . What kind of impact does that have on the community? . . . Suddenly it becomes a major centre in Western Europe for banking [and] finance . . . and because building technology is such that you can get these things up in five years you've got historically a new situation — the social change it just erupts like a volcano. It doesn't happen gradually. . . . Very very difficult to envisage what will happen because it's got so many things happening at the same time — you've got the ethnic thing, you've got a weak local authority you've got all sorts of problems. (One of the first "newcomers" who had lived on the Island for twenty years.)

For all the viciousness and immediacies of the battle for public housing on the Isle of Dogs and the racism that was a feature of it, the Island had already become a different place, both physically and demographically, by the time of the Bengalis' arrival. Yet, despite the sheer scale and pace of the development, the many changes that had already occurred, and the inevitability of further changes to come, many established residents were perceived to be living in a "time warp". "It's only quite recently, if indeed it's happened", a newcomer said, "that people have really taken on the fact that the Island is going to change. This is a different place altogether [but] it's taken quite a long time for people to realize the really radical changes that are bound to happen here."

The once predominantly white working class area that people had rarely heard of, where no one really wanted to go, and which one newcomer said was "like going back in time", almost overnight had been transformed into an emerging city with a diverse and fragmented residential population (poor, affluent and ethnically diverse) with an estimated 48,000 people working in the offices surrounding the once thriving docks, on the Isle of Dogs in 1998, and 85,800 across Docklands as a whole, with predictions of double that number by 2014 (LDDC 1996:6). The development had, as an Islander said, put them "on the map". Between 1981 when the development began and 1991, the population in the two electoral wards covered by the Isle of Dogs neighbourhood area increased by 20 per cent (1991 Census), when average growth in other inner London areas was 3 per cent (Lupton 1997:5). By 2011 the residential population of the Isle of Dogs is anticipated to grow by almost 60 per cent to approximately 30,000 (LDDC 1996:15, see Table 8.1) with

Table 8.1 Docklands population estimates

	Isle of Dogs	Wapping/Limehouse	Surrey Docks	Royal Docks	UDA total
1981	15,472	5,226	9,055	9,676	39,429
1986	15,800	7,000	9,000	12,900	44,700
1991	18,471	8,798	15,548	17,943	60,760
1996	22,092	10,490	21,403	22,864	76,849
2001	25,692	15,146	26,213	29,345	96,396
2006	28,604	16,753	27,841	33,517	106,715
2011	29,421	17,155	28,283	34,924	109,783

1981 and 1991 estimates as at April, all other figures are mid-year estimates.

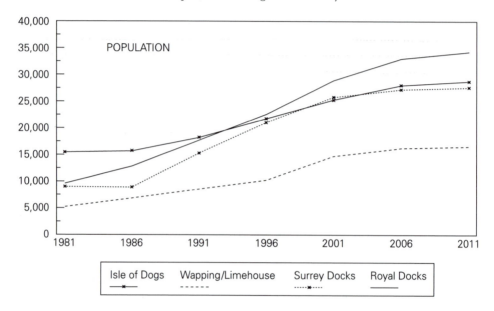

Source: 1981 and 1991 Population Censuses, OPCS and LDDC population estimates in LDDC facts and figures 1996. Reproduced with the permission of CNT as owner of the LDDC archives. © Commission for the New Towns.

increasing numbers of young people and the elderly (Lupton 1997:5, Inform Associates 1997:20).

This chapter describes the reflections of new and established residents on the impact of such rapid and dramatic urban change as it became manifestly evident that there could be no insulation from it. "The Isle of Dogs was rather like a little self-contained family unit, this is how it used to be", a businessman explained:

> You had a house on the Isle of Dogs and you worked on the Isle of Dogs . . . I don't think the locals actually had the wider vision of exactly what the LDDC were up to and how suddenly . . . thousands and thousands of people would want to be coming into Docklands to do a day's work. And I think to a certain extent [they were] quite naive about it. It was a sort of rethinking process that they were gonna have to go through to understand that the days of the fifties and the sixties when they were in

charge of the docks, what they said went, all those days they suddenly realized were going, but they couldn't accept it.

Although it is questionable whether local people were ever "in charge", certainly the activities, symbols and images with which the area was once imbued and to which some residents were very attached had little significance in the newly emerging postmodern landscape. As one Islander, commenting on the changed atmosphere following the redevelopment, said: "When we used to go in the docks it was all dockers . . . that was rough, that *was* the East End. But now you go in there and it's unbelievable innit what they've done." "If you look at the statistics", said another,

> it's shocking really. Talk about change it's unbelievable. You've now got more people on the Island who didn't know the Island before 1980 . . . and they're in the majority now . . . The sad thing is that . . . they could rightly say that they've got the community feel, the LDDC, because most of the people that are here now have been brought in by the LDDC or the Borough.

"You don't have to be a great mathematician" an activist commented "to work out what that's going to do to the demography and nature of the community". Some, especially in the light of its more negative characteristics, felt the development would, as a local Labour councillor said, finally "dispel the insularity of living on the Isle of Dogs in that people won't be able to avoid the outside world because it'll be on their doorstep" and welcomed this. Others felt, inevitable though it was, that this would rob the Island of its special character and feared that it would become like anywhere else. "The sad bit for me", a clergyman commenting on the changes in "community" pre and post development said, "is that I came into a community which very clearly believed in people":

> It did it naturally . . . Now . . . people are accepting and buying into a system that they don't have naturally but in effect is the one that's swamped them where people are secondary . . . [and] the prime thing is the prag- matics of your own survival . . . The pressure of the power, money, politics thing . . . left people buying into a system almost at times unconsciously and changing . . . I feel more sad than angry because I see something that's very valuable, and very precious that didn't grow overnight disappearing all but over night.

Once more the picture of change and its impact was a conflicting and con- tradictory one in which some had come to accept the necessity and inevitability of change while others lamented that so much had been lost for so little gain. A MORI survey in 1996 revealed that a quarter of Isle of Dogs residents, particularly the poorer and more established, still felt that the local community had received little benefit from the LDDC's activities and almost half of those surveyed (42 per cent) reported that neither they nor their families had benefited (Lupton 1997:2). However, almost half of the sample from the same survey felt that the LDDC took local people's views into account (Lupton 1997:3).

Whatever their view, as this chapter describes, the Isle of Dogs was a commun- ity in transition caught between its former working class and industrial past that still retained some symbolic power but no longer corresponded with the kind of place that the Island had become both visually and structurally or was to be in the future, and the distanced but powerful commercial might of the new development. "Now it's all here," one established resident said, people needed to ask "where are we? . . . Where can we fit in? And it's [about] people making that jump across."

Individualism versus community

> The big thing that struck me coming in [19]84 was the village community . . . there was a definite sense of the community being the community and there was no space for empire builders . . . that was not just totally inappropriate but was just unacceptable because there was a sense of look this is one community and we're in it together . . . but one of the things that certainly has happened through the regeneration is that everyone is paddling their own canoes. The perception is that you can't afford the luxury of doing it together because what you're doing in most cases is competing for the same resources in order to survive. (Church minister)

Although the "traditional" white working class on the Isle of Dogs was shrinking throughout the course of the development, and increasingly was associated with the area's past rather than its present or future, there remained a core of people for whom it would always be special whatever happened there. But those who were part of what was now seen as "the old Island" were an ageing population all too aware of new communities of interest emerging. "What we call the Island Mafia now is not the same group of people that it was ten years ago", a resident observed; "that's changing." "There's a different community coming along", a seasoned campaigner agreed, and "there are some elements in it that I find deeply worrying." Self-interest rather than "community" was perceived to be the order of the day, summed up by one activist as a "dog eat dog . . . situation" where care and compassion is regarded as of no consequence and . . . individual self-interest is here".

The most visible aspect of the transition on which many focused was the declining significance of the established white working class who, it was generally felt, by one means or another, would inevitably be driven out by the development. "There will be a huge pressure . . . for the area to be gentrified", a campaigner for local people's access to housing explained,

> and I think that's sad because I believe very strongly in trying to maintain historic bonds of family and kinship in any area. I think it's the most vital and precious thing to protect, but I think it's very very difficult on the Isle of Dogs because you'll have the most enormous financial up-market development, vast wealth sitting cheek by jowl with relatively poor people . . . After one or two generations I don't know how many of the old [families will be left] cos there are families who've been here literally for generations.

... Unlike other areas where you could maintain ... the identity of the community, I think the Isle of Dogs in the long run ... it may take two or three generations but I think that it will gradually change it.

"Areas like this ... are really villages in the inner cities ... [and] the place of sons and daughters in the community are vital to the continuing sense of having identity rather than just being like anywhere else in the suburbs, which are pretty anonymous". He continued:

> Everybody in the suburbs tends to complain about 'wouldn't it be nice if we had a sense of community'. We've got it here historically but unless you are going to keep people or at least provide the means by which people who want to stay can stay, then you're gonna destroy that and you end up with the worst of all worlds. (In Foster 1996:164)

Some, though, clung to a belief that the "community" would never be defeated. This was exemplified by the remarks of an activist who said, when asked whether commercial and more affluent interests would eventually swamp them: "That would have meant that we would have lost ... I think that we will struggle on to stop that happening and I think that we can win ... I can't see us being swamped ... even in my darkest moments." Such words, though well intentioned, now seemed to have a hollow ring about them. Yet he continued:

> Sometimes ... people [say we're] being ... pushed out ... but ... I never actually ... use that as an argument ... 'cos I never believe it and ... I don't want to encourage people to begin to ... believe that. ... We haven't been pushed out and to try to get people to ... sort of struggle on the fear that they might be pushed out is a negative one. I think what we say is ... we're never pushed out but we may be sort of pushed to the bottom, what we've gotta do is [say] ... we're gonna stay and we're gonna fight, nothing else we can do you know. This is the place where we go.

But increasingly the Island was beginning to feel like a place where some residents did not "go" and where even in local authority housing what they perceived to be "their place", as we saw in Chapter 7, was increasingly contested space. People were also tired of the struggle that, in the cold light of day, had achieved very little. "There is a community spirit here, there's a fighting spirit," a newcomer observed, "but there's almost a, it's not a defeatist [attitude], but I think a lot of us feel that we're really knocking our heads against a brick wall."

Many did perceive themselves to have been pushed out both literally and symbolically, which was reflected in the feelings of young and old alike. "The Island is like a village", a teenager explained, where "people don't like other people coming in. It's very complicated you see ... People who do not want to go ... people who've lived here so long, they've actually been forced out." This was a perception [though inaccurate] reflected in the comments of this pupil who said: "For every one of them [yuppies] who moves in one of us have to move out." An Islander fifty years this pupil's senior confided: "I honestly don't think we'll be here [in ten years' time] ... I think they'll force us all out" or "we'll just get that fed up ... and we'll be gone. . . . This is actually what they want isn't it." The insecurity of many was aptly expressed by another established resident, also in her sixties:

> You do wonder whether the council estates will go and it will all be private development. . . . They've done it before on [the estate demolished for the road] so there's nothing to say they couldn't do it here. [If that happened] I'd probably move right away I don't think I would stay. . . . I don't think I'd want to stay.

A former LDDC executive denied that local people were being forced out and suggested that it was hostile Conservative policies not the LDDC itself that caused the problems:

> There's no earthly reason in principle why . . . blocks of council dwellings shouldn't cohabit perfectly happily in that part of London as they do in any other part of London. . . . but if successive governments continue to be hostile towards local authority housing and to increasingly turn the screw to persuade local authority housing to dispose of their stock then [local people may be forced out] . . . because the houses have value and that's what will have changed . . . Ten years ago [1982] a council tenant on the Isle of Dogs probably wouldn't have wanted to buy his own house, it wasn't worth anything . . . now . . . he would be more likely to buy his own house. . . . And then it's lost to the council stock. So . . . there is a danger . . . but I think that's all about central government and local authority policies. There's not much the corporation or the people themselves could do about it.

Certainly the increasing competition for local authority housing was important but so too was the LDDC's emphasis on owner occupation, which undermined the position of some of the poorer sections of the white working class and pushed them out towards the margins. Nevertheless, some of the poorer sections of the "community" did benefit from these housing policies and were able to exploit the "value" created by them. An Islander who bought her pre-war council house in a highly prized area jokingly predicted that in years to come "we'll probably be like some little cul-de-sac in Chelsea or something . . . Bijou cottages . . . they might be sayin' half a million for a little bijou cottage on the Isle of Dogs!" She wanted to stay. But others who exercised their right to buy now had choice and some, like this couple, chose to sell and move away. "We intended to move when we got married. I've gotta be honest."

> We said we'd make the break then but finances wasn't permitting. . . . We always tried to own our own place and you know get on, but, I think definitely the building has helped us. The way they've developed it. Because my husband he's a builder so obviously . . . loads of work. So that has given us a step on the ladder really. . . . We brought our council place for £26,000 and sold it for £79,000 and brought this [house on a private development] for £123,000. It was the boom.

Their next move was to a large detached house off the Island. A woman whose sister also moved off said: "They're bettering themselves. They were lucky enough to sell [their council flat] and buy [a house]. It is a really nice semi-detached, three bedrooms . . . a garage and a nice big garden back and front. We can't possibly get that on the Island. It's only the yuppies [who can afford that] . . . the prices are right too heavy for us."

For those who had lived on the Island all their lives and had extensive family and kinship links, moving away was not a decision taken lightly because, as one pointed out, "our roots are here aren't they". Neither were the advantages of the extended family underestimated:

> Everyone said . . . "Won't you be sorry to leave here?" . . . It is a big move really. I mean, I'm minutes from my mum, five minutes from my mother-in-law. If I have to go to work they come and mind the children so we'll be losing all that. But then you've got to think about 'em [the kids] growing up. It's a difficult decision to make. . . . It's a shame we can't keep this and buy another house and then if it don't work out we can come back.

Some did move back and those who moved away often retained their attachment. "I think I'll always feel close to the Island," a young Islander planning to move said. "I don't think there is anywhere like the Island". This feeling is shared by the worldwide membership of the *Island History Trust*, as one of its members explained:

> I don't know whether it's the community spirit but we are definitely, I suppose you'd say, a breed apart. I don't know, there's something about 'em [Islanders] cos that's what brings them back you see and they're all so proud of being born on the Island. This is what gets you. When they write in from Australia places like that. . . . Still very close ties. It is such a closeness. You can't explain it.

"There's always that kind of feeling when you meet an old Islander, we call them old Islanders", one elderly women said who discovered, much to her delight, that the first generation of her family to move away had neighbours who were also Islanders:

> Immediately there were no barriers . . . I said "What is your name then?", so he said "—". So I said "Oh I knew your mother and your father and your aunts and your uncles and Nan." Ever since then we've been friendly . . . It's a funny sort of feeling you know what I mean. It's just there's no barrier once you've lived on the Island. . . . There's no barriers . . . there's just a closeness.

That is, if you were white, working class and had lived in the area for many years. Others, as we have seen, had to struggle harder to attain that kind of closeness and some never experienced it at all. But its existence was universally recognized.

There was a sense, particularly among many within the Corporation, that extended family networks were "outdated" and that all the talk of community did not have a place in the 1990s, as this LDDC executive aptly demonstrated. "Is there a concept of community anywhere? What does community mean?

> I think we have this sort of romantic notion put around by things like *The Archers* and *Coronation Street* and we say, ah that's the community — but you tell me anywhere real nowadays, anywhere in Britain, including in villages that lives like *The Archers* or *Coronation Street*. That's a romantic notion of community. You tell me what you mean by community and I'll tell you whether there's any sense of it round here.

I suggested a sense of belonging, an inter-relatedness, particularly prevalent among those who had lived in an area for a long time and had extended family. There is "heaps of that", she replied:

> To that extent the East End because of its traditional deprivation is still sort of many years behind the rest of Britain. Where in the rest of Britain do you still have extended family patterns? Nowhere. So yes there are some extended families still as no doubt there are in other places, but here as in other places the concept of the extended family as equalling community — people don't have extended families anymore and that's becoming true here as well. However, having said that, I do think the Island is one of the last bastions of it . . . local people that I talk to have got mums on the council estate here, the daughter's living in a self-build scheme there, her cousin's bought a house on Clipper's Quay and someone else's grand daughter's out in a house in Beckton — there's a bit of that in the area still. But I have to say it's gonna go . . . as it's gone everywhere else because extended families are not now a norm.

One seasoned community campaigner interpreted this response as part of the age-old class analogy. "We're always struggling against the class attitude", he said,

> because . . . the middle class for some reason . . . see it as their life's mission to convert the rest of the world to their way of thinking . . . The middle class, collectively, the attitude is very much about individuals, it's about getting on in society, acquiring an education, getting there you know . . . and the extended family . . . the community . . . that . . . sense of belonging, the sense of having roots . . . is not so important to them.

Yet even when the development was well under way, almost a quarter of the population in one survey were born on the Island, and 64 per cent of the sample had relatives living there (Wallman et al. 1987). Although these numbers were rapidly dwindling as older Islanders died and their younger counterparts moved away, and the concepts of "family", "neighbourliness" and "community" were themselves oversimplified, to some extent idealized, and changing in ways unrelated to the development, but established Islanders' influence and historical association with the area remained a potent force. As one resident observed: "It's a divided community that you've got now [but] there is still quite a strong community feeling down here." It was this strength that led some to believe that, despite all the forces for change, Island people would survive:

I really think that there's this, I don't how to explain it, but there's a tenacity and a survival instinct of people in this area which will go beyond all of this [development] and in the final analysis if [the local authority is] not giving them services . . . and the LDDC are a waste of time, they do it themselves . . . I think there is that strength and kind of focus and family networks. There's real webs of families here that keep together. But there is an argument I suppose with housing that maybe that's what will unstick it. The fact that there won't be the homes for local people's kids. (Local authority officer)

Although the sale of council housing represented a threat to the established Island community, in theory at least, it also provided a means of maintaining kinship ties. "The ones that are in fortunate housing, and there's a reasonable amount of local authority housing," a middle class resident pointed out, "and . . . have exercised their right to buy . . . They're sitting on a huge asset . . . If . . . they hand it on to their children then . . . those families will continue. But if they sell, "gradually we'll get an exodus and dilution". However, as the public housing stock diminished so did the opportunities for getting on to the local housing ladder. "Three of my children are the lucky ones", one local explained, "because they've exercised the right-to-buy, so they've just scraped in by the skin of their teeth."

My 14-year-old hasn't. Where will he get a mortgage to buy any property on the Isle of Dogs as a first-time buyer? . . . I've bought, I've exercised a right-to-buy . . . I don't agree with that because all that's happening is you're getting less and less rented accommodation available [but] . . . I could envisage a situation where one day, . . . you're left no choice, they tell you where you're going, you're the tenant and that's where you go. I wanted choice and the only choice I would have is if I exercise the right-to-buy. Okay I wouldn't, be able to buy on the Isle of Dogs if we sold, but at least I'd have choice where I did want to buy . . . and so would my kids.

There were, then, both winners and "losers" and without more social housing, which, as this activist said, "holds the community together because then your sons and daughters . . . can actually carry on living 'ere, . . . even if they can't afford to buy", many would find themselves unable to acquire housing locally. As we have seen, such requests largely fell on deaf ears, something that may not have been so much a rejection of the principles involved but the fact that on the Island the housing issue could not be separated from the insular and exclusive nature of the established white working class and the racism that was a feature of the battle for public housing there, or from broader political agendas locally and nationally. Yet the situation on the Isle of Dogs, as this community development worker argued, suggested reasons for optimism as well as pessimism:

People say that there's terrible racism down on the Island and yes there is, but there's terrible racism in Stepney and Spitalfields and everywhere else. Somehow because it's an Island they do see it as a different place and they do kind of exaggerate the problems, but they don't particularly exaggerate the advantages like I think there's a better sense of community in some ways, it's not always used in a positive way but the fact that people do have a sense of community is a good thing because in lots of areas that's gone. That's something that can be built on even if some of the attitudes that are expressed through that community are not particularly helpful. At least they're people who talk to each other, which is much better than working in an area where everybody keeps themselves to themselves. (Foster 1996:164)

It's not for me anymore

The community structure is changing. Those that . . . have [the] ability . . . are getting up and out because they don't like what's now happening . . . If you repeat that pattern with enough people a large part [of the] indigenous community are going to disappear. (Clergyman)

A lot of the old Islanders now don't like the Island, they wanna go, they don't wanna stay here. . . . Everybody sort of feels that the atmosphere's changed more than anythink else. It's sort of lost its, you know [community]. (Island woman)

After more than a decade of development in which they were marginal and from which little direct benefit was derived, established residents, some of whom had lived on the Isle of Dogs all their lives, questioned whether they *belonged* there anymore. "It's completely different" a young woman in her thirties argued: "Everything's changed from the beginning to the end of the Island. Nothing really is the same. I think it is good, you know . . . but it's not for me any more." She continued:

When we was growing up everybody knew everybody else . . . Families knew families, but that has gone now. You don't know anybody really. All the old families are moving off. . . . a lot of people have panicked and gone so that way you've lost friends and families . . . where like, say, older people died, like my grandparents, the grandchildren and the children they're not staying any more. They're going. So there's no big community any more. . . . It must be the development that has done that because I know a lot of people don't like it . . . A lot's been lost . . . I think it's happened too quick . . . I mean six or seven years ago when I had my first baby I used to walk round to see my mum and the docks wasn't developed at all and within seven years it's changed, everywhere you go through the Docklands there's another building. Or the start of another building. And I think it's happened too quick for the people that live here.

Despite the familiarity and security borne of many years of association with the same place, the Island felt as if it was increasingly populated with strangers. "I've always lived here so I know a lot of people," a young woman said, "but I walk round here now and I don't know nobody. I just walk past people now." "You don't know the people anymore," said another, "they're strangers", expressing the view of many established residents. A 14-year-old, too young ever to remember the "good old days", was equally adamant about the isolation: "You sometimes walk around and you feel isolated . . . because you don't know who to say hello to and who not to." Yet, talking in another context, an Islander emphasized that even now "you always know someone, if you don't know someone you know someone who knows 'em, do you know what I mean? I could be talkin' to someone and say oh yeah I know her she lives round by me." There were of course echoes of similar feelings after the traumas of post-war redevelopment and ironically the exclusive boundaries that Islanders drew between themselves and "outsiders" contributed not only to others' isolation but also to their own as their networks diminished.

The fact that the Island no longer felt like a place they wanted to be was not, it was argued, related to change *per se* but to the *nature* and type of change. "It's always been a story of change, but there is something about the nature of the change that's currently happening that [makes people say] I don't want to stay here anymore", a man closely involved with the established community said. "What's happening isn't meant for us. . . . There is nothing anymore that makes this place feel like the place that [we've] grown up in." In fact, one Islander simply asked: "Who is the Isle of Dogs for now?"

Certainly in terms of its commercial activity it did not seem to be for them and many found it difficult to imagine that they would ever work there. Despite its enormity and close proximity, the office developments surrounding the docks remained distant — another world entirely. "People here [are] feeling particularly alienated and isolated", a man who worked closely with the established island community said:

> I suppose I wonder why there aren't more demonstrations and protests and I wonder if on the Island ten years ago people would have been bowled over quite as easily as they are now. . . . People are beginning to say "I don't think this development's going to be too bad" . . . [but] it's another world.

Although the dock walls no longer existed physically, symbolically they remained firmly in place and, the clergyman continued: "The feeling of alienation [they] produced . . . carried on with the new developments." As Ted Johns wryly remarked: "The LDDC say, you know, 'we've pulled down the dock walls which formed a barrier' . . . so they pulled the dock walls down which was probably about 18–20 feet high and put a bloody office block there that's about 100 feet high . . . so you're even more cut off from the water!"

"One of the things . . . that makes it difficult for local people perceiving of Canary Wharf and . . . the other developments as being for them", an Olympia and York executive said, "is that they're in areas that have always been forbidden areas. They refer to Canary Wharf as being "inside" and they mean inside the dock walls. . . . The Island is the southern half of the Island not the area that is part of the Enterprise Zone."

"I don't know that local people will ever feel as though it's their space", a local clergyman said:

Even people that I thought of as mobile locals who, you know, did quite a lot of travelling around and looking around, the explorers, said "Where do you get to the other side of the dock at Millwall? What's at the back there?" And they didn't know because they weren't the people who worked in the docks when the dock was around and they're not people who go into the new development now because it's not for them. It's not their space. So I suppose I feel a bit sort of pessimistic about whether that invisible wall around the docks will in fact ever go. I suppose it will because these things do in a way but it's still this sense it's not for us.

Resentment concerning the primacy of commercial interests was still evident in 1996 when an IRA bomb exploded in the South Quay Plaza office development, killing one man and causing extensive damage in the central business and adjacent residential areas (see Figure 8.1). A report in *The Guardian* in the wake of the bombing observed: "After the IRA blast, counsellors . . . found traumatised families angry that their needs seem to rank lower than those of commercial victims" (*The Guardian*, 24 February 1996:9). An "us" and "them" was still manifestly evident partly because, as this resident explained, they felt distanced from the development: "Although it's physically very near," he said, "I really feel that I'm not part of it and it's nothing to do with me."

Rose-tinted spectacles

Although the development might have played its role in disrupting or accelerating the movement of families and was sometimes blamed for destroying the community, it was one factor in an ongoing process of change. People's perceptions of change, and the reasons for it, are as we have seen, often simplistically presented when the underlying processes are complex. On the Isle of Dogs the networks of established Islanders had gradually been diminishing for forty or fifty years but it was not perceived to be a gradual process. Instead, perceptions of change were linked, as I described in Chapter 1, with particular symbolic events, the war, slum clearance and now the development. Furthermore, although events on the Island were, as an LDDC employee said, "accentuated" and "thrown into neon lights" by the enormity of the development, many of the changes to community patterns were not unique to the Isle of Dogs even though the development offered an immediate and convenient explanation for them.

Establishing the extent of change was also difficult because the experience of "community" evoked different images for different people at different times (Crow & Allan 1994:1–13). It also operated on a variety of levels simultaneously and was inevitably accompanied by nostalgia and contradiction. For example, while some Islanders spoke on the one hand about the strength of community even in the contemporary context, following the development they grieved the loss of the pre-war "community spirit", a sentimental and emotive attachment expressed even among those too young ever to have experienced it. "I'd like to just get back to . . . how it was — without the war", a 14-year-old said:

Figure 8.1 IRA bomb damage, Janet Foster

Cos like . . . I went [to] Island History [a trust dedicated to documenting and archiving Island history] with me Nan and that and . . . in school I done a project on the Isle of Dogs . . . and you look at . . . Canary Wharf and all them ugly buildings, you don't realize [what there was before] . . . But when nanny like . . . told me [about how the Island was] you think oh, you know, it would be much better . . . So that's what I wish, knock down all the buildings and get back to the community.

Her grandmother had similar sentiments but a different experiential base for them:

When you say what would I wish for really and truly there's nothing I would wish for except the community spirit, that's all it is, innit, the community spirit. . . . I don't think any money or anybody could make the community spirit and that is what's lacking but not only here it's every-where, innit.

These perceptions, according to P. Cohen et al. are part of an "imagined community" that has grown up on the Island in the post-war years:

Certainly for the self-proclaimed Islanders, the Isle of Dogs is a very special place, whose uniqueness is bound up to a large extent with the dockers' sense of being a race apart, and the backbone of the nation. The pre-industrial values and forms of solidarity which informed the culture of costermongers and dockers were in fact constantly invoked, and not only by those who might claim to have actual memories of this vanished world. What we are dealing with here is an invented tradition, a certain construction of local history and geography, which attempts to recreate something which never existed in this idealised form in the first place. (Cohen et al. 1994:6)

There is some truth in this statement, as was demonstrated by an Islander in her thirties whose experiences of growing up on the Isle of Dogs strongly resonated with the descriptions of those two or three generations removed:

It was always like a close community . . . We was all born at 'ome except me older sister. There was always one of the neighbours in when she had her children. Her mum was there with her and it was that sort of com-munity and it always has been. . . . The whole street knew each other. They stayed actually wherever you went right up to the last few years, that's probably the one big change, the fact that the community's were probably split up or got smaller because people were either forced out or their children [were] . . . You see what happened before, we've been quite lucky, whereas I was born on the Island I've left 'ome and I'm still on the Island. Well that always happened. Children was just down the road so families were altogether but obviously now with no council 'ouses being built, prices that no one can afford, you know, locals, the families are split up.

This account appears as timeless as the descriptions of people bought up on the Island in the early decades of this century. In a factual sense the focus on the development breaking up the community is not accurate and many of the images she presents could be contradicted. But does this make her account any less valid? Is she presenting a cohesive view of community "up to the last few years" (i.e. before the development) in order to reinforce the disruption it caused? Her description is not merely recreating something that never existed. It is relating individual experience but in an oversimplified or "glossed" manner influenced in part by nostalgia but also to emphasize a point. As A. Cohen (1989) highlights, "In the public face internal variety disappears or coalesces into a simple statement" (that is, that community was characterized by togetherness, everyone knowing each other, and so on) but "in its private mode, differentiation, variety and complexity proliferate" (A. Cohen 1989:74). Furthermore, whether perceptions are founded in reality or not, what people believe to be true is real in its consequences, which was illustrated most fundamentally by the attitudes about race and public housing.

> *thesis*

Ted Johns used an analogy to describe the nature of community that seemed most apt for the Isle of Dogs. He suggested the need to differentiate between communities of interest or alliances, of which there had been several during the development itself, and a "oneness" or community unity often invoked but inaccurate. "At its most pragmatic", he said, community can be described as

> people that collide together from time to time and overlap. There are kind of pools . . . [and] as you get particular rainfalls, that pool will run into that pool, and that pool will run into that pool, and then they'll just drift apart . . . and then it'll come together again. . . . I think that communities are like that. In a sense a community is a collection of people that have a common purpose on a particular occasion. There is in communities an underlying unifying thing of being an Islander, being an East Ender, . . . that . . . brings them together. But . . . you can't talk about the East End community as being like a oneness because they're of a oneness on different occasions for different purposes.

Individuals also have different levels of investment, at different points in their lives and for different reasons (see Suttles 1972). Suttles suggests that the best way to conceive of communities is using a concept of "limited liability", which strongly parallels Ted John's comments above. "People tend to live in more than one community of limited liability and have many different adversaries or partners in maintaining more than one corporate identity." Furthermore:

> Since there are often two or more competing communities of limited liability, a resident frequently finds his (sic) interests divided among several adversaries or advocates and sometimes ones that are different from those of nearby co-residents. Participation in the community of limited liability then, is a voluntary choice among options rather than one prescribed on the basis of residence alone. (Suttles 1972:59)

In the late 1980s Hobbs described the East End of London as a place integrally bound to its history and culture, with fluid boundaries, "not directly aligned to any specific coagulation of concrete, steel or tarmac. They are defined by the inhabitants as an alignment of commonly held strategies" (Hobbs 1989:86–7). Hobbs suggests that the East End is characterized by a "one-class society" in which "east of the City of London you are either an East-Ender, a middle-class interloper, or you can afford to move sufficiently far east to join the middle classes of suburban Essex" (Hobbs 1989:87). However, in Docklands at least, and on the Isle of Dogs in particular, this monolithic and long-established pattern was breaking down, and the divisions were no longer so clear cut.

While they would always be a distinct group, newcomers had been absorbed into a fluid Island "community" in which the long-established residents on the Island remained dominant, and some of their accounts resonated strongly with those of more established residents. "It takes a while for Islanders to decide whether you're gonna belong or not," a businesswoman said, "but once they've decided . . . they're gonna accept you, then they accept you and it is like living in a village and that's lovely and there is a huge advantage in that." One woman who had visited the Island in the 1970s described it as "the sort of place I wouldn't have got out of the car on my own". Another recalled thinking if her husband "really loves me why is he bringing me to this place?" Yet both women, who moved to the Island deeply conscious of its different roots, sense of history and place, came to see these characteristics as part of what made it special. One newcomer with East End roots said: "This is the only place that I've actually settled that I really like. It's almost like coming home."

> We just felt like we'd moved back home with our own type. . . . it's just, you talk to an Islander and where we've all been bought up in, like, the dregs if you like, the dustbin of England innit they call it, the East End. We all know 'ow each other feels. Although [we] have moved on quite a bit, I'll never ever forget where I started from and how it was and what a struggle it was.

"I do feel part of a community", another newcomer said who did not have a hint of the nostalgia of the East End or Island pre-war and pre-redevelopment; but as we have seen, the "community" to which they belonged was in fact several communities, the limited neighbourly networks on the new developments, their involvement in communities where old and new came together, the local churches, the pubs, the play groups, and to a lesser extent the schools, and well beyond the Isle of Dogs in a mobile and sometimes global "community" too.

Two views of development: tradition or change

Although they were frequently regarded as uniformly anti-development, two very distinct views about the impact of the development existed among established Island residents: those who regretted and resented change, and those who sought to accommodate it. "Traditionalists" often perceived the changes as threatening. "Everything they have belongs in [this] small area", a community worker said: "It is just not in their make-up . . . to think of moving, not just out of the city, outside of their estate. Their mother lives there, their sister lives there, brother lives over there and to move out of that, they haven't anything." Others, also with extensive kinship ties, and subject to exactly the same forces of change, seemed more optimistic, and embraced change: "You've gotta move with the times," an old Islander pragmatically said,

> you can't get back what you've had, once it's gone it's gone. With life innit, that's how it is. Me meself I say oh God I'd love to get away from here . . . but then again I don't think I could. I don't think I could leave the Island because I've been brought up here.

"I'm rather glad I'm here and [can] watch it all happening", a fellow Islander explained. "I think it's great really . . . there's something to look forward to", said another.

But the very same places and changes were perceived quite differently by others. "Someone . . . asked me if I thought it was better now than it was", an Islander said,

> and I said "Well from my point of view, no it ain't" . . . From the purely personal point of view [I felt] more or less . . . you've got everything, you've got yer family round you, you got work in there [in the docks] . . . I thought I've cracked it . . . I suppose life's like that, when you think you're there you turn round and then it's all been . . . tak[en] away from you.

"I'm a great believer in tradition," he continued,

> improving the past but still based on the past and that's the way I am. . . . there's quite a lot of people . . . the same . . . We're not saying that . . . the past was brilliant 'cos it wasn't, but it had a place . . . and so what you try to do is you try to do away with what was bad 50 years ago but keep what was good . . . People . . . say "Yeah but people move on, life moves on". Yeah, but . . . there's a difference in moving on voluntary and being slung out or starved out so you have to look at it in perspective.

Another said:

> A lot of this talk is taken up with nostalgic looking back . . . I don't sort of contemplate the past and say well that's how the future should be. The future has it's own thing, . . . things change around. What I'd like to do is to hold on to the best of the old and graft in the best of the new . . . and that's where the past has it's place, it has it's lessons.

However, there was a fine dividing line between tradition and resistance to change, as this younger man who had lived on the Island most of his life explained: "They're still trying to motivate their force to stand against what is happening. They seem to want to live in the past . . . I think it's probably fear, fear of moving on. They want to maintain the status quo because they're frightened."

Although using it primarily in relation to race, P. Cohen (1996:193) characterized a "culture of complaint" that aptly described the negativity of some established residents. "They did absolutely nothing for the locals, only if they did it was an after thought or because it was benefiting the yuppies", an Islander in her twenties said bitterly. When I asked if there was anything good about it from her point of view, she replied: "Well I've not used anything that they've built." A businessman whose company had been on the Island for more than a century had this to say about the attitudes of many Islanders:

> You find people who've been here a long time get terribly cynical [about] the Isle of Dogs, and the people who are here — the biggest load of moaners on the face of this earth . . . I have a theory that people who leave are the people with energy and enterprise . . . if you studied islands and island people you'd find that the more remote the island the more introverted and awful they are.

A newcomer accustomed to the established Island population perhaps more charitably and wryly commented:

> If you took this development away, how long before the same people will be saying "How dare they leave us to live like this . . . have you seen that lot in such and such an area in London where they're getting all this new development and all these new compensations." How can you win in a situation like that? People are such difficult creatures aren't they?

However, as a newcomer who had close associations with some of the established Island families said, changing attitudes were even discernible among the negative people. "The whole area's been lifted up", she said,

> and now they actually want things out of life. They've seen a bit of the other side, where you've got what they call yuppies, move on the Island, they've got money, they've got property, they're well dressed. I think it's probably brought the whole Island up . . . [The development's] still a sore point but I think they're actually — whereas when I first came here [people said] "That railway's not been built for us, so we're not gonna use it" — they're actually doin' things on the Island that I s'ppose they swore "We'd never do that", oh no, and the Arena, no not going to the Arena no . . . and they've all used it now. So I think it's just where at the beginning it was a very sore point with 'em now they're actually beginning to use everythink that's here.

Some of these perceptions were reflected in a MORI survey (1996) that revealed that both improved transport and the newcomers who had moved into the area were viewed positively (quoted in Lupton 1997:2). Across Docklands, 61 per cent of respondents felt that the area had improved, but the least favourable view, as it had been almost a decade earlier, came from the Isle of Dogs (MORI 1996 in Lupton 1997:2–3).

Even if traditionalists wanted to recreate the past, the futility of this had become increasingly evident:

We're desperately trying to get back what we had before and you know it hurts me to say so but I don't think we will . . . I mean in pockets you can [find it] . . . Sometimes on a Friday night you get sort of 30 or 40 people that have been here for ever, as they say, and you can forget what's happening out there, or you can go to a party or you can go to a pub with the same sort of atmosphere, but you then come out into reality . . . So you ain't gonna get that back whether you like it or not, you ain't gonna get back the same involvement, the same community spirit that you had then.

The pain and loss in this ex-docker's account contrasted with the matter-of-fact feelings of the Islanders below who accepted that the Island community as they knew it was unlikely to survive:

We've more or less been absorbed into London, haven't we, rather than being a separate community . . . I don't think it will ever be like it was . . . there might be little pockets . . . yes little pockets sort of here and there [where] . . . they're born and bred on the Island, but I think they will probably be the last.

The strength of the feeling among some established residents, however, convinced a few that "Islanders" would survive:

All generations create their own style so that, you know, I'm as different from my father as he was probably from his grandfather . . . but in very many ways we're very much similar. So I think the underlying similarity will remain. On the surface there'll be plenty of changes going on but . . . there will be "Islanders", you know, in the future, I mean my daughters and my sons regards themselves as Islanders . . . They're still talking about Islanders.

But the Islanders of the future would be a diverse group including the children of affluent newcomers and Bengali families, with contrasting experiences and cultural influences, whose "differentness", and the threat they posed, had already resulted in their being "cold shouldered" and harassed. As one teenager talking about the tensions on the Island put it, "It's always like who's on top and who's going down." Another youngster said that the different communities on the Island had "different needs . . . they don't all sit together and discuss what they really want as a whole . . . They don't actually try to come together at one point and sort the thing out". When I asked what the communities had in common, he replied "The only thing the communities have in common is that there's rich and there's poor that's all there is."

Embracing change

I've often found that there's perhaps an excess of talk, exclusive talk about those who were there already and very little talk, if any, about the

incomers . . . and I think you've got to look at the two together. You've got
to say there must be a future for both. For those who were there, their
children and those who are coming in . . . Many's the time at a meeting
when all the discussion has been on "what are we gaining?" "what's in it
for us?" . . . there was always this absence of discussion about the newcomers, the incomers and welcoming them and making them a part of the
living community and getting them involved. . . . I think that will happen
and that will heighten ambitions and perceptions of the local population.
(LDDC Board Member)

We can see from a village community a new city emerging and I think
in due time it will be seen to be worth the struggle. (Community
representative)

The transformation of the Island from a village to a city was often discussed in
terms of loss, loss of a way of life, loss of employment and industry on which a
century of cultural heritage had been established, a process that was very disorien-
tating. However, as Massey points out, change is not simply about loss. It can also
involve recomposition. The global forces that had such a profound impact on
Docklands fortunes were not only negative. They also offered an opportunity to
liberate traditional perceptions of place and space. "Surely", she writes,

we should . . . be actively promoting a conceptualization . . . of place which
is precisely about movement and linkage and contradiction. A sense of
place which is extraverted as well as having to deal with and build upon an
inheritance from the past. That is surely the meaning of the joint existence of uniqueness and independence. . . . Why should the construction
of places out of things from everywhere be so unsettling? Who is it who is
yearning after the seamless whole and the settled place? A global sense of
place — dynamic and internally contradictory and extraverted — is surely
potentially progressive. (Massey 1994:142–3)

While Massey may be right that globalization offers new and exciting possibilities for redefining place and space, and there was evidence of a newly emerging
sense of identity, albeit one far more complex, contested and conflictual, than the
one that existed before, it is understandable that not all Isle of Dogs residents were
rushing to embrace change or even able to conceptualize change in this way, especially when, as a former docker said, history had taught them that they were most
frequently the losers rather than the winners. The reluctance to embrace change
was readily recognized by this Olympia and York executive:

When you do something new in this world, you generally do it to some
degree of scepticism no matter what it is. . . . It would be uncharacteristic
of human behaviour if everybody was applauding . . . At some point, I
don't know when it'll be, London will put their arm around Canary Wharf
[along with Docklands] . . . and expand the definition of what London is
to incorporate this . . . I believe perceptions lag reality . . . It means you
have to create the finished product and it has to be there and people have
to experience it.

Whether they liked the development or not it was important for local people to carve out a new place for themselves within this rapidly changing environment, something that Peter Wade, who had fought long and hard against the development, realized only too well. "You'll either be part of it or it will completely destroy you and you'll be embittered", he said:

> There are things there for you to gain from but you've got to think in [the] long term and it's your kids and it's your grandchildren because they quite honestly are not gonna give a toss for what you've done. They see it today and they're living it today with their children and their children and that's where you've gotta be.

Canary Wharf has become a "beacon" on the London landscape and well beyond. As one business person said, it made "a statement that here is the Isle of Dogs, here is Docklands" and with it, even before it was completed, came signs of a fusion between past and future. "This used to be the commercial centre of Europe," an Islander told an executive from Olympia and York on a visit to the site arranged by the company, "let's see if you can make it that again." "That was a good comment", she said, "because she had come there very anxious about it, was this what we wanted? This was being imposed on us from outside and yet when she was there she felt proud that it was on the Island."

The Isle of Dogs had become somewhere. While its poor and villainous East End status was difficult to shed, other more positive images were emerging. "Years ago, if you said you lived on the Isle of Dogs people didn't have a clue where you lived", an Islander said. "If you say you live on the Isle of Dogs now, people really think ooh, you know ... it seems to impress people all of a sudden if you say you're from the Isle of Dogs or you're from Docklands. They don't ignore you anymore." But the attention of the world's media was also a cause for concern and led to the need to defend their place once more. This time, ironically, defending the very development they had once sought to oppose, a situation aptly demonstrated in a meeting between the LDDC and the Association of Island Communities in which Ted Johns said they were "fed up with the negative press campaign being waged against Docklands and their portrayal of the area". "That [was] a most significant statement", an LDDC executive said. "First of all it was a recognition that this is a good place to be and secondly they are fed up with it being slagged off ... They said 'We're not saying we support the London Docklands Development Corporation, that's another issue. But we do support the London Docklands'."

"Getting through old Docklands into the regenerated Docklands"

> I think in 15 years' time ... it'll be great ... I think it's going through the period of getting through old Docklands into the regenerated Docklands [which] is causing the problems ... You've got a changing society, you're going to get changing attitudes and I think it'll be quite positive down there. I think they've got a problem of blending old and new, mind you ... I

think you're going to get even in the next 10 years, maybe even more than that, you're going to get problems of the sort of brand new housing estate and the old schools and the old types of shops and the old pubs. When I go down there now that's what hits me you know is that you can spend £250,000 on a lovely house and just across the road, I won't describe what I might think about it, that is actually as I perceive what the problem is at the moment. (Former resident)

You see the mock-ups, the artist's impressions and it looks fantastic. I wish it was here now because . . . I don't know if I'm going to be around in those years. . . . You just feel so torn because you don't want to leave here because of what's going on, you think I'm gonna miss all the best bits if I leave. All right so we're putting up with the not so good things now, the lack of facilities and whatever, but it's going to be fantastic in a few years' time so you're almost loath to give it up . . . I think it will be marvellous. (Affluent newcomer)

Without the sentiments and experiences of the Island past, and because the development was their reason for moving into the area, affluent newcomers wanted to look forward. Although some expressed doubts about the development and the problems it caused, most were optimistic because, unlike their poorer counterparts, they had a greater degree of choice.

The LDDC's push for private development, intended at the outset to create a more "balanced" community as I described in earlier chapters, backfired when intense pressure on land during the boom led developers to build flats rather than family housing, creating justifiable concerns about the long-term viability of the "community" in the area. A senior Corporation figure suggested that "market forces" would rectify this situation and that

we'll see people buying the adjoining one-bedroomed apartment and putting them together and hopefully having good families. One sees that happening in Chelsea today . . . there is a terrific shortage of the bigger number of bedroomed flats like that . . . I think things will work out. . . . I can see developers responding to the market.

Ironically, the long recession following the collapse of house prices actually generated a degree of stability and forced newcomers who had not initially planned to be on the Island for any length of time to stay. Some, much to their surprise, as I described in Chapter 5, found they liked the area and wanted to remain and as one newcomer observed: "You're actually seeing them integrate a bit more now." However, their long-term commitment to the area continued to be problematic because of the schools, as this newcomer explained: "If we had a really good school we'd think twice about leaving, that's how much we like it here. . . . We'd like a good school . . . , not a private one particularly, we'd like . . . a good state school . . . but we can't stay [here] long term because of the schools." Another said: "There are quite a lot of families that don't want to move, quite happy, but they're not happy about the schooling . . . The standards are appalling . . . And when you think about it, the schooling is probably the most important thing for your children." Although, as I discussed in Chapter 3, newcomers' concerns were justified, with only 26 per cent of school leavers gaining five or more GCSEs (the figure nationally is 45 per cent), class sizes higher than the national and Borough average, truancy rates treble the national average, and over a third of pupils going into secondary education deemed below average in literacy and numeracy skills (Inform Associates 1997:7), things were improving (see Lupton 1997:42). But it would be naive to think that education would not continue to influence people's decisions to move away. Race was an issue in this too, for established Islanders as well as newcomers, as growing numbers of children with English as a second language came into the area. "If the schools was better I wouldn't go," an Islander explained:

> it's just the school. There's nothing outside of school for 'em to do. Everything's off the Island that you have to travel to . . . and also I'm not, well I s'ppose this might sound prejudiced, I'm not against anyone's colour but there's a lot of Bengali children and I feel they're holding my children back. Everywhere you go everythink's in, is it Urdu their langauge? Everythink's written in that. I think they're catered for more now than what we are — than what our children (who) are born here are.

Without getting back into a discussion about how misplaced such stereotypes are, as long as these views prevailed and people continued to move off when their children reached school age this would mean, as one newcomer said, that the Island would "always be missing . . . families with children from five to eighteen" without whom "it may never settle down properly". Consequently, as Suttles (1972:99) observed, "schools . . . occupy a crucial position in the redevelopment of the area".

Part of the appeal of the traditional aspects of Island life for newcomers was its colour, a stark contrast with the anonymity and impersonal emptiness of the commercial development. Some newcomers feared that the transient and soulless atmosphere this often exuded would permeate the private residential developments around the Isle of Dogs too. "I hope it doesn't become like the City, nine to five, Monday to Friday", a newcomer explained. "The managing agent for this development says he reckons the trend will be towards renting here. I thought it was going to be the reverse, that more and more people would start being here permanently, but he can see companies buying or renting these places for their employees . . . saving them from commuting, being here during the week and going home at weekends . . . which I think's very sad."

As with the commercial development, the organization and design of the residential developments were themselves detached and, as one newcomer said, "terribly self-contained" with, in some cases, "high walls, automatically opening railing gates" and security that gave "the impression leave us alone we're not interested in anybody outside our little community". This was yet another division reflected in the highly defined sense of "us" and "them" of Island youngsters, who described a place where "the rich" and "the yuppies" had everything to gain at the expense of the poor. A Bengali student highlighted the divided nature of the Island very well:

Isle of Dogs is small island. Most of its land are sold to pivet owners. There are place where public are restriked. Most of the building are office's. There is a difference between low class and higher class family's. Also with working parents and non-working parents. (Female Bengali student reproduced in her own words)

Conclusion

To me it's everything I imagined and hoped it would be . . . To be absolutely honest I can't believe how much they've done so quickly. (Affluent newcomer)

People are making more of an effort to find common ground. . . . [But] there is still a feeling that the development does not address the needs of local people sufficiently. (Clergyman)

In less than a decade the redevelopment of the Isle of Dogs dramatically altered the nature of the place where Island residents lived. The battles over the development had largely been fought by what one woman described as "old war horses" accustomed to struggles in many other guises. The young occupied a different world. They could never remember ships coming in and out of the thriving docks and perhaps found it difficult to imagine what the Island looked like without high-rise office blocks. Their sense of identity was not linked with employment in which work passed from father to son. They had not participated in the struggles of the working class and campaigns to get better working conditions. Their experiences were different. In fact, the comments made by one elderly Islander about those moving into the area were equally true of the young as well: "It's just another place to them. It's got no links with the past."

For the youth, the Island in the 1990s represented contested space. Whether they were white, asian, or black, Islanders or non-Islanders they had no security of place. Their futures were characterized by uncertainty. Most did not expect the development to bring them jobs and they were acutely conscious of the endemic conflict that existed between different groups competing for scarce resources. The anger and resentment of the adult white working class was passed to the next generation whose sense of exclusion and of "being pushed out" was more prevalent than any sense ownership or pride — the development was not for them. Yet the sadness of the fragmentation and conflict was, as one of the "old war horses" observed, that whoever was fighting the battles, and however they manifested themselves, the underlying problems that confronted many of the communities had not changed at all: "The . . . pressing issues . . . about housing, jobs . . . the quality of life, health provision and all that sort of stuff, . . . hasn't changed, . . . all that happens is that the other side just continually move the goal posts."

Other continuities remained. The Island might have changed but it was still perceived by many to be "friendly" and to have a sense of "community" and in the contradictory and curious nature of human interactions even some of the Bengalis whose position was marginal and threatened found a sense of friendliness there too, as this young Bengali woman wrote:

> I like living in the Isle of Dogs . . . I think it's a nice place to live even though there are racist people [who] cause problems. But there not all racist. I [k]now people who are very nice friendly. Most of them are white. . . . There are some unfortunate people who don't like other colour people. They don't even try to get to [k]now people . . . Even though we are different colour, different religion . . . they should remember . . . we are all human being no matter where we come from or who we are. (Bengali pupil, reproduced in her own words) (In Foster 1996:158)

If all of those living in the area were mindful of this pupil's words, life on the Island might have been very different and combined "pools" of residents, whatever their origin or ethnicity would, as a Bengali teenager said, have been a "powerful force". But the fact remained that some were manifestly "not welcome on Island", as a young Bengali man said, "no matter how much I try to integrate".

It was not simply the divisions within and between the poorer sections of the Island that gave cause for concern, however. It was that "vast wealth" was, as one newcomer put it, "sitting cheek by jowl with [the] relatively poor", a potentially problematic cocktail. As I describe in the final chapter, this situation alongside the "market forces" that had rocked the Island's economic fortunes in the past, and that had facilitated the 1980s development in its early stages, made the future of the Island, just like its past, very fragile indeed.

Chapter Nine

Making sense of it all:
The global and the local ← *thesis*

I've always said that whether or not Docklands is a success . . . will be . . . dependent on whether the people actually living in Docklands, the old and the new, have assumed ownership, if not authorship, of what exists. If they do, on balance, it could be registered as a substantive success. If they do not assume at least partial ownership, if they continue to resent what has happened and oppose its on-going development — then it will have been a social failure. (Reg Ward, January 1998)

[The development] hasn't produced the regeneration . . . that we would certainly have been looking for, or I think what in the long term [is best]. . . . I think in the future when somebody . . . writes a book about Docklands I think they will say what a terrible waste it all was and . . . clearly I believe that they would be right. It was just a wasted opportunity, so much could have been achieved and wasn't and . . . so much was achieved that should not have been achieved. . . . It's a tragedy really. Docklands is a tragedy — the Isle of Dogs in particular. (Ted Johns, community activist, Isle of Dogs)

The preceding chapters have described the regeneration of the Isle of Dogs in London's Docklands through the words and experiences of those involved in shaping the development, those who moved to the area because of it and those who sought to oppose it, and over a period of time how ordinary Island residents came to accommodate, if not accept it.

However, the account would not be complete without a discussion of the broader structural factors that impinged on the development process, not only on the poor and powerless, but on the financiers and affluent residents too. This final chapter, therefore, moves to a different level of analysis and, drawing explicitly on the macro processes that provided the context within which the development occurred, assesses to what extent the development was a "success" or a wasted opportunity and what lessons might be derived from "the Docklands experiment" (DCC 1990) for future urban regeneration. These questions are simply phrased, but the solutions to them are very complex indeed.

Global forces: providing the context

Although there are important lessons to be learnt from the "Docklands experiment" it would be a mistake to believe that there was one solution or "master plan". None of the organizations that embarked on the regeneration of London's dock areas got it "right". The local authority-led development in the 1970s was widely criticized for its lack of vision and realism about market conditions. The LDDC at the outset, in its haste to create and exploit market opportunities, and hampered by government in its ability to move into social regeneration, neglected the needs and interests of local people and paid the price in terms of their negative image and mounting criticism even from groups not usually concerned about the plight of the poor or deprived; while later, despite the commitment to social regeneration in parts of the organization, the Corporation came to be viewed by the end of the fieldwork as a local authority, with increasing bureaucracy and no entrepreneurial edge.

However, there were also positive features of each wave of Docklands regeneration. The local authorities focused on the immediate needs for public housing and employment of residents in their areas, which might not have been "visionary" or exploited London's global city status but would as a bottom line have provided far more social housing for those most in need of it. The physical regeneration of Docklands in the early years of the Corporation's life was impressive and Reg Ward's passion for Docklands development, and his dreams for it, transformed entrenched views about the East End of London and led to development the likes of which no one had at the outset ever envisaged; while under Michael Honey the community rightfully became an integral part of the development process, if only for a relatively short period.

Yet the development was rarely viewed as a process in which there might be positive *and* negative features. For critics the "failure" of Docklands regeneration was perceived in absolute terms: it was a lesson in how not to develop: it did not benefit local people; it lacked adequate planning and neglected the need for social housing. The supporters were equally adamant that the ability to capitalize on and respond to market forces, the advantages of a more flexible approach, the sheer pace of development and its opportunities were self-evident. Yet change is rarely uniformly "bad", as the Docklands critics often saw it, or uniformly "good". In fact, at one level "bad" development might reap rewards just as "good" development might not succeed in achieving its objectives. As one of the key figures in the early days of the Corporation explained, the only sure thing was that "the way of life of the generality of residents in the area . . . has to change. Some ways for the better, some ways for the worse — change is like that isn't it".

Change is also part of a complex process operating simultaneously on a variety of interconnected levels, involving individuals, groups, areas and nations. Although the relationship between the "global" and the "local" is often difficult to comprehend fully, globalization has rapidly increased the pace of change, created similarities in experience for people in cities tens of thousands of miles apart and even changed the role of cities themselves (Lash & Urry 1994). These macro societal transformations also have a profound influence on individuals who occupy a world of uncertainty "in which everyday life is affected by international markets

and in which decisions are being spatially and temporally removed from the context in which they are eventually experienced" (Dickens 1990:4). Giddens sug- *def.?*
gests that this situation results in "'ontological insecurity,' in which people have
precious little understanding of the processes affecting their inner selves, their
daily lives, the sense of who they are and where they belong" (in Dickens 1990:4),
which accurately summed up the feelings of confusion that many poorer Islanders
experienced.

Such profound changes, however, go far beyond the Isle of Dogs and Docklands, and yet making sense of the development and the process of transformation
there requires a framework that encompasses them. Cooke (1989:18–19) suggests
that Britain's economic structure and free market policies result in our "experiencing the extremes of global restructuring, both negative and positive". London,
like New York and Tokyo, has become a global city (Sassen 1991) with opportunities unparalleled in other areas, something those at the forefront of development
in the 1980s realized could be exploited. But the flip side of global city status, as I
describe later, is vulnerability. So as the next century approaches, we live in a world
with increasingly complex sets of opportunities and constraints (Savage & Warde
1993:191).

Unfortunately, as the development itself revealed, opportunities are not equally distributed. "The well-resourced select the best means of minimizing the risks
and maximizing the potential of modernity" (Savage & Warde 1993:192), placing
them in an ideal position to extol its virtues, but leaving the poorer sections of the
community to experience maximum disruption and uncertainty with few rewards,
something that on one level was exemplified by the Docklands story.

The complexities of globalisation also make the task of establishing which
factors are most important in determining change highly problematic, which has
led to "two opposing tendencies": a determinist one that focuses on singular explana-
tions, and another that draws out a multiplicity of factors (Fainstein 1994:12–13).
Fainstein believes that "economic and spatial structures" are most important because these "restrict choice . . . to further economic well-being. Everything may
matter", she argues, "but not equally" (Fainstein 1994:12–13). Yet her own work
demonstrates the importance of understanding *both* structure, which is strongly
gendered as well as economic and political, *and* the individuals involved in the
development process (agency). The difficulty with structural primacy is that it fails
to recognize the diversity of responses and differential impact of the same economic forces in different localities (Lash & Urry 1994:212 and Taylor et al. 1996:
303–6), and that human actors (agency) are important too, because as Goodwin
(1993:149) emphasizes: "It is of course people who create and destroy these urban
geographies, even if they do not always do so in freedom from others or in conditions of their own choosing."

The precise nature of the balance between local complexity and global fragility may be very difficult to establish (see Hill 1994:193), and some factors may be
more important than others. However, the principle should be that in trying to
make sense of urban change we need, as Keith & Cross (1993:24) argue, to consider the local and global simultaneously.

Giddens' (1984) structuration theory offers another way of thinking about these
issues that is most apt in the case of Docklands. He suggests that the complexity
and interdependency of late modern society should encourage us to conceptualize

change more imaginatively than the rather simplistic divide between macro (structural) or micro (local or individual) factors and which of the two should have primacy (Giddens 1984:142). Neither should be given "priority over the other" (p. 139), Giddens argues, since it is the interaction *between* them that is important in shaping and understanding social change that occurs simultaneously on a variety of different levels. The micro and macro are "co-present" (p. 142). Although Giddens does not employ the terms micro, meso (intermediate) and macro to explain the differing levels of interaction, because structure is important to all types, Bottoms & Wiles (1992:29–30) suggest that these terms are useful for explaining the way in which actors and structures interact and influence social processes. In this final chapter, therefore, I have sought to place the detailed ethnography of the Isle of Dogs into a broader framework offered by structuration theory to discuss some of the political, economic and structural limitations that impinged on the development process and inevitably influenced the ways in which it was implemented and how it should be judged.

A failure or a success?

> The children who are growing up on the Island . . . will have enormously more diverse opportunities and better opportunities than their fathers had and despite the great disadvantages that the parents are going through . . . that is so positive that you can't possibly neglect it. (Affluent newcomer)

> I would say . . . whilst not every single aspect will be beneficial to every single resident, every single resident will benefit from at least one aspect of the development. (LDDC employee)

Assessing the relative "success" or "failure" of Docklands regeneration is a challenging task aptly illustrated by Sir Christopher Benson reflecting on his time at the Corporation. He said it was a "wonderful place and wonderful opportunity to try and do something better . . . That's where one says . . . did you do well?" His answer: "I don't know" — an honest reflection perhaps of the complexities of the development process, which was not just about how many offices and housing units were erected but far less tangible factors that are difficult to measure and about which there was considerable controversy.

After years of decay and decline, many argued that irrespective of the particular form the development took it *had* to benefit the locality. However, the extent to which "success" could be attributed to the development was rather more complex than many involved in the regeneration process acknowledged or understood. How much of the change was merely physical, and perhaps largely superficial, and how much was structural? As a community division employee said: "At the end of the day it doesn't matter how nice an environment you've got and how many local facilities you've got if you haven't got any money . . . if you can't get decent health care and you can't get your kids into a school, or you're racially harassed and you can't set foot outside your door." But she also recognized that "a lot of that is dependent on what happens nationally in terms of health policy . . . education, people's access to employment", issues that lay with central government policy not the LDDC. Furthermore, a belief that "things *had* to be better" was not always supported by the available evidence on the impact of regeneration in Docklands or in other areas that experienced similar difficulties. As Fainstein explains:

> The sponsors of the regeneration programs of the eighties claimed that they had achieved a remarkable reversal in the trajectory of inner-city decline. Numerous studies, however, have characterized this growth as extremely uneven in its impacts, primarily benefiting highly skilled professionals and managers and offering very little for workers displaced from manufacturing industries except low-paid service-sector jobs. (Fainstein 1994:9)

So how can we judge the "success" of regeneration? Bailey (1995:206), assessing the effectiveness of "partnership agencies", identifies three criteria that are most apt for Docklands. Was it "an adequate response", did it "achieve urban regeneration" and "what qualitative and quantitative impact" did it have? I want to discuss the first and third of Bailey's criteria.

"Was it an adequate response?"

> We may still be at the stage where in response to "what is the answer to ICR [inner city regeneration]" we may still need to ask "what is the question?" This is not as quirky as it sounds. Are we seeking to reduce unemployment in some areas at the expense of others? Are we intent on removing blips and blind spots? Are we, like the Germans, committed to a game of economic equilibrium between the regions? Are we committed to market freedoms with only the worst externalities attended? Or what? . . . What an examination of the recent record of ICR makes clear is that we have not yet adequately addressed these issues in Britain. (Lewis 1992:60)

The controversy about the type of regeneration suitable for London's Docklands led to over a decade of conflict before the LDDC was ever established. Not surprisingly, therefore, whether the regeneration was perceived as an "adequate response to the task" was equally contested and depended on how the "problem" was defined.

The demise of London's docks, the loss of related industries (estimated to have cost 18,000 jobs across Docklands; LDDC 1988c) and the consequent unemployment that blighted the area was the "problem" the LDDC was set up to tackle. The fact that Docklands was in the middle of a capital city with global status gave it "national" importance and provided the opportunity to exploit its location and justified an entirely different approach to regeneration. As this LDDC Board member explained:

> In trying to regenerate the area you are looking after the needs of something very much wider than the geographical area you are operating in,

because the immediate interest for the people living in [Docklands] . . . was to have blue-collar jobs . . . Those jobs just don't exist and there's no indi-cation that they will ever exist on the scale that they used to 20, 30, 50 years ago . . . There is after all an opportunity in Docklands to do a great deal for Britain and for London . . . but it's not in the first place and in the very short term what the existing inhabitants of the area . . . want to achieve.

It was not, as Fainstein (1994:237, 248–9) argues, that the choice of commer-cial property development in Docklands was the wrong mechanism for change but the absence of balance that was the problem because it increased the development's fragility and exposure to global forces. As in many other areas of Britain blighted by de-industrialization where the work had disappeared, in order for those "blue-collar workers" in Docklands to have any hope of meaningful new employment in a rapidly changing market place, heavy investment in education and training, and the development of new skills were required from the outset, but this was not the priority of either the government or the Corporation.

For their part, the LDDC and most business people saw the development as certainly more ambitious and effective than other urban development pro-jects that would bring jobs and hopes of a better future. In 1996, it was estimated that in excess of 36,000 people were working on the Isle of Dogs (LDDC, 1996:6). Employers who moved to the area had "prospered", with over a third having increased their growth in spite of a long recession. Optimism about the future was high, with 50 per cent of employers predicting growth in the years between 1994 and 1999 (LDDC Employment Survey 1994/5, quoted in Lupton 1997:30).

However, these signs of "success" were, as critics of the development and much of the research evidence suggested, paralleled by other factors that questioned the "adequacy of response", taking the failure to tackle underlying social and eco-nomic problems generated by de-industrialization and the inadequacies of "trickle down" as the core of their criticisms. As Bailey (1995:227) suggested: The ability of "urban policy . . . to improve the economic and social well being of localities and to reduce urban deprivation remains at best uncertain and at worst requires a negative conclusion."

Donnison (1987:277–9) identifies a number of areas that urban regeneration should seek to address including: enhancing the quality of life, creating employ-ment, provision of services, a more diverse population profile and better housing provision. Although the development certainly encouraged a wider population profile by attracting new, affluent residents to the area, which in conjunction with other aspects of the development improved transportation and other facilities (many, for example, saw the light railway as having "a liberating effect" by making other areas of London more readily accessible) and were positive factors reflected in a MORI survey of the area in 1996, the LDDC did less well in the other areas.

Although people talked of "success", little or no attempt was made to define precisely how "success" should be measured at the outset. Neither did the emphasis placed on the virtues of "trickle-down" result in mechanisms to assess its veracity. When I asked an LDDC executive whether the trickle-down philosophy worked, he replied: "Golly that's a very difficult one! It depends how you measure it. My own experience is it works in certain sectors, and also over certain time spans. . . . we certainly as a Corporation believe in the trickle-down effect but we've never done any measuring to see how effective it is."

Even if efforts had been made to "measure" trickle-down, this was not a straightforward task, as another LDDC employee explained:

> I think trickle-down does work but it takes a lot longer to feed through than you can expect people to be patient and wait for. I mean, trickle-down will be manifesting itself from now until long after we're [LDDC] gone. The area *has* changed the people *are* changing but they change slowly. . . . so you will get trickle-down and you will get all sorts of impact and it *must* be happening it *has* changed people's lives but for them to notice and . . . [believe] this is good is gonna take a lot longer than the Corporation's operating in.

"I can't believe that in pumping in the billions of pounds" that had been invested in Docklands, an LDDC board member said, "that it doesn't represent a pay-off for the local people." Yet as Logan & Molotch note:

> The mere *circulation* of wealth between cities and countries does nothing in itself to benefit disadvantaged people and places; appropriate *institutional* forces are always necessary to direct the flow toward the disadvantaged. . . ., and without them the trickle won't be down. (Logan & Molotch 1987:284) (original emphasis)

Furthermore, the vehicles to "pump-prime" regeneration in Docklands further empowered those already possessed of every advantage, an irony that was not lost on this local Labour politician:

> The LDDC have spent a lot of money on . . . some of the richest organizations in the world [big business and developers] . . . where the people who live here are amongst the poorest families in Europe, they've got very little out of it. And it seems a very strange way, you know, the sort of trickle down, if you give money to the rich it will eventually trickle down to the poor but it never seems to quite get there!

Yet as Lupton (1997:3) argues, "the very success of the LDDC's work in addressing its economic regeneration brief may even have brought new and different social needs and tensions" that were largely neglected.

The Corporation's focus on physical regeneration and altering image, driven by Conservative ideology from central government and the *gut* instincts of business philosophy, not only neatly evaded the complex and seemingly intransigent problems that already existed (see Gilroy & Williams 1991:80) but was politically expedient. Social regeneration is not particularly appealing to politicians as its benefits are reaped in the long rather than the short term. Politicians and their civil servants have different concerns: maximum impact in minimum time with minimum expenditure, as this senior LDDC figure said on the basis of his own experience: "Ministers and civil servants don't like to take decisions before they have to", which largely produces unimaginative approaches and means "the emphasis is on trying to be just in time rather than prophetic." Furthermore, he continued, "the cycle of parliamentary elections" and "civil servants whizzing about from job to job at huge speed" results in little continuity:

> There's nobody there who you agree the thing with [who] is gonna to be there the following day. The civil service are extraordinary in this way, . . . and they do this I'm jolly sure deliberately . . . the politicians are electoral cycle and they're all playing their game of musical chairs and the civil servants know very well that they're not gonna be there when the consequences of their efforts are matured.

"The United Kingdom is perhaps notorious for muddling through," Lewis (1992) observed, "for reacting to events, for crisis management. Planning in any real sense seems to be an anathema to our system of government" (Lewis 1992:50), which results in an *ad hoc* system, where central government has vested interests in keeping things as they are. This ensures they retain control even if the result is fragmented (see Carley 1990, quoted in Lewis 1992:58) and less successful inner city regeneration, a scenario that all of those linked with the Corporation during Michael Honey's era knew only too well, and that Reg Ward sought to circumvent by trying to move forward as far as possible before consulting with the LDDC Board and the Department of Environment.

The fragmented nature of the British approach to regeneration and the divisions and sometimes antagonisms that exist between government and the different ministries within it, and then the local authorities and the private sector beyond, have been the subject of much discussion (Jacobs, B.D. 1992, Chap. 6, Audit Commission 1989, Bailey 1994:222, Lewis 1992:65). Brunskill (1989:26) suggests that "the weak overall structure of economic development means that the lessons are not learned and we have to keep reinventing the wheel" with a "complex, inefficient and fragmented" approach characterized by "increasing competition between similar agencies" (Brunskill 1989:26) and different localities (see Bailey 1995:223, Savage & Warde 1993:173, Robson 1994:40).

Bachtler (1990, quoted in Lewis 1992:64) suggests that the best means to achieve a co-ordinated strategy for regeneration is to have different layers of approach: a national level that sets the agenda; an intermediate level that develops strategies according to regional needs in line with national objectives and with local representation; and finally a local level to allow the development of "bottom-up" approaches. Lewis (1992:64) describes this approach, (similar to the French system (see Le Gales & Mawson 1994:68)) as a "policy structure involving the 'top-down' establishment of a framework, and 'bottom-up' implementation", an idea that is rather appealing since it answers the need for an overall strategy without permitting the kind of excesses exemplified in Docklands during the height of the development when the market was left to dictate despite the medium- and long-term consequences. Some of these ideas have now filtered their way into the thinking of the new Labour government, elected in May 1997, in which partnership, local accountability and a more integrated approach are very much on the agenda.

In search of jobs: "quantitative impact"

Bailey's (1995:206) third criterion for judging effectiveness was qualitative and quantitative impact. Here I deal with the quantitative before moving on to the qualitative in the next section. The prospect of new employment was frequently seen as the most definitive indicator of the development's "success". Many in the Corporation and the business community perceived the opportunities to be so plentiful that if locals chose not to benefit it was no one's fault but their own, a view that was no doubt influenced by optimistic LDDC estimates that by the "end state", calculated to be the year 2014, "when all sites are developed and occupied" there would be 175,000 jobs in Docklands the majority of them in the Isle of Dogs (LDDC/NOP Employment Survey and Lute 0395 Employment Forecasts, 1994, in LDDC 1996:6).

However, actual and anticipated levels of employment claimed by Urban Development Corporations (UDCs) have often been at variance. As Robinson et al. (1994:327) argue: "The record of the UDCs on job creation is very difficult to establish. All publish employment targets and outputs . . . but these have been bedevilled by inconsistent definition." The National Audit Office highlighted different recording practices for job estimates in which some Corporations "measured jobs gross, others net" where "jobs claimed were often based on estimates when projects were first appraised rather than actual jobs created" or were "calculated on the basis of density assumptions applied to available floor space rather than actual jobs in place" (National Audit Office, 1993:11–12). Confident predictions were not always realized in practice (see Loftman & Nevin 1994:312–14). In Sheffield, for example, the World Student Games did not bring the employment originally envisaged (see Goodwin 1993:150–55, Darke 1991), and in Birmingham the levels of employment anticipated from "prestige projects" were greatly overestimated (Loftman & Nevin 1994:312–16). There was a similar story in Tyne & Wear and Teesside in which not only fewer jobs than anticipated were created but "cost-per-job" relative to investment seemed "very high indeed" (Robinson et al. 1994:335), a finding supported by Imrie & Thomas (1993) in other Urban Development Corporations.

There were questions not only over the volume of jobs created but also over the "poor quality" of many jobs (Loftman & Nevin 1994:314–15). In Docklands, for example, between 1981 and 1994 there was more than a hundred per cent increase in "distribution, repairs" and "hotel and catering" jobs employing approximately 4,000 in 1981 and almost 9,500 by 1994 (LDDC *Key Facts and Figures*, July 1995:5) most of which will have been low paid. While it is certainly better to have some employment than no employment, as Lowe (1993:225) highlighted, the "*quality* of employment provided" is as important as the raw numbers of jobs created. Widgery (1991:219) concluded that even newly created "menial" jobs "will have been immensely expensive judged against the tremendous cost to the public exchequer of the LDDC as a whole".

The trends observed here are by no means just a British phenomenon. In New York the concentration on finance, insurance and real estate (FIRE) resulted in 400,000 job losses in a five-year period (Fitch 1994, quoted in Taylor et al.

1996:303). A comparative analysis of London and New York observed that the emergent "'flexible' labor markets" result in increased female employment, dwindling male employment and "a polarized distribution of earnings" between skilled and unskilled (Harloe and Fainstein 1992:240).

The pattern of Docklands employment reflected these wider trends. Between 1981 and 1994 the numbers employed in "banking, finance, insurance, leasing and business services" increased from approximately 1500 to over 20,000 (1981 Department of Employment, 1994 LDDC/NOP Census of Employment, in LDDC 1995:5). The manufacturing base in Docklands improved on its 1981 figure, but most of the jobs were "based on printing and publishing rather than on the previous traditional Docklands manufacturing" (LDDC 1992, LDDC *Key Facts and Figures*, July 1995:3). Yet, despite these gains, in April 1996, fifteen years after the development began, the LDDC's *estimated* number of unemployed was greater than when the development began (3,533 in the LDDC area in 1981; 4,673 in 1996; LDDC 1996:14; see Table 9.1) even though in this period the number of "employees working in Docklands" more than doubled (LDDC, 1995:3). In January 1997, the unemployment rate for the Isle of Dogs neighbourhood was 12 per cent in Millwall ward and 19 per cent in Blackwall (compared with an average of 10 per cent across the Urban Development Area as a whole (LDDC Executive Office quoted in Lupton 1997:36).

Whyatt (1996:267) highlights the importance of differentiating between "economic performance" and "labour market position" and using this distinction it is possible to understand how development that may have been successful in attracting inward investment had little impact on those in most need. Beynon et al. (1989) provide an excellent example in Cleveland, which "has become a model of the new industrial Britain" but "still has unemployment problems as serious as almost anywhere in the country" (*Financial Times* 1981 quoted in Beynon et al. 1989:272). The same is the case in East London where "unemployment remains a long standing East London problem ... for most" local residents (Church & Frost 1992:149/151). "Whether Canary Wharf and the Isle of Dogs" are successful "may be ... irrelevant", according to Church & Frost (1992:149/151), as "few new recruits to the office developments will be drawn from the ranks of the unemployed and, particularly, the long-term unemployed." This is borne out by Lupton (1997:v) recent analysis. She who found that on the Isle of Dogs itself "81 per cent of workers ... are in white-collar jobs". Yet between 23 per cent and 29 per cent of those resident in the two electoral wards that cover the Isle of Dogs neighbourhood are unskilled or semi-skilled (Lupton 1997:v), a mismatch that the regeneration process exacerbated rather then redressed. "The work brought to the East End had virtually nothing to offer local unskilled and semi-skilled workers", an East End GP said. "Among one's patients especially the over-fifties and the chronically sick, unemployment is a way of life and I have, with time, had to alter the question 'What is your job?' to 'When did you last work?'" (Widgery 1991:44).

Although, as Bailey (1995:218) notes, it is not feasible to expect "exclusive" access to employment for local people in development areas, especially in the light of economic restructuring (Bailey 1995:218), those affected by the closure of the docks and related industries, which had caused massive unemployment and to which the Development Corporation was tasked to respond, deserved something better. Now 37,000 people work on the Isle of Dogs and predictions are that this

Table 9.1 Unemployment in Docklands 1981–1996

Date	Estimated number in LDDC area	London Borough of Newham	London Borough of Southwark	London Borough of Tower Hamlets	Total number in three Docklands boroughs	Percentage for Greater London	Percentage for Great Britain
			Unemployment numbers and rates (seasonally adjusted)				
Jun 1981	3,533[a]	11,952	12,968	12,282	37,202	–[b]	–[b]
Apr 1982	4,108	14,221	16,258	14,268	44,747	7.1	9.7
Apr 1983	4,637	15,957	17,156	15,598	48,711	8.1	10.6
Apr 1984	4,769	16,332	18,675	15,413	50,420	8.6	10.9
Apr 1985	4,896	16,952	20,696	16,131	53,779	9.0	11.2
Apr 1986	4,912	17,465	21,776	16,102	55,343	9.4	11.3
Apr 1987	4,669	16,530	20,447	15,630	52,607	8.7	10.7
Apr 1988	4,063	13,499	17,700	13,576	44,775	7.2	8.3
Apr 1989	3,056	10,455	13,415	10,303	34,173	5.2	6.3
Apr 1990	2,658	10,321	11,541	9,601	31,463	4.6	5.4
Apr 1991	3,770	14,365	15,995	12,504	42,864	7.6	7.5
Apr 1992	4,749	18,296	20,334	15,610	54,240	10.2	9.4
Apr 1993	5,343	20,624	22,409	16,965	59,998	11.7	10.3
Apr 1994	5,109	19,706	21,821	16,536	58,063	10.9	9.5
Apr 1995	4,923	18,438	20,164	15,179	53,781	9.8	8.2
Apr 1996	4,673	16,877	18,837	14,528	50,242	9.2	7.7

[a] The LDDC unemployment figure is for July 1981.
[b] Compilation of unemployment statistics changed in 1986. Revised seasonally adjusted figures for 1981 are not available. Figures through to 1984 for Great Britain and to 1985 for Greater London are annual averages.

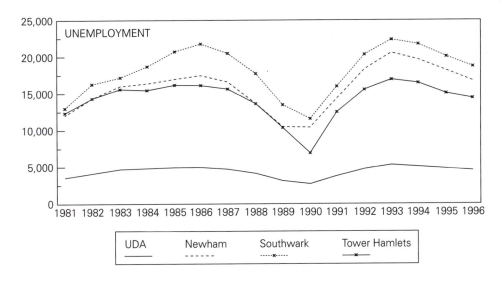

Source: Central Statistical Office Monthly Information Notice, Employment Gazette and LDDC Estimates in LDDC facts and figures 1996. Reproduced with the permission of CNT as the owner of the LDDC archives. © Commission for the New Towns.

figure will reach 60,000 by the year 2000 (LDDC 1996:6). Yet "2,500 people, more than one in ten of the total resident population" are unemployed (Isle of Dogs Community Foundation 1996/7:5), as many as a decade ago (Inform Associates 1997:4). Although the rate of unemployment has, like national trends, declined, one of the two Isle of Dogs electoral wards has a high rate (16.3 per cent) when compared against the inner London average (Inform Associates 1997:4). Ethnic minority groups are "three times more likely to be unemployed" than their white counterparts (Inform Associates 1997:4) and almost 50 per cent of those unemployed on the Isle of Dogs "have been out of work for more than a year" (Inform Associates 1997:4).

The extent of the development's influence on levels of unemployment will be clearer in the 2001 Census. However, it would need to buck a significant trend from its predecessors in 1981 and 1991, and the LDDC's own figures on unemployment in their designated area, to show any impact. The last two censuses suggest that in the East End of London the number of men joining the ranks of the unemployed are growing (see Rix 1996:47) and the development, it seems, had not stemmed unemployment or compensated for the loss of the traditional economic base of the East End or helped to alleviate disadvantage, especially for minority ethnic groups living in the area.

Lupton (1997:37) argues that "although employment opportunities in the area have increased significantly, they do not seem to have impacted on the long term unemployed" who form "a core . . . who are increasingly unlikely to get work. As time progresses, it becomes less likely that these are people who used to work in the dock industries" although, as she points out: "They may . . . be people from dockers' families" Lupton (1997:37). The most vulnerable group of all are the young: "27 per cent of all unemployed people" in Tower Hamlets "are aged under 25"; "the proportion of Isle of Dogs school leavers who become unemployed (15 per cent) is higher than the Borough average" and "take up of further education is low, particularly among young white people" (Inform Associates 1997:4).

The 500,000 people who commute into London, a significant proportion of whom come to Docklands, rob the city of investment put into regeneration and take, Whyatt estimates, about £10 billion back to the suburbs with them, contributing to unemployment that is "concentrated consistently in core areas of the city" in which ethnic minority settlement is high and "creating a central crescent of decline" (Whyatt 1996:272).

This very depressing picture of the development's failure to generate employment in the way envisaged by its architects was not readily accepted by supporters of the regeneration because it seemed counterintuitive. In the face of the evidence on unemployment rates, many questioned the statistics on which these observations were based. For example, this senior figure in the LDDC suggested that official statistics on employment were not representative of economic activity in the area:

> You've got to be realistic in assessing these chaps. There's a lot of money there. If you analyse building societies, the Docklands branches then and now great depositors of money. . . . There's an unknown amount of money in socks and under floorboards, these people and their ancestors have been cheating and pilfering for centuries and they . . . included a large

number of minor capitalists and entrepreneurs. If you ask a building employer out there ... you'll find they can attract labour and skilled wage rates from places like Guildford and Woking and Hertfordshire, St Albans perhaps, that are wages the local people unemployed won't work for. . . . There's a lot of black economy here and people don't pay any tax, either income tax or VAT. . . . Published statistics are not very meaningful.

A local teacher concurred with this view:

I'm not awfully impressed by the statistics produced on the unemployed. I get back from various sources what my kids are doing after they leave school and very often the information [statistical] is inaccurate. There are always going to be a percentage of my kids who are going to be in the black economy because their family has a small business and we know that some of the kids have three or four jobs and claiming unemployment benefit. So there's all that aspect of it that gets hidden. We know kids who've gone off to further education and are claiming unemployment benefit as well, they register in the figures as unemployed. [Others] create the impression of course that they are unemployable so I'd digest those statistics with a pinch of salt.

Although it was certainly the case that there was a thriving hidden economy in operation (and that the development opened up new opportunities for its operation), hidden economic activities are arguably greatest in areas with least opportunity for legitimate enterprise for a significant proportion of the population or where structural disadvantage has been a feature of an area's history over a long period (see Hobbs 1989, Foster 1990). Furthermore, the existence of a hidden economy does not detract from fundamental structural shifts in patterns of employment that have led to the increasing marginalization of the unskilled and poorly educated, many thousands of whom resided in Docklands and understandably wanted and needed the development to deliver something better. Urban Development Corporations were simply not the most appropriate means to achieve this goal (see Robinson et al. 1994:326).

Ironically, given all the emphasis by those driving the development on facing up to the loss of "traditional industries", Robinson et al. suggest that continuing support for major industries like ship building in Sunderland might have proved more cost-effective than Development Corporations:

Sunderland has seen millions spent on establishing an Enterprise Park but few jobs have so far been created. Would the money have been better spent on saving the local shipyards — which, far from being a "sunset" industry were amongst the most modern in the world? Similarly would it not make more sense — economically as well as socially — to subsidise and save Swan Hunter's yard on the Tyne and the region's remaining coal mining capacity? Even conventional regional policy looks a better bet, creating or safeguarding 39,705 jobs in the North East for an expenditure of £181 million . . . If these figures are reasonably accurate they compare very well with the two UDC's total of 16,116 jobs at the cost of £384 million. (Robinson et al. 1994:335)

In an evaluation of urban policies conducted for the Department of the Environment, Robson et al. (1994) found that although some areas in Britain had improved they often did so at the expense of others and that in "the most deprived areas . . . problems grow increasingly severe" (Robson et al. 1994:x, in Bailey 1995:227). Bailey concluded therefore that

> urban policy initiatives have a negligible impact on the overall level of economic activity or household income at a national level, but may have the effect of displacing jobs from one group of relatively deprived residents to another. Other non-measurable benefits may arise from the improvement of housing conditions, the provision of social facilities and environmental improvements, but again benefits may be reallocated between areas so that one gains at the expense of another if the total amount of resources is not increased. (Bailey 1995:229)

Despite all the structural constraints which miltitated against a better outcome for the urban poor in Docklands, the benefits of development were frequently individualized. It was assumed that there was open competition and all were equally qualified to exploit these opportunities. Those who failed to take advantage were often regarded as simply work shy or resistant to change. "For god's sake", a newcomer urged, "learn to live with it as quickly as you possibly can and take advantage." "The essence", another explained, is "that . . . people have hope and opportunities and won't see it." While a businessman said that some locals "felt that they had been deprived and the world effectively owed them a living". "*If only* they could adopt that philosophy of not seeing things as a problem but being positive and seeing it as an opportunity," another newcomer remarked, "then I think everything would be a lot different. . . . But you see . . . they're blinkered."

Just as recent research suggests that even the long-term unemployed are not work shy and still express a strong desire to work (Kempson, 1996) the ability to grasp opportunities was structurally constrained. For example, the harsh realities of life for some of the Bengali families on Island made the development a peripheral issue — their concerns were rather more immediate.

> From this flat where we looked to this high before [two-storey buildings and dereliction]. . . . Now it looks to us a dream, somethink like a dream within three years. Within three years it absolutely changed and it is really an agreeable, exceptional place. But most of the Bengali peoples does not like it because they consider themselves "Oh we are very poor and helpless with this" . . . so they are not thinking how to utilize them . . . they are not thinking about that . . . they are not really understanding what is going on . . . [The Island is] not a safe and happy place . . . we are far, far behind all this.

Racism was seen as a major exclusionary process for young black and Asian youngsters. Even work experiences geared towards easing the transition from school into employment were viewed negatively by some black and Asian students I interviewed. They described offices on the Island in which there were few, if any, black and Asian employees, which they perceived as direct evidence of discrimination and a favouritism for employing whites. One black youth said that he "didn't see black people in suits and that getting on where he lived, involved dealing in drugs" (Research notes).

Solomos' observations at the end of the 1980s seem sadly to mirror the situation a decade later:

> The options for most young blacks are, if anything, becoming more narrow as a result of the social and economic restructuring of the major urban conurbations. Hence the promise that more opportunities for employment and self-achievement lie ahead will do little to change the current realities unless it is linked to a programme of action at national and local levels; a programme involving black youngsters and their communities in deciding on their futures.
>
> The paradox, however, is that as black youngsters become the personification in neo-right ideologies of a "dangerous enemy within" such policies are becoming less and less likely. Political agencies and institutions seem only to become interested in doing something about the social conditions faced by young blacks at times of crisis and disorder. (Solomos 1989:242–3)

In a report setting an agenda for dealing with some of the Isle of Dogs' outstanding problems, employment — and the needs of the ethnic minorities in particular — are highlighted (Inform Associates 1997:4). "The Isle of Dogs has unique difficulties with regard to unemployment", the report said:

> Action needs to be specific and tailored, to address the difficulties that Isle of Dogs residents experience in finding employment as result of a lack of appropriate skills and knowledge, educational qualifications and language difficulties; the particular difficulties faced by young people and people from the ethnic minority communities; the specific problems experienced by the long term unemployed; the reasons why more Isle of Dogs residents are not taking the opportunities that are available locally; and to prepare residents for a further influx of job opportunities, both on the Island and further afield. (Inform Associates 1997:4)

In addition to that provided by the statutory agencies and Labour's welfare to work, the report identifies the "need for specific local initiatives, which may be community-led" (Inform Associates 1997:4), including subsidised work placements for school leavers, a co-ordinator for "employer–school links", a "forum for networking between employers and local community agencies", "after-school programmes" and "training courses" (Inform Associates 1997:4/5). These types of initiatives seem to be at the heart of regeneration. That 17 years later they had still not been adequately addressed is a catastrophic failure in human and social terms.

However, given broader economic changes beyond the LDDC's remit, was it realistic to expect the development to have had a more significant impact on patterns of employment and unemployment? Levy & Arnold (1972:95) suggest that in the United States at least "the only jurisdiction that should be concerned with the effects of its policies on the level of employment is the Federal government.

Small jurisdictions do not have the power to effect significant changes in the level of unemployment" (quoted in Logan & Molotch (1987:91). Logan & Molotch (1987:89) conclude, therefore, that: "The reality is local growth does not make jobs: it only distributes them. . . . The number of jobs in this society . . . in any . . . economic sector, will therefore be determined by rates of return on investments, national trade policy, federal decisions affecting the money supply, and other factors unrelated to local decision making".

Changing hopes and aspirations

Few would deny that the opportunities open to residents in Docklands prior to the development were restricted or that something needed to be done to enhance them, even though this was, until the Honey administration, commonly perceived to be an indirect benefit rather than a specified aim. The difficulty was, as we have seen, that the existing skills base of local residents did not match those of the emerging market place and the barriers to poorer sections of the community accessing it were not fully appreciated. An LDDC Board member said:

> We're building a new city . . . and there's no question that the jobs will be there but it's a two-way thing. The filling of those jobs is essentially up to us and the local authorities to raise the educational standards and the training standards and it's also up to those without jobs to seize those opportunities because they are there . . . But we've also got to persuade the existing adult population to . . . see what has been created as an opportunity for their children . . . to encourage their children to take the further education and the training . . . because there's an incredible opportunity for their children to take themselves right forward in terms of the types of jobs that they can fill.

In terms of the statistics at least, it seemed that youngsters across the East End were responding to new opportunities, with record numbers of young people moving into higher education and with increased examination performance (Albury & Snee 1996:359). The hopes and aspirations of the young Bengali quoted below seemed representative of the wider East London picture: "I want a good job in a managerial area, that's what I want. I don't want people telling me what to do, I wouldn't like that. A really good job which is well paid in Canary Wharf specially." Ethnic minority students were most successful of all, outperforming their white colleagues in examinations (Policy Studies Institute in Albury and Snee 1996:359, Lupton 1997) and choosing to move on to higher education in larger numbers (see also Rix 1996:45–6, analysis of census data, Lupton 1997:49). Recent research (O'Brien & Jones 1996) in Barking and Dagenham, another East End borough, suggested that the "occupational and educational aspirations" of the predominantly white working class are also changing with "ambitions towards a professional/managerial" employment and "some form of further education" (O'Brien & Jones 1996:75).

In order to assess the impact that the development might have played in shaping young people's aspirations, I conducted some group discussions and a written exercise with a group of 14- and 15-year-olds in a local school (see Foster 1992:176). Their wish to continue into further education was fully endorsed. A Bengali student, for example, wrote: "I am going to college next year to study and to make something of my life." However, the relationship between these aspirations and the development was less immediate than I had anticipated and in fact it was the barriers they perceived to their achievement rather than an ability to exploit new opportunities that was sometimes most evident. They did not expect the development to bring them jobs and were acutely conscious of the endemic conflict that existed between different groups competing for scarce resources. "There is a lot more jobs about locally," a student wrote, "but many of them are for yuppies." "Canary Wharf hasn't brought anything for the young people", said another. Perhaps indicative of their ambivalence was the fact that at the Island's secondary school in 1995 59 per cent of pupils moved on to further education or stayed on at school. The figure for the borough as a whole was 69 per cent. Twelve per cent of school leavers moved straight into employment but more (15 per cent) "went straight into unemployment". "The proportion going on to further education was particularly low" (34 per cent compared with 47 per cent) across the borough (Lupton 1997:49).

When I asked a local head teacher whether he felt that there was a sense of ownership evident among pupils in the school, he said:

> I think that's . . . a process that at the moment is incomplete. We've got some indication of that through . . . some of our sixth-formers and their attitudes and they were mixed I have to say. Some still felt a residual resentment, others were very much more positive and felt that they themselves would benefit either in being able to get a job which could allow them to purchase property . . . I would say that the majority of them are fairly positive about what's was going on cos they see things in it for themselves, it's self-interest and that's useful. I think you can get an indicator from the fact that the kids who fulfil the goals of Compact, probably about 70 per cent of the school population, and the majority of them want to stay on to further education . . . and that's associated with ambitions which are also to do with the development.

However, achieving the mobility offered by the development was not without impediment. Almost 30 per cent of youngsters in the Island secondary school "have special education needs" (compared to a borough average of 20 per cent); almost 40 per cent of those going into secondary school on the Island in 1995 had "below average reading, writing, spelling and numeracy skills" (Inform Associates 1997:7); and although the level of GCSE attainment is higher than average for Tower Hamlets it is well below the national average (26 per cent gaining five GCSE grades A–C in 1996, against a national figure of 45 per cent). Tower Hamlets was fourth from bottom in the league table of education authorities on this criterion (Lupton 1997:46), though "its performance at all levels is improving rapidly" (Lupton 1997:vi). For almost half of the school pupils on the Isle of Dogs, English is not their first language, and about 25 per cent of bilingual students leave school with only basic English. However, the vast majority who overcome the language problems "gain better results" than their white counterparts and "are more likely" to stay on at school post 16 (Lupton 1997:vi).

If the younger generation are influenced by parental attitudes, an important part of the process of aspirational change might come from more positive attitudes about the development among parents themselves — a process that was evident particularly among young women, as a teacher explained:

> The difference now is quite dramatic . . . A lot of the women that I speak to are very resentful about their own lack of academic career. They were more or less forced to leave school at the first opportunity . . . [and] had no formal qualifications whatsoever. Many of them are very bright, articulate people . . . If they'd had employment at all . . . it was low-level employment. They now re-enter the labour market as mature personnel. They often get employment as receptionists then gravitate from there to high-tech switchboards . . . from there into . . . computerized offices . . . and so on. They're now mixing with people with different social backgrounds. They're now aware of the opportunities that were missing for them. They themselves are very ambitious and . . . they have high expectations for their children. That's been very good in shaping local people's perceptions because they pass that on.

Changing aspirations were also detected by this newcomer:

> Instead of sittin' down there in the doldrums and lettin' it all wash over 'em . . . they've actually got up and done somethink now and I think most of the girls between, well mainly the ones in their twenties, are actually doin' somethink with their self now and tryin' to make somethink. A lot of 'em are actually buying their own flats, . . . and then they've moved . . . off. Quite a few people have moved off now. So they're actually doing somethink with theirselves and if they can't actually buy their house or flat where they're livin' they're actually doin' somethink like they're buyin' theirself a decent car and it's just the whole [thing] has just been lifted.

Church & Frost (1992:150) in an analysis of employment patterns on the Isle of Dogs suggested that in "the long term . . . female clerical employees . . . might well make up 50 per cent or more of the total workforce of the area". A trend reflected in the gender composition of the labour force more widely. However, in the context of employment patterns across the East End, "despite the Docklands development and the creation of many non-manual jobs, by 1991 Tower Hamlets had the lowest proportion of women in the labour force of all London boroughs (57 per cent)" with a decline in full-time and part-time female employment in the borough between 1981 and 1991 (Rix 1996:50). One step forward two steps back?

An employment survey conducted for the LDDC 1994/5 revealed that over 80 per cent of work on the Isle of Dogs is full time and just 14 per cent part time (nationally the figure is 25 per cent, HMSO 1995) which may, as Lupton (1997:33) suggests, be one of the reasons why "women are under-represented in the workforce in the Isle of Dogs", comprising a third of the Isle of Dogs workforce as opposed to

45 per cent nationally (Lupton 1997:33). Lupton (1997:34) also suggests that, as census data indicates Asian women "tend to have lower rates of economic activity", this in part explains the lower participation in part-time work.

Despite the promising signs of aspirational change, and acknowledging that different levels of benefit might accrue over different time spans, some hard questions need to be asked about *exactly* what the impact of the development for local people was. In the final analysis, it is not the aspirations but the realization of them that is most important.

The growing divide

> You see history repeating itself here again and again and you get a tiny bit depressed ... What one mustn't forget is that this is not the first regeneration in Docklands ... The regeneration before came from when it was nothing until the docks came in. And you've created the problems then of the housing and whatever and you're doing exactly the same thing again. You know, ten years to fifteen years' time we're gonna have the same problem. ... but nobody wants to know ... You had a classic Dockland community here where the people and their work lived together ... Yes they got a brand new city but at what cost? If you're not careful you'll have the have-nots of society round the wall. (Businessman, Isle of Dogs)

> Recent trends and events in cities throughout the world, from London to Los Angeles, Liverpool to Lima, Manchester to Manila, have made clear the realities — and prospects — of increasingly uncertain, generally unsatisfactory and dangerously unequal ways of living in a majority of world cities in the 1990s. An accelerated trend towards "private affluence and public squalor" has been accompanied by a decline in the standards of both amenity and civility, with a "transnational" elite and a "marginal" underclass competing for available space and co-existing by necessity tempered by antagonism rather than conviviality. (Safier 1993:34)

The sentiments expressed in the two quotations above demonstrate the complexities of urban change, highlighting the need for an historical as well as contemporary perspective, the necessity of appreciating global trends and structural forces (not only have we been here before, but we are not alone, and may have limited ability to tackle it), and the fact that social life is characterized at a local and global level by increasing inequalities with potentially troubling consequences. For all the superficial glitz of Docklands regeneration and its rapidly changing postmodern landscape beneath the surface, the East London of the 1990s still had parallels with its late nineteenth-century counterpart, as Murshid described:

> East London generally, ... [is] one of the poorest and most deprived parts of the country. ... the periods of prosperity seem to pass it by. The periods

of depression seem to affect it worse than other places. So you have this contradiction . . . all this money that's been invested into the Docklands . . . none of that development has really bought any substantial changes to the local communities. . . . [which] deepens its sense of alienation . . . because somehow the rest of society doesn't invest in these poor areas . . . What it does do is it invests in very spectacular ways, under their noses, in other things, and in other people. . . . that is a kind of alienation, that kind of dispossession, because effectively land is taken away, opportunities are taken away, and people are left looking at the signs of prosperity but not being able to participate in any manner in that process. That creates the basis for a deep sense of grievance. (Murshid 1993:18)

The contrasts between affluence and poverty, inclusion and exclusion, brought into sharp relief by the development contributed to a growing concern, even among those not normally given to thinking about the poor and deprived, about the long-term future of the Isle of Dogs. There is "this obvious comparison", a businessman said: "The LDDC has encouraged developers to put expensive houses here, which makes the difference between across the road there and there dramatic. . . . It is always gonna be a problem." Another said: "The only question still in my mind" is ". . . what's going to happen? What you've created in fact is effectively a new city in a city" but "you've got vast areas of deprivation, dereliction, appalling houses and social conditions" all around it — which many likened to Hong Kong.

> . . . In Hong Kong you have these spectacular developments and they're surrounded by shanty towns. . . . My fear is that the council estates on the Isle of Dogs will become Canary Wharf shanty town where it'll be low-paid workers living in low-grade housing. There would be a constant supply . . . of cleaners, tea ladies, shop assistants who would all be undervalued and underpaid and they will live in the shanty town of the Isle of Dogs on the crumbling council estates. (Labour councillor)

This situation may get worse, Sassen (1991:329) suggests, because "the growing inequality in the bidding power for space, housing, and consumption services means that the expanding low-wage work force that is employed directly and indirectly by the core sector has increasing difficulty living in these cities". To demonstrate this point, one needs to go no further than the following statistics: Approximately 70 per cent of Isle of Dogs residents have an income of less then £20,000 (Inform Associates 1997:1); 1 in 3 are in receipt of Income Support (Lupton 1997:17); 40 per cent of children live in households where there is no earned income (Inform Associates 1997:1). "A typical rent for a council one-bedroom flat is less than £40 per week". Yet, "68 per cent of new social housing tenants can not even afford this level of rent" (Inform Associates 1997:1, see Figures 9.1 and 9.2).

Some of these housing poor were caught in the "poverty trap", unable to move into low-paid employment because they would lose valuable benefits (for example, housing benefit) that enabled them to survive (Lupton 1997:25). Lupton (1997:25) suggests that this trap "may account in part for the fact that despite a great improvement in the number of jobs available, there are still more people unemployed

Figure 9.1 "Live in tomorrow's world", impossible for many to attain, Janet Foster

Figure 9.2   The flip side, Janet Foster

than there were in the late 1980s": a problem the Labour government is seeking to rectify by radical tax changes and an overhaul of the welfare system.

On the Isle of Dogs, the obvious contrast between the 4,000 new homes (LDDC 1998:30) erected for private sale and the paucity of affordable housing for poorer sections of the Island community (which was exacerbated by the right-to-buy legislation, producing a net loss of council housing) was bound to create difficulties. "The close proximity of extreme wealth", as a recent report points out, "exacerbates the sense of disadvantage for those who are poor, and can create social stress" (Inform Associates 1997:1), graphically demonstrated by the battle over public housing on the Isle of Dogs, the racism that was an integral part of it and the manipulation of that crisis by the British National Party, as I described in Chapter 7.

Despite the BNP's failure to be elected in the May 1994 elections, the conditions that led some to support their extreme politics have not disappeared and are likely to get bleaker still. Increasingly scarce local authority accommodation, across the country, is allocated to the most desperate and needy households, creating a residualization of public housing (see Forrest & Murie 1988) and an increasing concentration of poverty and other social problems in particular areas of our cities. In the capital as a whole, 25 per cent of tenants are dependent on welfare benefits, and "were twice as likely as others to be unemployed" (London Research Centre 1996:72). "70 per cent of London's lone parents are in social rented housing", as are "half of its black and ethnic minority households, half of its pensioners who live alone and half of its disabled people" (London Research Centre 1996:72). Crime also disproportionately affects these very same areas (see Hope 1996).

On the Isle of Dogs, as Lupton rightly argues, "high levels of council-renting . . . are increasingly indicative of poverty and social exclusion" but reside alongside more affluent households and very wealthy households — "a community with increasingly different lifestyles, needs and opportunities" (Lupton 1997:21). They were indeed occupying "different worlds", as one of the new affluent residents commented in Chapter 5, and those contrasting worlds were for the time being insulated to some extent, a luxury the Bangladeshi community did not enjoy from the taunts and aggression of poor white racists. Nevertheless, even for the affluent, though armed with crime prevention technology and occupying a different space, the underlying and real potential for more serious forms of crime and conflict existed very close to their doorsteps.

Social exclusion breeds crime and increases the likelihood of disorder as those, especially the young, with little stake in conformity find alternative means to acquire the status symbols and consumer durables where conventional routes have failed or opportunities are circumscribed. During the fieldwork, many residents (with the important exception of the Bengalis) on the Isle of Dogs said that the Island felt safer than other areas of London, although a MORI survey in 1996 (quoted in Lupton 1997:66) suggested that a quarter of respondents were more worried about crime than any other issue. Limehouse Division (in which the Isle of Dogs falls), as I mentioned in Chapter 5, has the lowest number of total notifiable offences (just under 10,000 in 1996 and approximately 10,500 in 1997; see Table 5.5) compared with surrounding inner divisions (Metropolitan Police Statistics 1996/7) — a similar picture to ten years previously (see Foster 1988). In that sense, despite significant social and economic change, for the time being the area seems to have retained an advantage.

However, recorded crime data are only a partial indicator of the total amount of crime and many, especially in poor urban areas do not report offences against them to the police because they regard them as too trivial, do not believe that anything can or will be done about it, fear retribution, or are not insured (Mirrlees-Black et al. 1996). Although the bulk of victims, as they have always been, will continue to be the poor, the potential consequences of social exclusion, especially in the context of such marked contrasts between affluence and poverty, are very worrying indeed, something exemplified by the British National Party's election victory in 1993.

The rather dismal picture painted by the academic literature and research focusing on social exclusion could not contrast more significantly with the LDDC's up-beat account of their 17-year reign in which social exclusion is not mentioned. An LDDC review of their housing strategy suggested

A new housing market was discovered in the inner city, which encouraged similar developments in other cities. More than 17,800 houses and flats will have been built for sale by March 1998, together with nearly 6,250 for rent or shared ownership by housing associations and the three local authorities. 4,800 council homes in older properties have undergone major refurbishment inside and out, with an environmental facelift, such as landscaping, new play areas and lighting, carried out to a further 3,100 properties. In all, nearly 80 per cent of local authority homes received attention. During the 17 years, the number of houses and flats in the Docklands' eight square miles has more than doubled. So has the population, which by 1998 was more than 83,000 . . . The process will continue — more homes, more people, more jobs — as schemes are completed and the potential of the area, including the Royal Docks, is even more fully realised. (LDDC 1998:3)

The trouble is that those who developed Docklands did not take into account the prevailing context. Social and economic polarization is a problem in many industrialized nations, especially those with free market economies (see Taylor et al. 1996:164) and for "major cities which are strategic sites of the global economy" (Kofman 1996:1, see also Castells 1994, Sassen 1991). In fact, Harloe & Fainstein (1992:264) identify "growing poverty and associated disadvantages as the principal policy issue created by economic restructuring" where "reliance on market forms of distribution produces increasingly inequitable results." "Increases in absolute and relative poverty", they argue, were "a legacy of the eighties." Although there is some disagreement about the impact of these changes on the structure of global centres, Kofman (1996) suggests that

> whatever theoretical framework is followed, the core idea is a restructuring of capitalism which expels the middle strata from the city and leads to the expansion of higher level professional and managerial classes, on the one hand, and increasingly precarious and informal activities at the lower end on the other hand, filled disproportionately by women and immigrants. (Kofman 1996:1)

It may be, as an LDDC executive told me, that "they cared", but perhaps they did not understand how crucial the broader social and economic factors as they impacted upon the poor were to the development of Docklands.

In Britain the gap between rich and poor has risen dramatically over the last 15 years. "About 14 million people — one in four of the British population — live in households with incomes that are below half the national average" (Kempson 1996:1), the vast majority of whom (almost 10 million) live on Income Support (DSS 1994, in Kempson, 1996:1). While the poorest 10 per cent of the population experienced real reductions in the level of income between 1979 and 1991, those at the top experienced a 62 per cent increase (*Households Below Average Income* (HMSO 1994), in Taylor et al. 1996:163). All this pain and hardship is not, according to Glyn & Miliband (1994, in Wolkowitz 1994:442), even profitable or "efficient" because economic and social inequalities actually "inhibit greater prosperity" (Wolkowitz 1994:442).

After more than £2,000 million of public funds and many times that in private investment, the two wards of Blackwall and Millwall which make up the Isle of Dogs local authority neighbourhood are "amongst the most deprived in the country" (Isle of Dogs Community Foundation 1996/7:4). Blackwall is among "the worst 2 per cent" and Millwall "the worst 5 per cent", statistics that hide pockets of even more extreme deprivation because "the averages for the Isle of Dogs, which are improving due to the influx of new better-off residents belie the facts of continuing real poverty and deprivation intensely concentrated in certain areas" (Isle of Dogs Community Foundation 1996/7:4–5). These areas "have multiple problems; high unemployment, low incomes, large proportions of single parents, high child density and overcrowded housing. Concerted social and economic regeneration is still needed in these disadvantaged communities" (Inform Associates 1997:1). These comments were not written prior to the redevelopment but after 17 years of it. As a person commenting on these figures said, "what an indictment" on the LDDC and its record.

The failure to address underlying structural changes and the problems generated by economic restructuring could have devastating consequences making the parallels between the late nineteenth and twentieth centuries far closer than might seem possible at a superficial glance. In fact, some commentators have not only made these linkages but suggested that in certain respects the situation might have been better handled at the end of the nineteenth century as it produced philanthropy and reform. In the early and mid 1990s there was fragmentation and *laissez faire*. The upshot of which, as Harvey suggests,

> is to leave the fate of the cities almost entirely at the mercy of real estate developers and speculators, office builders and finance capital. And the bourgeoisie, though still mortally afraid of crime, drugs and all other ills that plague the cities, is now seemingly content to seal itself off from all that in urban or [more likely] suburban and ex-urban gated communities suitably immunized [or so it believes] from any long term-threats. (Harvey 1996:41)

Widgery (1991:222) identifies similar themes focusing on Docklands itself:

> It is, some might say cynically, simply the oldest of city stories: the callous indifference to the social consequences for the existing population in the pursuit of profit. But when this process happened in London in the late nineteenth century, it was moderated by a sense of civic pride and social conscience and underwritten by an empire whose super profits could afford acts of charity and municipal contrition.

Widgery's passionate description of Docklands blighted by ill-health, disease and inadequately funded services documents the tragedies of contemporary London life just as Booth's *Life and Labour* did in the nineteenth century:

> The despoliation of our cities concerns me not just as a Londoner but as a doctor because it generates a great deal of ill-health, depression and family disruption. It is compounded by the reversal over the same decade of the century-long trend towards greater equality of income and the prising open of the already wide gap between rich and poor. The consequences of that economic process are presented in human terms, conveniently out of sight to the politicians, in our surgeries every day. Patients made sick by poverty, living an unhealthy, overcrowded existence which is exhausting them and making them ill. Mrs Thatcher's chosen monument may be the commercial majesty of Canary Wharf topped out only two weeks before her resignation in November 1990, but I see the social cost which has paid for it in the streets of the East End: the schizophrenic dementing in public, the young mother bathing the new-born in the sink of a B-and-B, the pensioner dying pinched and cold in a decrepit council

flat, the bright young kids who can get dope much easier than education, wasted on smack. (Widgery 1991:233–4)

The litany of poverty, ill health, poor education and lack of access to housing discussed in this chapter and contained in graphic detail in recent reports (Lupton 1997 and Inform Associates 1997) is a far cry from the glossy images of urban development promoted by marketing moguls (see Widgery 1991:161) that Harvey characterizes as a "carnival masks that diverts and entertains, leaving the social problems that lie behind the mask, unseen and uncared for . . . in which every aesthetic power is mobilised to mask the intensifying class, racial and ethnic polarisation going on underneath" (Harvey 1989b:21, quoted in Crilley 1993:249). Such divisions are reinforced by the architecture of cities like London and New York in which the type of commercial and "luxury" residential developments "encouraged developers to engrave the image of two cities — one for the rich and one for the poor — on the landscape" (Fainstein 1994:236–7).

Campbell's (1993) vivid account of "angry young men" on housing estates in Newcastle, Cardiff and Oxford whose hopelessness and desperation had no constructive expression were mirrored among those who sat uncomfortably on the margins on the Isle of Dogs whose simmering discontent, as I described in Chapter 7, the BNP exploited for its own ends.

It may be that without the development the deprivation and exclusion might have been much worse. Certainly I found the impact of economic marginalization far more evident in an area of the North East than in London Docklands because not even a glimmer of a brighter future existed there (Foster & Hope 1993, Hope and Foster 1992). At least on the Island, as one local commented, there were some opportunities: "There's a change here", he said, and "I just thank God for the younger ones and certainly the tiny ones that there is that opportunity there." However, accessing and being able to exploit the opportunities, as we have seen, was sometimes easier said than done as described in the following report:

> Poverty not only impacts on people's material quality of life, their health and their opportunities in education and work. It also impacts in many ways on their daily lives; restricting choice, limiting aspiration and narrowing horizons, damaging self esteem, creating stress and anxiety and putting strain on relationships. People who live in deprived areas are often multiply disadvantaged . . . [and] . . . tend to need more support from their local community than people in more affluent areas. (Inform Associates 1997:20)

The underlying trends for the poorest, and the ethnic minorities in particular, were very worrying indeed. Yet their needs will become decreasingly visible as the more affluent population grows (Lupton 1997:viii).

The fragility of the regeneration

We don't know how lucky we are to have an initiative going on down here . . . London has got to look to its laurels to maintain its competitiveness with everywhere else in the UK . . . we're in the age of regionalism. . . . That's why I go nuts with all the people here who are so parochial that they can't even see how vulnerable they are. They're making themselves twice as vulnerable by not realizing just how very fragile an infrastructure, work, and everything [are]. The interrelated [nature of everything] — the whole scene, it's a very fragile business. (Newcomer, Isle of Dogs)

Docklands' fortunes have always been integrally linked with national and global trends, but economic restructuring and the increasing pace and impact of globalization have produced an alarmingly uncertain and fragile economy. Savage & Warde (1993:62) aptly summarize what we have come to know about cities and how they work, suggesting that "we should recognise the inherent impermanence of [their] economic foundations . . . and the multiple roles of cities in the world capitalist economy". The notable absence of any mention of fragility in the brash confidence and (as it turned out) overconfidence of property developers, big businesses and government about the development and its potential impact is rather disturbing. Cooke's observation (1989:305) that "government policy . . . is too restrictive in a downward direction and too liberal upwards and outwards to a global economy which is subject to increasingly alarming fluctuations" is one exemplified by the Docklands story.

In her excellent analysis of the development industry, Fainstein (1994) suggests that although it is commonly portrayed (and portrays itself) as market responsive — that is it builds in response to demand — in actuality the development process is dictated by rather less rational criteria and

> constructs and perceives opportunity through the beliefs and actions of its leaders operating under conditions of uncertainty. Real-estate developers participate in a dynamic process in which they sell themselves to governments, financial institutions, and renters, combat their opponents and estimate their competitors' intentions. They do not merely react to an objective situation, but operate within a subjective environment partly of their own creation. Often they build projects with little chance of success and press for governmental policies that may not be in the best long-term interests of their industry. Because personal rewards are not wholly tied to the ultimate profitability of projects, individuals within both government and the industry often succumb to wishful thinking in pushing for ever more, ever larger development. (Fainstein 1994:18)

Fainstein suggests that this approach is further compounded by the propensity to continue even in the face of adverse economic indicators because developers have short memories. "For most of the eighties", she argues,

> the constant fanfare trumpeting new development projects and the army of building cranes punctuating the London and New York skylines did appear to herald progress, whatever its imperfections. The visibility and hopefulness of new construction tended to override the caveats of critics. Community representatives who railed against the overwhelming effects of large projects on their neighborhoods were derided for standing in the way of progress. Despite soaring office vacancy rates in other American

cities as the eighties progressed, New York developers continued to pro-
pose ever larger projects. And in London, memories of the property-
market collapse of the mid-70s faded, as banks ratcheted up their real-estate
investments. (Fainstein 1994:26)

The developers and bankers Fainstein interviewed readily identified with this
"irrationality" and were less than complimentary about their industry and the
people who operated within it. "Leading developers, officers of lending institutions, property consultants, and public officials" told her that "project commit-
ments had continued in the face of mounting evidence of their fragility" because
of "herd instinct" and "stupidity". "Most people in the business have a pretty
impoverished level of intellectual capacity", she was told. "The market is not driven
by experience or technology but by emotion" (Fainstein 1994:64). The depressing
conclusion we might draw from this is that the much-vaunted public–private partnership promoted by the Conservative government for a new kind of development
amounted to a free for all for not very bright developers who continued to develop
despite the precariousness because everybody else was doing it and because the
banks were willing to finance it!

Docklands had the greatest advantages of any urban regeneration project,
offering, as a Board member explained, opportunities for development that were
"probably unique in the United Kingdom" because of London's position in world
financial markets. "It's closeness to the City . . . is a very significant part of our
success", he said, "because much of what has come into Docklands in the way of
employment is City of London type jobs, financial services . . . and that's something which is unique to London." If concerns were expressed about Docklands'
future in the light of all its advantages, then we might assume that the fate of
other locations without these advantages would be even more problematic. As one
businessmen commented: "If urban development doesn't work here, God help
Scunthorpe or Liverpool or anywhere else." But after 17 years of development
Docklands' future was very fragile in an absolute and symbolic sense and its global
city status may in fact have made its fortunes more precarious.

Whyatt (1996:270) presents a depressing picture of the "challenges which face
London" in order to maintain its competitive edge in increasingly competitive
global markets. She argues that its role is changing in ways that will affect not
merely the city itself but the whole of the South East satellite and beyond that the
rest of Britain. Despite many advantages, she argues, London will "have to work
harder than almost any other region in the UK if it is not to fall back even further
from its present position" because the concentration of service sector employment
is too great in London at a time when manufacturing "is likely to be one of the
fastest growing sectors in the UK economy", and employment in service sector jobs,
generally regarded to be the "substitute" for lost manufacturing, has been declining in London (Whyatt 1996:272). An indicator of its potentially precarious position is reflected in patterns of employment: 29 per cent of Isle of Dogs employees
work for just five companies employing more than a 1,000 people and a further 64
per cent of people work for only 4 per cent of employers (LDDC Employment
Survey 1994/5, in Lupton 1997:31).

Other cities, for example Sheffield, have developed less vulnerable strategies
and created employment linking "subsidy programs for businesses to job training

efforts thereby reducing the mismatch between displaced manufacturing work forces and new industries." "Neither London or New York has worked out a sectoral strategy to lessen its dependence on the financial services industry; in the meanwhile large-scale lay-offs in this industry point to the riskiness of such dependence" (Harloe and Fainstein 1992:267). In order to generate less insecurity, Fainstein (1994:248) suggests that "assisting the non-profit sector" (education, hospitals, etc.) provides a means to encourage more "stable growth" and offers greater prospects for employment. Such activities are also "much more insulated from global economic competition than multinational corporations": a lesson that was finally learnt and encapsulated in plans for the Royal Docks, also part of the LDDC's area, which included a university (see Albury & Snee 1996) and mixed commercial and housing development — a more people-orientated plan that local residents and community groups in the area had long campaigned for.

However difficult its market place might be to retain, exploiting London's global city status and the market opportunities it presented was heralded as a "success", especially during the boom years, and the rapidity of the physical regeneration was certainly breathtaking. However, the "property crash" in the late 1980s that left thousands of square feet of office space unoccupied, private housing developments unsold and the banks overextended led many to question the efficacy of this approach (Fainstein 1994:14). Just as the LDDC's nineteenth-century predecessors who developed the docks exploited markets, which led to oversupply and eventual decline (see Hardy 1983a:9), so too huge office developments in a previously unknown part of London may themselves become surplus to requirements, but far sooner as Barras (1994:6) explained:

> the technological needs of "Big Bang" were the trigger for the late 1980s development boom . . . However, what happened was that this demand requirement produced a magnified supply response, and one that may prove quite transitory in its utility. . . . the latest developments in network technology . . . mean that . . . many buildings considered technically obsolete just a few years ago now seem quite serviceable . . . without the need for redevelopment. Here is a particularly acute example of the general disjunction between the rapid processes of innovation affecting urban functions and the much slower development processes which can change urban built forms. In this case, a particular technological requirement erupted and then receded within the time span of a single building cycle.

One problem solved, another generated, as another LDDC Board member reflected in the early 1990s:

> The sheer volume of [the development] has coincided with what we can now see is the worst recession really in my lifetime. . . . It's not just another blip on a cycle, it's far more serious than that, so it's coincided with vacancy rates through central London, I think, of seventeen, eighteen per cent and you need ten million feet of new space like a hole in the head at this particular time . . . For landlords it's bad news because they're carrying voids in their properties all over the place partly attributable to Docklands. I don't know the precise numbers but Canary Wharf is only

half of what's happening in Docklands. There's another ten million feet elsewhere and so that's, if you like, good for Docklands but a problem for London, even a problem nationally.

In the wake of recession and a collapse in commercial and residential property in Docklands, concerns about its future were frequently expressed. As by the businessman below who outlined two very different fates for the area demonstrates:

> I suppose the whole catalyst will be whether Canary Wharf and the commercial development takes off. I mean, if that takes off then I think this area will become like a second City, office workers being very busy during the day and then tending to die a bit at night. The luxury housing will pick up again, you know, the housing market would pick up as people want to live close to the offices where they work and I think pressures for housing will become far tougher. Alternatively, if you have a collapse in the commercial office market . . . I think this area could almost become as desolate as it was when the Docks closed, you know, just boarded up office blocks where again nobody wanting to live here and I think there's a very real danger of that happening so, you know, what will happen in 10–15 years — I think either one of those will happen but I'm not sure which.

Even the financial might of Olympia and York could not weather the recession unscathed and although many local people took some satisfaction in their financial problems and those of the development more generally, they were also deeply concerned: "If they can't sell their centre piece what chance have they got of selling the rest of the office space?" an Islander explained. Fainstein (1994:212) shared this view: "The failure of office development will not cause it to be replaced in the foreseeable future by a more community-oriented scheme and borough residents will probably lose whatever benefits they had achieved through development agreements and trickle down."

When I asked a senior director of one of the large house builders what his vision of Docklands future was, he said:

> I think it's a bit scary. It's a little bit on a knife edge. Somebody put it to me the other day that the epicentre of the boom was Docklands and that died first and then the market died slowly like, you know, you drop a pebble in a pond. . . . Therefore the resurgence in the market is going to come back the same way and the last place it's going to get to is Docklands . . . so . . . it's a very great worry and until we know which way the economy is going to go the last place people want to live is in Docklands.

None of this is very reassuring after the investment of £1,744 million of public funds, expenditure of £2,014 million by the LDDC (between 1981 and March 1996) and £6,227 million "by the private sector" (LDDC 1996:4, see Table 9.2). Inform Associates (1997:2–3), in "a broad costed agenda for social regeneration in the Isle of Dogs, on which to base the fund raising efforts of the Isle of Dogs Community Foundation", which included tackling unemployment, improving educational attainment, improving access to good housing, improving health and, reducing crime

Table 9.2 Total expenditure by LDDC, 1981 to March 1996

Areas of expenditure	£ million
Land acquisition	185
Land reclamation	145
Utilities	155
Roads and transport	648
Environmental	141
Docklands Light Railway	311
Social housing	163
Community and industry support	108
Promotion and publicity	22
Administration	163
Total	2,041

Source: LDDC Finance Department Key Facts and Figures (1996:19).
Reproduced with the permission of CNT owners of the LDDC
archives. © Commission for the New Towns.

and its impact, suggested a little over £4 million for a three-year programme, just
over half of which would be raised through the Community Foundation. A drop in
the ocean compared to the enormity of the expenditure on the development.

However depressing the messages emanating from much of the work on
London and New York, Harloe and Fainstein (1992:251–2) advise against an overly
pessimistic view, highlighting the need to differentiate between "cyclical and secu-
lar changes". "Overpessimism in the 1990s may be as misplaced as the overoptimism
of the 1980s proved to be." Much of London's future depends, they suggest, on
whether the city retains its "core" function or whether trends towards decentraliza-
tion accelerate. Perhaps with their advice in mind, this LDDC executive expressed a
more optimistic view:

> It's not going to be easy to retain the advantages that have been created
> . . . But . . . Docklands is . . . not your local regeneration project, it's very
> much tied into government policies. And I think one has to be optimistic
> about the country climbing out of recession, interest rates going down,
> Europe being seen as an attractive place to locate in, Docklands within
> that, also having a number of advantages to attract companies to move in.
> And for that purpose the jobs will continue to come. (LDDC executive)

And they did. By 1997, "1.4 million square metres of commercial and industrial
space [had] been completed" (Lupton 1997:iii) and building was underway on a
new phase of the Canary Wharf development. Memories were short. Yet the underlying structural concerns remain.

An agenda for urban regeneration

After 15 years, and many new initiatives, surprisingly little has been achieved.
Given the record so far, it is difficult to have much confidence in more of

the same or to feel at all hopeful about the future prospects for deprived urban areas. (Willmott & Hutchison 1992:82, quoted in Bailey 1995:227)

The preceding discussion has demonstrated how problematic it is to establish the "success" or "failure" of Docklands regeneration given wider and extremely powerful structural forces outside the control of individuals and agencies in Docklands. The juggling act that faces those involved in urban development is aptly described by Hill:

> The demands of . . . global . . . restructuring . . . make providing a stable local economic environment very difficult. The local "branch" economy becomes vulnerable to regional and international movements of investment, and also to the counter-urbanisation that inward investment fosters . . . In this climate, balancing proactive urban initiatives between a multiplicity of local actors with demands of community responsibility and responsiveness is a major task. So, too, is balancing social with economic objectives. (Hill, 1994:193)

Making the connections between macro economic changes, with the alarming predictions about globalization, and an individual development programme and its impact on local residents is very problematic. Indeed, it is often difficult not to feel overwhelmed (and depressed) about the extent of change that can be achieved in any individual locality given underlying and powerful structural forces (see Savage & Ward 1993:174, Sassen 1991). But this suggests that individuals or groups can do nothing to ameliorate or stem the flow of processes that originate from external forces. Individuals *are* important in making and shaping processes both positively and negatively (the latter being exemplified by more than a thousand electors who voted for the British National Party in September 1993). What the fragility of the economy does demonstrate, however, is the importance of striking a balance in development, which should not be entirely driven by market forces, especially given the imperfections of the market mechanism and the "irrationality" of developers and financiers who wheel and deal within it (Fainstein 1994:64).

Furthermore, given the problems generated on a global scale, identifying what a local urban regeneration programme should look like is highly problematic because of the conflicting interests that have to be accommodated as Fainstein explains:

> It is an extremely difficult task to devise an appropriate system for land use and economic development planning that takes metropolitan area-wide considerations into account, operates efficiently and effectively, involves citizens in reviewing development proposals without succumbing to the "not in my backyard" syndrome . . . and responds to initiatives emanating from urban neighbourhoods. Because planning must cope with genuine conflicts of interest, trade-offs between long- and short-term considerations, and considerable uncertainty over the results of any project, no process will produce a fully satisfactory result. (Fainstein 1994:252)

During the course of researching and writing this book, I have come to appreciate the complexities of urban development, and come to understand therefore that the solutions are far from straightforward. I have attempted to outline below some of the key principles that should form part of any urban regeneration agenda, none of which is new:

- a clear notion of *what the problem is*;
- the need for *a co-ordinated strategy with clear aims and objectives in the short, medium and long term* with adequate but flexible planning;
- the strategy needs to be *locally and globally responsive*, based on some understanding of what works;
- the necessity of *partnership* between government, development agencies and local authorities, and between the public and private sectors with adequate consultative processes to ensure that policies can be responsive to local needs where this is appropriate;
- the *importance of "bottom-up" approaches* that have an enabling function.

I have discussed the importance of the first three areas in earlier parts of this chapter. I now want to make the connection between the macro and micro and discuss how better partnerships can be developed and the way in which individuals can influence processes and encourage grass-roots initiative.

Partnership

Regeneration is . . . dependent on the construction of alliances and coalitions between parties with differing agendas and dominant concerns. Successful regeneration requires the development of a unifying vision which articulates the needs and aspirations of the institutions and individuals concerned. (Albury & Snee 1996:355)

[The LDDC] were genuinely creative down here and I think the local authorities have learned from what they've done in terms of developing their assets. Certainly Tower Hamlets council have learned and is in my view able to handle and deal with their development sites much more effectively than they were. (Liberal councillor)

During the early years of the LDDC's existence the uneasy and, in some cases, hostile relationship, between government and the local authorities made any notion of genuine partnership highly unlikely (precisely at the time when strategies for community gain could have been developed and later extracted). Emotions ranging from ambivalence to hostility existed on both sides. By the end of a decade of development the wider agenda for urban regeneration had changed and the notion of partnership was at its core (see Bailey 1995:1, Lewis 1992, Scottish Office 1993, Bailey 1994). As this LDDC executive described, the initial belief, that what was required was a regeneration machine that drove development through irrespective of the local authority and community, changed fundamentally by the end of the 1980s:

> My view of sort of working in Docklands now for the last ten years is that I suspect this non-elected agency set up by Heseltine was perhaps on reflection an overreaction ... He said, "I want something sharp, action orientated", and I say it was perhaps an overreaction because of course there were no arrangements made to look at the software areas and we've come around to that middle ground in effect from a sort of rather inef-fectual local authority model about the regeneration to a government model which is now from it's early beginnings taken on board to a certain extent, [the] economic and social consequences of what it's doing ... And I suspect in a way the new UDCs (Urban Development Corporations) set up after the eighty-seven election actually reflect that middle ground. They do actually have from day one community departments, community officers. They have tried from day one to have a better working relation-ship with the local authorities.

This is an approach reflected in plans for the "urban village" in the Royal Docks, another part of the area covered by the LDDC (see Osborne 1993:12–13, LDDC 1998:24) in which the scheme has a tripartite structure of LDDC, private and local authority involvement and is mixed use.

Imrie and Thomas (1993) suggest this shift is typical of other Urban Develop-ment Corporations with moves towards closer working partnerships with local authorities and more responsive partnership modes with communities as well as business. This, according to Mackintosh, produces gains on all sides as the respec-tive interest groups attempt to pull one another closer to an understanding of their perspective:

> The private sector is seeking to bring private sector objectives into the public sector, to shake it up, get it to seek more market-oriented aims, to work more effectively in its terms. The public justification offered is that this will be in the long-term public interest. The public sector, conversely, is trying to push the private sector towards more "social" and long term aims, justifying this in precisely the same terms. (Mackintosh 1992:216, quoted in Bailey 1995:33)

While "a meeting of minds" is desirable, many accounts of partnership pro-jects demonstrate the difficulties of achieving this. Moore (1995:14), for example, described the poverty programme implemented with partnership arrangements in Liverpool, where the experience was far from "a real meeting of minds". Given the political context and culture in Liverpool, the limited funding that was avail-able for the programme, and a number of other factors, the experience was

> an important indicator of the problems of partnership in an intensely conflictual inner city context. . . . Over a period of five years . . . No one

had been prepared to put him or herself into another's position, to see the institutional and interpersonal constraints within which the other works. Putting yourself in the other's shoes is an essential element both of partnership and of developing a strong bargaining position. Neither was achieved in the Granby Toxteth Community project. Pre-judgement was the order of the day. (Moore 1995:14)

It is one thing, then, to argue for the principles of partnership and another to execute them especially where there are significant and important differences in power. Whether Southwark, Tower Hamlets and Newham would have derived more for their residents if they had formed a joint power base on the LDDC Board over the most prosperous phase of the development is difficult to establish. What is clear however is that other Labour-controlled local authorities like Sheffield did take the entrepreneurial road, working in partnership with the private sector throughout the 1980s, although the results were not as promising as many hoped (see Darke 1991, Cochrane 1991, Goodwin 1993:150–52).

Had there been a spirit of partnership in Docklands, the redevelopment would almost certainly have been different and would have occurred largely without the concentrated, intense and spectacular commercial development as local authorities and community groups were strongly opposed to these aspects of it. Public consultation and involvement in the planning process would inevitably have changed the mix of land usage and led to greater emphasis on affordable housing and more investment in refurbishment. Even with the kind of large-scale commercial development there would have been a more concerted attempt to exact direct community gain from private developers.

The problem prior to the LDDC was that there was much talk and little action. The principles involved in partnerships are sound and the need for community (in its many and conflicting forms) involvement are important. So too is negotiation within and between different agencies and organizations. But with so many agendas this is not at all straightforward; problems that the single-minded Corporation in its early days did not suffer. Furthermore, although accountability is an important aim, it is not always very successfully executed, as Bailey argues:

Our findings suggest that they [partnerships] represent relatively closed institutions by which dominant local stakeholders collaborate in order to achieve partial definitions of the public interest. These definitions are largely determined by technocratic rather than democratic processes where dominant interests (which in some cases might include the local authority and sections of the local community) structure the boundaries of the debate in their own interests. (Bailey 1995:220)

Returning to the wider structural issues and the way in which these impinge on partnerships, Jacobs (1992) observed:

The globalisation of policy issues and their seriousness place a heavy burden on public and private agencies which seek to address the problems of cities through partnership. The approaches to participation . . . rely upon the consent of those involved and the harmonisation of interests. However, globalised issues are not easily interpreted in terms of consensus. There are inherent conflicts in such issues between those who see economic development as a necessary and desirable way of improving the cities and those who regard growth and its consequent problems (pollution, congestion, etc.) as detrimental to the broader social interest. The corporate private sector increasingly favours socially responsible community partnerships, while local interest groups remain unconvinced that the companies are genuinely concerned about the community. (Jacobs 1992:257–8)

On balance, whatever the problems, partnership allows combined skills and resources to be maximized (Haughton & Whitney 1989, quoted in Bailey 1995:11), and according to Bailey (1995:11) "will bring a new sense of urgency to local problems", exploiting the entrepreneurial skills of business to minimize bureaucracy. Partnership, he continues, also marks a change in the approach of local and central Government to regeneration issues where:

at the local level central government has broadly shifted its position from the New-Right free-market model to the centrist social market, [and] many of the more interventionist-inclined urban authorities have, voluntarily, or under duress, moved from an interventionist stance to the centre. (Bailey 1995:227)

Enabling people as well as property

I have retained throughout the research for and writing of this book a feeling that the poorer sections of the Island "community" in its broadest sense should have derived more from the development than they did. Consequently, it should come as no surprise that I regard "local" benefit, in its most inclusive definition, as a crucial ingredient of regeneration and that it should involve "bottom-up" as well as "top-down" approaches. Empowerment was not a term that readily described the experience of poorer sections of the "community" on the Isle of Dogs; their emotions were dominated by powerlessness. The development was imposed upon them, and even in its more benign guises was still quite autocratic in style. As one senior LDDC figure observed: "Our communication really left a lot to be desired because we didn't talk, we didn't talk things through, which means we didn't talk to people. We tended to talk at them."

Local residents never really gained a voice or participated in the process at all. Yet as Hill suggests:

Local people have to feel that they are stake-holders in the process [of urban renewal], and that they have something to contribute as well as receive . . . Balancing proactive urban initiatives between a multiplicity of

local actors with the demands of community responsibility and responsive-
ness is a major task. So, too, is balancing social with economic objectives.
(Hill 1994:193)

Despite the difficulties, Hill (1994:193) is convinced of the need for "the enabling
authority . . . to be concerned with . . . focusing on a bottom-up, not a top-down,
approach to urban renewal".

An example of the impact that small-scale but focused development can achieve
is provided by Lewis (1992:25) on a deprived Newcastle housing estate, where,
following an estate "audit" and one of businesses in the immediate locality, over a
hundred residents obtained employment and twice as many began training programmes with the aid of a resident-led, but publicly and privately funded, "Community and Enterprise Centre . . . with a full-time community worker . . ." which
acted as job-shop and advice centre. Decentralized housing management, like that
promoted by the Priority Estates Project has also had an empowering role (see
Power 1987a, Foster and Hope 1993, Foster 1995). It is often, as one of the LDDC
consultants commented, the small things that are important to people on the
ground, yet they frequently seem more difficult to achieve. One tenant wryly
remarked, for example, that it "took exactly the same time to build [the] Canary
Wharf tower" as it did to get entry phones promised by the local authority installed
on their block of flats — that is real demoralization. Yet, as a community worker
argued, "they could have made a spectacular impact on the community at very
little expense".

Lewis' example demonstrates that even with economic restructuring and global
change, positive changes within a locality are possible and can have an important
impact, but agencies need to work *with* people rather than *for* them, to generate real
ownership. Michael Barraclough, the man responsible for getting the self-build
housing scheme under way on the Island (another small but important scheme for
those local people who took advantage of it), said:

Neither the left or the right have anything to offer the inner city because
. . . all that the left will do is . . . impose another bureaucracy . . . [and]
the right is equally bad. [They] . . . think that the way to change the inner
city is to get business in the community, to get entrepreneurs there.
The point is . . . what they're doing in Docklands is totally irrelevant. . . .
The right has to . . . understand that big business are actually overwhelming. They actually deny local opportunity. . . . big money coming in just
buys out.

Furthermore, what the Corporation, Government and many businesses failed
to understand until it was too late was that ameliorating local people's distress was
in fact in everyone's best interests. "There's no reason why local people should not
have benefited as much as business people from [the] development", a community worker pointed out. "If you're trying to persuade somebody to buy into Canary
Wharf do you really want them to see the kinds of places you let people live in just
round the corner?"

Figure 9.3 All that remains, Janet Foster

Facing the future

> Perhaps the chief sin of the twentieth century was that urbanization hap-
> pened and nobody much either cared or noticed in relation to the other
> issues of the day judged more important. It would be an egregious error
> to enter in upon the twenty-first century making the same mistake. It is,
> furthermore, vital to understand that what half-worked for the 1950s will
> not be adequate for the qualitatively different issues to be fought over the
> next civilization in the twenty-first century. (Harvey 1996:58)

It is seven years since I began the research on which this book is based. As I finish,
the redevelopment of Docklands has been in progress for over 15 years. I have
constantly been reminded during the course of researching and writing this book
that there is an historical continuity that flows through the Docklands story but
one that, superficially at least, might not be apparent to the visitor any more.
Comparing London Docklands with Liverpool docks, also developed in the 1980s,
one senior LDDC figure was concerned that London Docklands had forgotten its
roots (see Figure 9.3). In Liverpool, he said: "They'll remember their heritage,
whereas here, people will think we've dug these docks one day. They won't realize
they were dug in 1804." Docklands' roots, however, were not only about the physical landscape but the people too — the thousands who lived and worked there
before the developers and estate agents came.

When I asked a community activist what he thought of Reg Ward's remarks
about local people "owning" the development (see p. 313), he replied:

> When you can start influencing their lifestyle by this development, i.e.
> good jobs, yes they will. They'll become very very protective of that . . .
> look at us with our docks. We won't let that go because that was our
> living . . . [we] didn't want to lose it . . . [but] it's all gone. There's a change
> here . . . and we couldn't alter it in any way, shape or form . . . It may not
> be in terms of its height, its mass and maybe lots of other things all what
> we would have wanted but my grandchildren don't give a toss about that,
> it's their future, it's their market place . . . As far as I'm concerned the
> single most important issue is a job . . . It's got to be providing the source
> of their income because if it don't that ain't for them, that ain't for us.
> (Foster 1992:180)

The Island certainly had changed and Peter Wade was right that his grandchildren would inherit a different market place. But would future generations of
poorer Island residents have a place and employment in this dramatically altered
space as their predecessors for many generations before the development had
done? The only certainty about the future was, as we have seen, its uncertainty and
fragility. Although some believed, and many people hoped, the future would be
better, too many, through processes well beyond their control, still remained on
the wrong side of the now symbolic dock walls looking in on another world that
did not seem to be for them at all.

Postscript

Figure 10.1   Canary Wharf and the Millennium Dome, Janet Foster

On a chilly December morning in 1997 I returned to the Isle of Dogs five years after my field work was completed. The Docklands Light Railway meandered out of Bank station in the heart of the City of London with passengers absorbed in their newspapers, conversations or thoughts, largely oblivious to their surroundings as the run-down areas of the East End went past outside, row upon row of council flats, visible through the murky windows of the train. With few exceptions, these still looked neglected and forlorn, and I was surprised by how little seemed to have changed.

As the train rapidly progressed towards the Isle of Dogs, the Canary Wharf development, the "beacon" of the new Docklands, still takes my breath away — perhaps because of its quality, but also because of the stark contrasts with the

Figure 10.2   Memories of recession left behind, Janet Foster

deprivation surrounding it. I am reminded of the comments of a consultant I interviewed who, reflecting on the enormity of the sums of money involved in the development said: "Sometimes when I come out of my office and I look at Canary Wharf, you can't help but be attracted to it . . . But then . . . I think of the four thousand million pounds going into Canary Wharf, what could that four thousand million do for the starving people of the world?" The contrast is perhaps a little unfair, but I understand his sentiments. The size of the development is so impressive and overwhelming and, not for the first time, I think it is "another world", geographically so close to the poorer sections of the Island population and yet so far away for those who have no part in it.

Despite the pessimistic predictions that Canary Wharf would be a white elephant, it is thriving with activity and those working or simply visiting get out in large numbers here. As the train pulls out of the station new buildings in the second phase of the speculative development are well under way (Figure 10.2): memories of the recession and its aftermath seemingly left behind (Figure 10.3). The damaged buildings in the central part of the Island are a stark reminder of the 1996 IRA bombing (Figure 10.4) and the Audi car showroom (Figure 10.5) is perhaps an indicator of just how much some aspects of the Isle of Dogs have changed.

Yet, beyond the central business areas, as I walked the short distance from Mudchute station to the Docklands Settlement in an "old" part of the Island, past a quiet street of pre-war housing, little appeared to have changed (Figures 10.6, 10.7 and 10.8). "I thought you looked familiar", a woman I had interviewed five

Figure 10.3 Another decade of building, Janet Foster

years before said, as she made me a cup of tea and we chatted about the intervening period since we last met. "There's three communities now," she explained, "the established, the business/affluent and the Bengalis." The development was still alien and divorced symbolically and spatially "over there". Without my asking, she continued: "The BNP episode was not a proud part of the Island's history" but reflected the anger and frustration that people felt about the housing situation at the time — anger that had been replaced by resignation and, according to another, by "acute demoralization".

******

Sitting in a cafe bar of a central London hotel, Reg Ward talked fondly of his "love affair" with Docklands. I asked what he thought of it now. He reflected for a moment and replied: "It's not bad. It's not as good as it could be. It's not complete." He wanted the record set straight about the community and social housing. The Department of the Environment, he said, had not given the Corporation the powers to get involved in these areas and, in any case, social housing was part of the difficulty of the area because of its concentration and low demand when the Corporation began work. He also emphasized the importance of not being too focused on the Island. The LDDC's task was greater than this. Indeed it was, and I commented that I was probably guilty of taking on the parochialism of the Island. Reg believed ample social housing was built in other areas of Docklands, both south of the river in Bermondsey and in Beckton to the east. This of course did not help those Island people who wanted to remain there and could not gain

Figure 10.4 IRA Bomb damage, Janet Foster

Figure 10.5   Audi car salesroom, Janet Foster

Figure 10.6   Pre-war Island housing, Janet Foster

Figure 10.7   The "old" in the shadow of the new, Janet Foster

Figure 10.8 An "old" Island street, Janet Foster

affordable housing locally. He reinforced too the "constant dialogue" he maintained with the community and did not feel this really came out in the account.

******

Perhaps confirming Reg Ward's point, Ted Johns rang shortly after and said in response to reading the section on the early days of the Corporation: "The one bloke who should have got a knighthood was Reg Ward. If he had stayed another two years we might have seen a different kind of Docklands. Reg went out on a limb. Reg went to the DOE and said 'We've got to build social housing'" but at that time the government would not hear of it. Ted explained that he became a Labour councillor again for a short while in an attempt to pacify some of the discontent on the Island that emerged as a result of the housing crisis. "Of course it wasn't enough", he said, "and the BNP got in." But, he continued, they were "soon dispatched back to political oblivion" and the shock of their election certainly focused minds on the frustrations and pressures.

After having read the manuscript Ted wrote and said that he was "very impressed" and had "nothing to add which would enhance the story — you have told the story of the LDDC impact on Docklands as it was, and by far (even to a biased old warhorse like me) the best and fairest analysis of what really happened from both sides of the fence that I have read". He enclosed a copy of a review he had written of Phil Cohen's work on race on the Island in which he highlighted that for all the focus on white working class racism, which he did not seek to support or to deny its existence, he wanted to add that "racism, in a more virulent form is endemic in the middle and upper classes of British society and always has been" and yet rarely reaches the gaze of researchers. "It is this society," he continues,

> long indifferent to the needs of the East End, and other working class areas — where many died of cholera, typhoid, rickets or sheer starvation on the streets of the East side of a capital city at the heart of the greatest empire the world has ever seen, which largely oversaw (and profited from) the workings of the British Empire. . . . Cockneys . . . are the backbone the nation. They have a culture and a background which they treasure and regard highly. When they feel this to be threatened or diminished they react, and, more particularly, when they are preached about being "racially correct" by those living in areas, untouched by the harsher realities of life, mainly middle-class and exclusively white!" (Ted Johns, *The Islander* July 1996:7)

As if personifying the competing visions and views on the regeneration's "success" or "failures", two contrasting reports were drawn to my attention by different people. The first highlighted the "continuing poverty and deprivation" after 17 years of the LDDC now hidden in the statistics by the general levels of improvement due to affluent settlement (Lupton 1997). The second, a review of the LDDC's work with communities (Hillman 1998) that presented a very different more positive picture:

> A MORI opinion poll of local people commissioned by the LDDC at the end of 1997 showed that over two-thirds of people involved in the survey thought the LDDC had done a good job and that prospects for the area were good. . . . 69% of those living on the Isle of Dogs were positive about future prospects, compared to 58% in a 1996 MORI survey. (Hillman 1998:28)

The report continues:

> By the end of its regeneration remit, the LDDC provided the means for widescale improvements to the lives of the original 40,000-strong communities. In the 1990s, progress became obvious with schools, colleges, social housing, parks and play areas all around the area on the ground. It is not surprising if the brickbats continued instead of bouquets. Local people were shrewd enough to realise that continual pressure right up until the end might secure additional benefit. (Hillman 1998:28) (see Figure 10.9)

May be it was that simple. My sense is that it was not. Ironically even the picture of children in a playground that accompanied these comments, taken from a different angle, reveals a play area strewn with graffiti — an altogether less pleasing image (see Figure 10.10).

******

Back in my office at Cambridge I spoke to Peter Wade. Peter remained pragmatic. Still in his community relations post despite the upheavals experienced by Olympia and York, he felt much had been done to support local education and employment but that the LDDC's record remained "lamentable". Recent partnerships between Olympia and York and the borough resulted in a skills match scheme to facilitate more local employment, and as the construction of homes and offices has increased so have jobs in the industry. On phase II of the Canary Wharf development, he explained, £18 million of local contracts were allocated with 224 local

Figure 10.9   Improvements to social housing, Janet Foster

Figure 10.10   Children's play area, Janet Foster

jobs. Investment in education also continued with a £200,000 early readers scheme at primary, and a literacy, numeracy and recovery scheme at secondary level. "A lot of good things are happening", he said, but in a broader context "there's not a lot left to win". There is little available land on the Island for housing and "the Isle of Dogs stands in some respects as a warning to others not to make similar mistakes". His advice: "Go for more measured development, with less reliance on a particular type of market and with community gain." "You talk about the old Island being in a time warp," Peter said, "the new development is in a time warp. It bears no relation to sanity."

The consequences of such intense development that inflated land prices were all too apparent as the younger generations exploited the opportunities of mobility offered by the right-to-buy scheme. Unable to move up the housing ladder on the Island and concerned about the quality of education their children would receive, established Islanders were moving out, Peter's family among them. "I'm very sad about it", he said, but he also recognized its inevitability.

Peter's verdict on the book: "Your book is the only one that has given a balanced and unbiased view across the spectrum of the parties involved . . . I found myself re-living the past 16 years as if I had written the book." "The only section I take exception to is the one relating to the influx of the Asian community to the Island." Here he wanted "balance". "If an outsider were to read this chapter (7) their impression would be that the Islander's were a bunch of racist thugs. The East Enders have always supported people under threat . . . all they ask for is equal treatment." "What is not being said", he continued, "is the escalating attacks on Island people." "The Isle of Dogs", he said, was "used, it was orchestrated." He admitted that there were a minority of hard-core racists on the Island both before and after the BNP victory, but "it was the grandmothers and grandfathers voting for their grand children's future" that allowed the BNP in. "It was a protest vote purely on the housing allocation system on the Isle of Dogs."

The Island is now infamous as the place where the first BNP councillor was elected and will always be linked in people's minds with this event. Yet Peter felt the area's notoriety was unjustified. Recorded levels of harassment have declined. In 1997 reported racial incidents were approximately a third (63) of their 1994 (180) level and the local police Superintendent confirmed that the area was "much quieter now". There had, as some predicted, been another twist in the tale, with the victims becoming aggressors and, as in other contested areas of London, white youths had been seriously victimized by Asian youths.

Newcomers who retained their involvement with the community continued to play an important role and offered, through their plethora of professional experience, invaluable help and advice. "I only wish we'd had these people here in 1981," Peter said, "then it might have been different", perhaps forgetting for a moment that they did have a few even at the outset, who had played key roles, but whose emphasis on pragmatism was ignored.

******

Just after Christmas, Reg Ward's comments on the manuscript arrived at my home. He said that he "thoroughly enjoyed reading it despite his increasing discomfort with the analysis". He confessed that he had considerable difficulty in recognizing the Isle of Dogs as he had experienced it over the period 1980–88 and as he found it on his visits in 1997. His discomfort, he said, could only be explained in two ways. That he had simply read the Isle of Dogs constantly wrongly. Or, over time, he had become like his "Docklands Crow", so enjoying the experience and looking down from far too high above the ground and from the disturbing reality. There is another explanation. I began this research after Reg Ward had left the LDDC. I never witnessed the 1980–88 period and the place that he knew so well and which his energies and enthusiasm had shaped. People *see* different things and *perceive* different forces depending on *where* they are located and *why* they are there — as we have seen, the different stories and perceptions of Docklands development were so diverse and in many senses could not be reconciled. I am conscious too that when you have striven, with a dedicated team, to transform a place that was so desperately in need of regeneration and succeeded in making your dreams real-ities, it must be deeply dispiriting to have a sociologist take apart that cherished world, in which you passionately believed and for which you fought hard and for so long, in very difficult circumstances. Furthermore, despite the insight I gained through my contacts with those who worked in the Corporation, the business com-munity and the developers, I am conscious that my sociological background and support for the underdogs has not deserted me and that at one level the Docklands story is an excellent example of the neglect of what a community activist called the "little people" by big business, government and developers. However, that was only one part of the story. Those vested with the Docklands regeneration, as Sir Christopher Benson reminded me, "were not experts in sociology"; and while their vision may have been different they were certainly committed and in some cases passionate about their task.

the author

******

In March 1998 the last of the areas within the LDDC's jurisdiction returned to local authority control. How did people feel about that? Concerned. New upheavals and new uncertainties, not least because the millions of pounds collected in busi-ness rates in Docklands flows straight into treasury coffers, providing no additional income to local authorities, and once more no direct redistribution of wealth to the locality.

******

After 17 years of dramatic physical social and economic change on the Isle of Dogs, the development continues. In fact the building may go on for another seven to ten years. "There's no precedent for that", said Peter Wade. But what there was a precedent for was the conflict and anger that it can cause. I leave the Isle of Dogs and this book with a sense of sadness. There was so much energy and excitement that went into the development of Docklands, and without which the outcome would have been less breathtaking. But it wasn't simply those who were vested with the area's regeneration who had dreams; those who lived there previously had dreams too, which were largely unfulfilled. The sentimentalist in me asks was it so impossible that their lives and those in other areas of the neglected East End of London could have been transformed too?

Janet Foster, University of Cambridge
January 1998

References

ABRAMS, P. (1986) in BULMER, M. (1986).

ADAMS, C. (1987) *Across Seven Seas and Thirteen Rivers: Life stories of pioneer Sylhetti settlers in Britain*, London: THAP Books.

ALBURY, D. & SNEE, C. (1996) Higher Education in East London: A case for social renewal, in BUTLER, T. & RUSTIN, M. (eds) (1996), pp. 353–69.

ANDERSON, B. (1991) *Imagined Communities: Reflections on the Origin and Spread of Nationalism*, London: Verso.

ANDERSON, J. (1982) Growth, welfare and conflict, in *Social Change, Geography and Policy* (Block 6 in the Social Sciences Foundation Course), Milton Keynes: The Open University.

ASSOCIATION OF ISLAND COMMUNITIES (AIC) (1982) Figures reported in *Thames Mirror*, issue No. 5, 18 March 1982.

ASSOCIATION OF ISLAND COMMUNITIES (AIC) (1990) *Into the Nineties*, London: AIC.

AUDIT COMMISSION (1989) *Urban Regeneration and Economic Development: The Local Government Dimension*, London: HMSO.

BACHTLER, J. (1990) *Issues in European Regional Development*, Glasgow: European Policies Research Centre.

BAILEY, N. (1994) Towards a research agenda for public-private partnerships in the 1990s, *Local Economy*, **8**(4), 292–306.

BAILEY, N. (1995) *Partnership Agencies in British Urban Policy*, London: UCL Press.

BARRAS, R. (1994) How Much Office Building Does London Need? Paper presented to ESRC London Seminar Series, London School Economics.

BASSETT, K., BODDY, M., HARLOE, M. & LOVERING, J. (1989) Living in the fast lane: economic and social change in Swindon, in COOKE, P. (ed.) (1989).

BAUMGARTNER, M. (1988) *The Moral Order of a Suburb*, Oxford: Oxford University Press.

BEAZLEY, M. (1993) NHS cuts, penthouses and Canary Wharf: the community and Docklands development, *Regenerating Cities 6*, **1**(4), 45–8.

BECKER, H. (1965) Whose side are we on?, *Social Problems*, **14**, 239–47.

BEDARIDA, F. (1975) Urban growth and social structure in nineteenth century Poplar, *The London Journal*, **1**(2), 159–88.

BELL, C. & NEWBY, H. (1971) *Community Studies*, London: George Allen & Unwin.

BEYNON, H., HUDSON, R., LEWIS, J., SADLER, D. & TOWNSEND, A. (1989) "It's all falling apart here": coming to terms with the future in Teesside, in COOKE, P. (ed.) (1989).

BOOTH, C. (1889–91) *Life and Labour of the People of London*, 3 vols, London: Williams & Norgate.

BOOTH, C. (1902–3) BLPES *Booth Collection Group A*, vol. 33, fol.3.

BOTTOMS, A. & WILES, P. (1992) Explanation of crime and place in EVANS, D., FYFE, N., HERBERT, D. (eds) *Crime, Policing and Place: Essays in environmental criminology*, London: Routledge.

BOYLE, R. (1989) Partnership in practice *Local Government Studies*, March/April.

BROWNILL, S. (1990) *Developing London's Docklands: Another Great Planning Disaster?*, London: Paul Chapman Publishing.

BRUNSKILL, I. (1989) *The Regeneration Game, A Regional Approach to Regional Policy*, London: Institute for Public Policy Research.

BULMER, M. (1986) *Neighbours: The Work of Phillip Abrams*, Cambridge: Cambridge University Press.

BURGESS, R. (1984) *In the Field: An Introduction to Field Research*, London: Routledge.

BUTLER, T. & RUSTIN, M. (eds) (1996) *Rising in the East?: The Regeneration of East London*, London: Lawrence & Wishart.

CAMPBELL, B. (1993) *Goliath: Britain's Dangerous Places*, London: Methuen.

CARLEY, M. (1990) Housing and neighbourhood renewal: Britain's new urban challenge, quoted in FREESON, R. (1990).

CASTELLS, M. (1994) European cities, the informational society, and the global economy, *New Left Review*, **204**, 18–32.

CENSUS OF EMPLOYMENT (1981) *Census of Employment*, Watford: Department of Employment.

CENTRE FOR LOCAL ECONOMIC STRATEGIES (CLES) (1990) *Inner City Regeneration: A Local Authority Perspective*, CLES Monitoring Project on Urban Development Corporations, Manchester: Centre for Local Economic Strategies; written by COLENUTT, B. & TANSLEY, S.

CHURCH, A. (1988a) Urban regeneration in London Docklands: a five year policy review, *Environment and Planning C: Government and Policy*, **6**, 187–208.

CHURCH, A. (1988b) Demand-led planning, the inner city crisis and the Labour market, London Docklands evaluated, in HOYLE, B., PINDER, D., HUSSAIN, M. (eds) *Revitalising the Waterfront*, London: Bellhaven.

CHURCH, A. (1991) Urban regeneration, economic restructuring and community response in London Docklands, Paper presented to the London and New York: Cities in Decay Conference.

CHURCH, A. & FROST, M. (1992) The employment focus of Canary Wharf and the Isle of Dogs: a labour market perspective, *The London Journal*, **17**(2), 135–152.

CHURCH, A. & HALL, J. (1989) Local initiatives for economic regeneration, in HERBERT, D. & SMITH, D. (eds) *Social Problems and the City: New perspectives*, Oxford: Oxford University Press, pp. 345–69; quoted in CROW, G. & ALLAN, G. (1994).

CHURCHES STANDING COMMITTEE FOR LONDON'S DOCKLANDS (1995) *Housing and the Regeneration of Docklands: Creating Sustainable Communities*, A report prepared for the Churches Standing Committee for Docklands.

CLARK, G. (1996) East London and Europe, in BUTLER, T. & RUSTIN, M. (1996), pp. 316–26.

COCHRANE, A. (1991) Restructuring local politics: Sheffield's economic policies in the 1980s, Paper presented to Eighth Urban Changes and Conflict Conference, Lancaster.

COHEN, A. (1982) A sense of time, a sense of place: the meaning of close social association in Whalsay, Shetland, in COHEN, A. (1982), pp. 21–49.

COHEN, A. (1982) Blockade: a case study of local consciousness in an extra-local event, in COHEN, A. (ed.) (1982), pp. 292–321.

COHEN, A. (1982) (ed.) *Belonging: Identity and social organisation in British Rural Cultures*, Manchester: Manchester University Press.

COHEN, A. (1989) *The Symbolic Construction of Community*, London: Routledge (originally published 1985).

COHEN, P. (1993) *Home Rules: Some Reflections on Racism and Nationalism in Everyday Life*, London: University of East London, The New Ethnicities Unit.

COHEN, P. (1996) All white on the night? Narratives of nativism on the Isle of Dogs, in BUTLER, T. & RUSTIN, M. (eds) (1996), pp. 170–96.

COHEN, P., QURESHI, T. & TOON, I. (1994) Island Stories: Race and class in the remaking of East Enders. Paper presented to ESRC London Seminar Series.

COLE, T. (1984) *Life and Labor in the Isle of Dogs: The Origins and Evolution of an East London Working Class Community 1880–1980*, Ph.D. Thesis, University of Oklahoma.

COLENUTT, B. & LOWE, J. (1981) Does London need the Docklands Urban Development Corporation, *London Journal*, **7**(2), 235–8.

COMMISSION FOR RACIAL EQUALITY (CRE) (1988) *Homelessness and Discrimination: Report of a formal investigation into the London Borough of Tower Hamlets*, London: CRE.

COOKE, P. (1989) The local question — revival or survival, in COOKE, P. (ed.) (1989), pp. 296–306.

COOKE, P. (ed.) (1989) *Localities: The Changing Face of Urban Britain*, London: Unwin Hyman.

COOPER, J. & QURESHI, T. (1993) *Through Patterns Not Our Own: A Study of the Regulation of Racial Violence on the Council Estates of East London*, London: University of East London, New Ethnicities Research and Education Group.

COOPERS AND LYBRAND ASSOCIATES (1980) *Docklands Urban Development Corporation, Draft Business Plan 1981/2*, vol. 1.

CORNWELL, J. (1984) *Hard-Earned Lives: Accounts of health and illness from East London*, London: Tavistock.

COWPER, B.H. (1853) *Historical Account of Millwall*, quoted in *Greater London: A Narrative of its History, People and Places* (c.1900), vol. 1, London: Cassell.

COX, A. (1995) *Docklands in the Making: The Redevelopment of the Isle of Dogs, 1981–1995*, published for the Royal Commission on the Historical Monuments of England, London: Athlone Press.

CRILLEY, D. (1992) Heat and dust from crumbling Olympia, in *The Higher*, 12 June, p. 18.

CRILLEY, D. (1993) Architecture as advertising: constructing the image of redevelopment, in KEARNS, G. & PHILO, C. (eds) (1993), pp. 231–52.

CROW, G. & ALLAN, G. (1994) *Community Life: An introduction to local social relations*, New York: Harvester Wheatsheaf.

CROW, G. & ALLAN, G. (1995) Community types, community typologies and community time, *Time and Society*, **4**(2), 147–166.

CYBRIWSKY, R. (1978) Social aspects of neighbourhood change, *Annals of the Association of American Geographers*, **68**(1 March), 17–33.

DAMER, S. (1974) Wine Alley: the sociology of a dreadful enclosure, *Sociological Review*, **22**, 221–248.

DAMER, S. (1989) *From Moorepark to "Wine Alley": the Rise and Fall of a Glasgow Housing Scheme*, Edinburgh: Edinburgh University Press.

DARBY, M. (1973) A local resident's view, in HALL, J. et al. The Docklands Study Revisited *East London Papers*, No. 15, 30–6.

DARKE (1991) Gambling on sport: Sheffield's regeneration strategy for the 90s. Paper presented to Eighth Urban Changes and Conflict Conference, Lancaster.

DICKEN, P. (1986) *Global Shift*, London: Harper & Row.

DICKENS, P. (1988) *One Nation? Social Change and the Politics of Locality*, London: Pluto.

DICKENS, P. (1990) *Urban Sociology*, London: Harvester Wheatsheaf.

DOCKLANDS CONSULTATIVE COMMITTEE (DCC) (1985) *Four Year Review of the London Docklands Development Corporation*, London: DCC.

DOCKLANDS CONSULTATIVE COMMITTEE (DCC) (1988) *Urban Development Corporations: Six Years in London's Docklands*, London: DCC.

DOCKLANDS CONSULTATIVE COMMITTEE (DCC) (1990) *The Docklands Experiment: A critical review of eight years of the London Docklands Development Corporation*, London: DCC.

DOCKLANDS CONSULTATIVE COMMITTEE, THE DOCKLANDS FORUM and CAMPAIGN for HOMES IN CENTRAL LONDON (1989) *Priced Out of Town*. Papers, reports and situation update from A Conference on the Future for Housing in East London, 15 November 1988.

DOCKLANDS CONSULTATIVE COMMITTEE AND THE DOCKLANDS FORUM (1995) *LDDC Exit Putting Housing on the Agenda*, September.

DOCKLANDS FORUM (1987) *Housing in Docklands*, London: Docklands Forum.

DOCKLANDS FORUM (1989) *Community Benefit?*, London: Docklands Forum.

DOCKLANDS FORUM (1993) *Race and Housing in London's Docklands*, London: Docklands Forum.

DOCKLANDS FORUM/BIRBECK COLLEGE (1990) *Employment in Docklands*, London: Docklands Forum.

DOCKLANDS JOINT COMMITTEE (DJC) (1976) *London Docklands; A Strategic Plan*, London: DJC.

DONNISON, D. (1987) Conclusions, in DONNISON, D. & MIDDLETON, A. (eds) *Regenerating the Inner City*, London: RKP.

EADE, J. (1995) Profiting from places: local residents and struggles over land in London. Papers presented to British Sociological Association Conference, University of Leicester.

EADE, J. (1997) Reconstructing places: changing images of locality in Docklands and Spitalfields, in EADE, J. (ed.) *Living the Global City: Globalization as Local Process*, London: Routledge, pp. 127–45.

ELIAS, N. (1976) Introduction: A theoretical essay on established and outsider relations, in ELIAS, N. & SCOTSON, J. (1994), pp. xv–lii.

ELIAS, N. & SCOTSON, J. (1964) "Preface" in ELIAS, N. & SCOTSON, J. (1994) *The Established and the Outsiders: A Sociological Enquiry into Community Problems*, 2nd edn, London: Sage.

ELIAS, N. & SCOTSON, J. (1994) *The Established and the Outsiders: A Sociological Enquiry into Community Problems*, 2nd edn, London: Sage.

ELLMERS, C. & WERNER, A. (1991) *A Dockland Life: A Pictorial History of London's Docks 1860–1970*, London: Mainstream Publishing.

EYLES, J. (1976) *Environmental Satisfaction and London Docklands: Problems and Policies in the Isle of Dogs*, Occasional Paper No. 5, Department of Geography, Queen Mary College.

FAINSTEIN, S. (1990) Economics, politics and development policy: the convergence of New York and London, *International Journal of Urban and Regional Research*, **14**(4), 553–75.

FAINSTEIN, S. (1994) *The City Builders: Property, Politics and Planning in London and New York*, Oxford: Blackwell.

FAINSTEIN, S., GORDON, I. & HARLOE, M. (eds) (1992) *Divided Cities: New York & London in the Contemporary World*, Oxford: Blackwell.

FIELD, F. (1989) *Losing Out: The emergence of Britain's Underclass*, Oxford: Blackwell.

FIELDING, N. (1981) *The National Front*, London: RKP.

FITCH, R. (1994) New York's road to ruin: explaining New York's aberrant economy, *New Left Review*, **207**(Sept–Oct), 17–48.

FORMAN, C. (1989) *Spitalfields: A Battle for Land*, London: Hilary Shipman.

FORREST, R. & MURIE, A. (1988) *Selling the Welfare State: The Privatisation of Public Housing*, London: Routledge.

FOSTER, J. (1988) Crime and community: an ethnographic evaluation, unpublished report of work in progress.

FOSTER, J. (1990) *Villains: Crime and Community in the Inner City*, London: Routledge.

FOSTER, J. (1992) Living with the Docklands' Redevelopment: community view from the Isle of Dogs, *The London Journal*, **17**(2), 170–82.

FOSTER, J. (1995) Informal social control and community crime prevention, *British Journal of Criminology*, **34**(4), 563–84.

FOSTER, J. (1996) "Island Homes for Island People": Competition, conflict and racism in the battle over public housing on the Isle of Dogs, in SAMSON, C. & SOUTH, N. (eds) *The Social Construction of Social Policy: Methodologies, Racism Citizenship and the Environment*, London: Macmillan, 148–168.

FOSTER, J. (1997) Challenging perceptions: "community" and neighbourliness on a difficult to let estate, in JEWSON, J. & MACGREGOR, S. (eds) *Transforming Cities: Contested Governance and New Spatial Divisions*, London: Routledge, pp. 116–26.

FOSTER, J. & HOPE, T. (1993) *Housing, Community and Crime: the impact of the Priority Estates Project*, Home Office Research Study No. 131, London: HMSO.

FREE, R. (1904) *Seven Years' Hard*, London: William Heinemann.

FREESON, R. (1990) Partnerships for urban renewal, *Search* (August), York: Joseph Rowntree Foundation.

FRIEDEN, B.J. & SAGALYN, L.B. (1989) *Downtown Inc: How America rebuilds cities*, Cambridge, Mass.: MIT.

GANS, H. (1961) Planning and social life: friendship and neighbor relations in sub-urban communities, *Journal of American Institute of Planners*, **27**, 134–40.

GANS, H. (1962) *The Urban Villagers*, New York: Free Press.

GANS, H. (1967) *The Levittowners*, London: Allen Lane.

GIDDENS, A. (1984) *The Constitution of Society*, Cambridge: Polity Press.

GIDDENS, A. (1991) *Modernity and Self-Identity: Self and Society in the Late Modern Age*, Cambridge: Polity Press.

GILROY, P. (1987) *There Aint No Black in the Union Jack*, London: Hutchinson.

GILROY, R. & WILLIAMS, R. (1991) Community involvement in neighbourhood regeneration: a British perspective, in ALTERMAN, R. & CARS, G. (eds) *Neighbourhood Regeneration: An International Evaluation*, London: Mansell.

GLASS, R. (1963) (ed.) *London: Aspects of Change*, Centre for Urban Studies, London: Macgibbon & Kee.

GLASS, R. (1973) The mood in London, in DONNISON, D. & EVERSLEY, D. (eds) *Urban Patterns Problems and Policies*, London: Heinemann.

GLYN, A. & MILIBAND, D. (1994) (eds) *Paying for Inequality: the Economic Cost of Social Injustice*, London: Institute of Public Policy Research.

GOLDTHORPE, J., LOCKWOOD, D., BECHHOFFER, F., PLATT, J. (1968–1970) *The Affluent Worker*, 3 vols, Cambridge: Cambridge University Press.

GOODWIN, M. (1993) The city as commodity: the contested spaces of urban development, in KEARNS, G. & PHILO, C. (eds) (1993), pp. 145–62.

GREATER LONDON COUNCIL (GLC) (1985) *The East London File — Mark II*, London: GLC.

GREATER LONDON COUNCIL (GLC) (1986) *Docklands Housing Needs Survey*, London: GLC.

HALL, P. (1980) *Great Planning Disasters*, London: Weidenfeld & Nicolson.

HALL, P. (1988) *Cities of Tomorrow*, Oxford: Blackwell.

HAMNETT, C. & WILLIAMS, P. (1980) Social change in London: a study of gentrification, *Urban Affairs Quarterly*, **15**(4), 469–87.

HARDY, D. (1983a) *Making Sense of the London Docklands: Processes of Change*, Geography and Planning Paper No. 9, Middlesex Polytechnic.

HARDY, D. (1983b) *Making Sense of the London Docklands: People and Places*, Geography and Planning Paper No. 10, Middlesex Polytechnic.

HARLOE, M. & FAINSTEIN, S. (1992) Conclusion; the divided cities, in FAINSTEIN, S., GORDON, I. & HARLOE, M. (eds) (1992), pp. 236–268.

HARVEY, D. (1989a) *The Urban Experience*, Oxford: Basil Blackwell.

HARVEY, D. (1989b) *The Condition of Postmodernity: An Enquiry into the Origins of Cultural Change*, Oxford: Basil Blackwell.

HARVEY, D. (1996) Cities or urbanization?, *City* **1/2**, 38–61.

HAUGHTON, G. & WHITNEY, D. (1989) Equal urban partners?, *The Planner*, **75**(34), 9–11.

HEBBERT, M. (1992) One "Planning Disaster" after another: London Docklands 1970–1992, *The London Journal*, **17**(2), 115–34.

HILL, D. (1994) *Citizens and Cities: Urban Policy in the 1990s*, Hemel Hempstead: Harvester Wheatsheaf.

HILL, S. (1976) *The Dockers: Class and Tradition in London*, London: Heinemann.

HILLMAN, J. (1998) *Learning to Live and Work Together*, London: LDDC.

HOBBS, R. (1988) *Doing the Business*, Oxford: Clarendon Press.

HOLCOMB, B. (1993) Revisioning place: de- and re-constructing the image of the industrial city, in KEARNS, G. & PHILO, C. (eds) (1993), pp. 133–44.

HOLLAMBY, T. (1990) *Docklands — London's Backyard into Frontyard*, London: Docklands Forum.

HOPE, T. (1996) Communities, crime and inequality in England and Wales, in BENNETT, T. (ed.) *Preventing Crime and Disorder: Targeting Strategies and Responsibilities*, Cambridge: University of Cambridge Cropwood Series.

HOPE, T. & FOSTER, J. (1992) Conflicting forces changing the dynamics of crime and community on "problem" estates, *British Journal of Criminology*, **32**(4), 488–503.

HOSTETTLER, E. (undated, a) *An Outline History of the Isle of Dogs*, London: Island History Trust.

HOSTETTLER, E. (undated, b) *Isle of Dogs Old and New Pictures 1900–1993*, vol. 1, London: Ilford Old & New Series.

HOSTETTLER, E. (ed.) (undated c) *Island Women: Photographs of East End Women 1897 to 1983*, The Photo-Library 5: Nishen Photography.

HOSTETTLER, E. (ed.) (undated d) *The Island at War: Memories of War-Time Life on the Isle of Dogs, East London*, London: Island History Trust.

HOSTETTLER, E. (1992) Dogs: the truth, *The Islander*, July 1992, p. 6.

HOUSE OF COMMONS EXPENDITURE COMMITTEE (1974–5) *Redevelopment of London Docklands*, PP, XXII HC 348.

HOUSE OF COMMONS EXPENDITURE COMMITTEE (Environmental Subcommittee) *Redevelopment of London Docklands: Minutes of Evidence*, PP 1978–9, XVII HC 99.

HOUSE OF COMMONS (1988) *The Employment Effect of UDCs*, Employment Committee Third Report HC327-1, London: HMSO.

HOUSE OF LORDS (1981) *Report of House of Lords Select Committee on the London Docklands Development Corporation (Area and Constitution Order)*, London: HMSO.

HUSBANDS, C. (1982) East End Racism 1900–1980, *The London Journal*, **8**(1), 3–26.

INFORM ASSOCIATES (1997) *Responding to Need in the Isle of Dogs An Agenda for Action*, prepared by Inform Associates for The Isle of Dogs Community Foundation.

IMRIE, R. & THOMAS, H. (1993) *British Urban Policy and the Urban Development Corporations*, London: Paul Chapman.

ISLE OF DOGS COMMUNITY FOUNDATION (1996/7) *Helping the People of the Isle of Dogs*, Annual Report. London: Mirror Group.

JACOBS, B. (1988) *Racism in Britain*, London: Christopher Helm.

JACOBS, B.D. (1992) *Fractured Cities: Capitalism, community and empowerment in Britain and America*, London: Routledge.

JAGER, M. (1986) Class definition and the aesthetics of gentrification: Victoriana in Melbourne, in SMITH, N. & WILLIAMS, P. (eds).

JANOWITZ, M. (1967) *The Community Press in an Urban Setting*, Chicago: University of Chicago Press.

JOHNS, T. (1996) Cockney trickery or common sense?, *The Islander*, July, p. 7.

JOINT DOCKLANDS ACTION GROUP (JDAG) (1982) *Employment in Docklands*, London: JDAG.

JOINT DOCKLANDS ACTION GROUP (JDAG) (1976) *The Fight for the Future*, London: JDAG.

KANE, F. (1987) The new eastenders, *The Independent*, 26 September.

KEARNS, G. & PHILO, C. (eds) *Selling Places: The City as Cultural Capital, Past and Present*, Oxford: Pergamon Press.

KEITH, M. & CROSS, M. (1993) Racism and the PostModern City, in CROSS, M. & KEITH, M. (eds) *Racism, The City and The State*, London: Routledge.

KEMPSON, E. (1996) *Life on a low income*, York: Joseph Rowntree Foundation.

KLAUSNER, D. (1987) Infrastructure investment and political ends, the case of London Docklands, *Local Economy*, **11**(4), 47–59.

KLEIN (1965) *Samples from English Culture*, London: RKP.

KNIGHT, FRANK & RUTLEY (1989) *Docklands Commercial and Residential Development*, London: KFR.

KOFMAN, E. (1996) Whose City?: Gender and immigrants in globalizing European cities, paper presented to ESRC London seminar series, London School of Economics.

LASH, S. & URRY, J. (1994) *Economies of Signs and Space*, London: Sage.

LASHMAR, P. & HARRIS, A. (1988) Anarchists step up class war in cities, *The Observer*, 10 April, p. 9.

LAWRENCE, E. (1982) In the abundance of water the fool is thirsty: sociology and black "pathology", in CCCS, University of Birmingham, *The Empire Strikes Back: race and racism in 70s Britain*, London: Hutchinson.

LE GALES, P. & MAWSON, J. (1994) *Management Innovations in Urban Policy: Lessons from France*, Luton: Local Government Management Board.

LEDGERWOOD, G. (1985) *Urban Innovation: The Transformation of London's Docklands 1968–84*, Aldershot: Gower.

LEVY, S. & ARNOLD, R.K. (1972) *An Evaluation of Four Growth Alternatives in the City of Milpitas, 1972–77*, Technical Memorandum Report. Palo Alto, Calif.: Institute of Regional and Urban Studies.

LEWIS, N. (1992) *Inner City Regeneration: The Demise of Regional and Local Government*, Buckingham: Open University Press.

LOCAL GOVERNMENT LAND AND PLANNING ACT (LGLPA), 1980, London: HMSO.

LOFTMAN, P. and NEVIN, B. (1994) Prestige project developments: economic renaissance or economic myth? a case study of Birmingham, *Local Economy*, **8**(4), 307–25.

LOGAN, J. (1990) From beyond the city limits: local impacts of the globalization of real estate, paper presented to Urban Affairs Conference.

LOGAN, J. & MOLOTCH, H. (1987) *Urban Fortunes: The political economy of place*, Berkeley: University California Press.

LONDON BOROUGH OF TOWER HAMLETS (LBTH) (1995a) *An Analysis of the London Reading Test*, London: LBTH Education Policy, Quality and Equality.

LONDON BOROUGH OF TOWER HAMLETS (LBTH) (1995b) *Statistical Bulletin* LBTH London: Education & Community Services.

LONDON BOROUGH OF TOWER HAMLETS (LBTH) (1997) *Draft Community Care Plan 1997/8*, London: LBTH.

LONDON DOCKLANDS DEVELOPMENT CORPORATION (LDDC) (1988a) *Report of London Docklands' Survey*, conducted by Gallop for the LDDC, London: LDDC.

LONDON DOCKLANDS DEVELOPMENT CORPORATION (LDDC) (1988b) *London Docklands Housing Review*, London: LDDC.

LONDON DOCKLANDS DEVELOPMENT CORPORATION (LDDC) (1988c) LDDC *Briefing: Jobs*, London: LDDC.

LONDON DOCKLANDS DEVELOPMENT CORPORATION (LDDC) (1990) *The London Docklands Household Survey*, London: LDDC.

LONDON DOCKLANDS DEVELOPMENT CORPORATION (LDDC) (1991) *Report of London Docklands Survey*, conducted for the LDDC by Gallup.

LONDON DOCKLANDS DEVELOPMENT CORPORATION (LDDC) (1992) *Briefing — Jobs*, London: LDDC.

LONDON DOCKLANDS DEVELOPMENT CORPORATION/NOP (1994/5) *Employment Survey*, London: LDDC.

LONDON DOCKLANDS DEVELOPMENT CORPORATION (LDDC) (1995) *Key Facts and Figures*, London: LDDC, Corporate Information Unit.

LONDON DOCKLANDS DEVELOPMENT CORPORATION (LDDC) (1995a) *Property Fact Sheet*, May, London: LDDC Press Office.

LONDON DOCKLANDS DEVELOPMENT CORPORATION (LDDC) (1996) *Key Facts and Figures*, London: LDDC, Corporate Information Unit.

LONDON DOCKLANDS DEVELOPMENT CORPORATION (1996b) *Local Communities*, prepared for the LDDC by MORI, London: LDDC.

LONDON DOCKLANDS DEVELOPMENT CORPORATION (LDDC) (1998) *Housing in the Renewed London Docklands*, London: LDDC.

LONDON DOCKLANDS STUDY TEAM (1973) *Docklands: Redevelopment Proposals for East London*, 2 vols, London: Travers Morgan and Partners.

LONDON RESEARCH CENTRE (1993) *Housing Statistics*, London: London Research Centre.

LONDON RESEARCH CENTRE (1996) *The Capital Divided, Mapping Poverty and Social Exclusion in London*, London: London Research Centre.

LONDON, B. (1980) Gentrification as urban reinvasion: some preliminary definitional and theoretical considerations, in LASKA, S. & SPAIN, D. (eds) *Back to the City: Issues in Neighborhood Renovation*, New York: Pergamon Press.

LONDON WEEKEND TELEVISION (1994) *The London Documentary: Councillor Beackon and the Battle for Millwall*, 17 July.

LOWE, M. (1993) Local hero! An examination of the role of the regional entrepreneur in the regeneration of Britain's regions, in KEARNS, G. & PHILO, C. (eds) (1993), pp. 211–30.

LUPTON, R. (1997) *The Isle of Dogs Development, Deprivation and Diversity, Phase One of the Isle of*

Dogs Community Foundation's Review of the Needs of the Isle of Dogs Area, prepared by Ruth Lupton for the Isle of Dogs Community Foundation.

MacCormac, R. (1993) Social Purpose, Image and Structure: Reflections on some recent architectural work, an interview with Richard MacCormac in *Regenerating Cities*, **6**, 22–6.

Mackintosh, M. (1992) Partnerships: issues of policy and negotiation, *Local Economy*, **7**(3), 210–24.

Marcuse, P. (1986) Abandonment, gentrification and displacement: the linkages in New York City, in Smith, N. & Williams, P. (eds) (1986).

Marshall, Sir F. (1978) *The Marshall Enquiry on Greater London*, London: Greater London Council.

Marshall, J.N. in collaboration with Wood, P. (et al.) (1988) *Services and Uneven Development*, Oxford: Oxford University Press.

Massey, D. (1994) *Space, Place and Gender*, Cambridge: Polity Press.

Martin, R. & Rowthorn, B. (eds) (1986) *The Geography of Deindustrialisation*, London: Macmillan.

Mayhew, H. (1860) *London Labour and the London Poor*, 4 vols, Griffin, Bohn & Co. (The section on casual labour in the docks first appeared in *Morning Chronicle* 26 October 1849.)

Mirrlees-Black, C., Mayhew, P. & Percy, A. (1996) *The British Crime Survey*, Home Office Statistical Bulletin, Issue 19/96, London: HMSO.

Moore, R. (1995) Poverty and partnership in the Third European Poverty Programme: the Liverpool case Paper presented to British Sociological Association Conference, Leicester.

MORI (1997) *Policing in the Community* (research conducted for Metropolitan Police), London: MORI.

Murray, C. (1984) *Losing Ground: American Social Policy, 1950–1980*, New York: Basic Books.

Murray, C. (1994) *Underclass: The Crisis Deepens*, Choice in Welfare Series No. 20, London: IEA in association with the *Sunday Times*.

Murray, C. (1996) *Charles Murray and the Underclass: The Developing Debate*, Choice in Welfare Series No. 33, London: IEA in association with the *Sunday Times*.

Murshid, K. (1993) Reversing a racist agenda: community experience and community politics in Tower Hamlets, *Regenerating Cities* 6, **1**, 16–20.

National Audit Office (1993) *The Achievement of the Second and Third Generation Urban Development Corporations*, London: HMSO.

Newby, H. (1977) *The Deferential Worker*, London: Allen Lane.

Nossiter, B. (1978) *Britain: A Future that Works*, London: Deutsch.

O'Brien, M. & Jones, D. (1996) Family life in Barking and Dagenham, in Butler, T. & Rustin, M. (eds) (1996), pp. 61–80.

OPCS (1993) *1991 Census, Ethnic Group and Country of Birth, Great Britain*, London: HMSO.

Open University (1982) *Social Sciences Foundation Course Unit 25, Social Change: Geography and Policy*, Milton Keynes: Open University Press.

Osborne, T. (1993) Beyond monolithic development: a London and national developer, an interview with Trevor Osborne, *Regenerating Cities* 6, **1**, 12–13.

Payne, G. (1993) The community revisited: some reflections on the community study as a method Paper presented to British Sociological Association Conference, University of Essex.

Payne, G. (1996) Imagining the community: some reflections on the community study as a method, in Samson, C. & South, N. (eds) *The Social Construction of Social Policy*, London: Macmillan, pp. 17–33.

PEARSON, G. (1983) *Hooligan: the history of respectable fears*, London: Macmillan.

PHILLIPS, D. (1986) *What Price Equality?*, GLC Housing Research and Policy Report 9, London: GLC.

POLICY STUDIES INSTITUTE (1993) *Review of Advice Services in Leicester*, Leicester: Leicester City Council.

POWER, A. (1987) *Property before People: the management of twentieth century council housing*, London: Allen & Unwin.

POWER, A. (1987a) *The PEP Guide to Local Management*, 3 vols, London: Department of Environment.

POWER, A. (1989) Housing, community and crime, in Downes, D. (ed.) *Crime and the City*, London: Macmillan.

RATCLIFFE, P. (1992) Renewal, regeneration and "race": issues in urban policy, *New Community*, **18**(3), 387–400.

RATCLIFFE, P. (1997) "Race", housing and the city, in JEWSON, J. & MACGREGOR, S. (eds) *Transforming Cities: Contested Governance and New Spatial Divisions*, London: Routledge.

RAUSE, V. (1989) Pittsburgh cleans up its act, *New York Times Magazine*, 26 November.

REES, R. (1967) The Port of London and economic change, *East London Papers*, X, 2, p. 121.

REX, J. (1986) *Race and Ethnicity*, Milton Keynes: Open University Press.

REYNOLDS, F. (1986) *The Problem Housing Estate: An Account of Omega and Its People*, Aldershot: Gower.

RIX, V. (1996) Social and demographic change in East London, in BUTLER, T. & RUSTIN, M. (eds) (1996), pp. 20–60.

ROBINSON, F., SHAW, K. & LAWRENCE, M. (1994) Urban development corporations and the creation of employment: an evaluation of Tyne and Wear and Teesside Development Corporations, *Local Economy*, **8**(4), 326–37.

ROBSON, B. (1988) *Those Inner Cities*, Oxford: Clarendon Press.

ROBSON, B. (1994) Assessing the impact of urban policy, *Regenerating Cities 7*, **2**(1), p. 40.

ROBSON, B. et al. (1994) *Assessing the impact of urban policy* (DOE Inner Cities Research Programme), London: HMSO.

ROGER TYM AND PARTNERS (1983) *The Potential of Future Docks Use of the Royals*, report for the Greater London Council.

ROGER TYM AND PARTNERS (1987) *The Economy of the Isle of Dogs*, prepared for Isle of Dogs Neighbourhood Committee, June 1987.

ROSE, G. (1990) Imagining Poplar in the 1920s: contested concepts of community, *Journal of Historical Geography*, **16**(4), 425–37.

RYAN, M. & ISAACSON, P. (1974) Structure planning in Docklands, *Political Quarterly*, **45**, 323–32.

SADLER, D. (1993) Place-marketing, competitive places and the construction of hegemony in Britain in the 1980s, in KEARNS, G. & PHILO, C. (eds) (1993), pp. 175–92.

SAFIER, M. (1993) Leading from the ground up: organisational landscapes, institutional innovations and shifting terms of trade in community-led urban regeneration, *Regenerating Cities*, 1(3/4), 33–6.

SASSEN, S. (1991) *The Global City: New York, London, Tokyo*, Princeton, NJ: Princeton University Press.

SAVAGE, M. & WARDE, A. (1993) *Urban Sociology, Capitalism and Modernity*, London: Macmillan.

SCHOEN, D.E. (1970) *Powell and the Powellites*, London: Macmillan.

SCOTTISH OFFICE (1993) *Progress in Partnership: A Consultation Paper on the Future of Urban Regeneration Policy in Scotland*, Edinburgh: Scottish Office.

SOCIAL TRENDS (1995), London: HMSO.

SOUTH EAST REGION OF THE TUC (SERTUC) (1989) *Wall Street on Water? Employment patterns in London Docklands*, Gavin Pointer for SERTUC.

SHORT, J. (1989) Yuppies, yuffies and the new urban order, *Transactions of the Institute of British Geographers*, **14**, 173–88.

SKELLINGTON, R. (with MORRIS, P.) (1992) *"Race" in Britain Today*, London: Sage/Open University Press.

SMITH, D. (1992) (ed.) *Understanding the Underclass*, London: Policy Studies Institute.

SMITH, N. (1987) Of yuppies and housing; gentrification, social restructuring, and the urban dream, *Environment and Planning D: Society and Space*, **5**, 151–72.

SMITH, N. & WILLIAMS, P. (1986) (eds) *Gentrification of the City*, London: Unwin Hyman.

SMITH, S. (1989) *The Politics of "Race" and Residence: Citizenship, Segregation and White Supremacy in Britain*, Oxford: Polity Press.

SPEARING, N. (1978) London docks: up or down river?, *The London Journal*, **4**(2), 231–44.

SOLOMOS, J. (1988) *Black Youth, Racism and the State*, Cambridge: Cambridge University Press.

SOLOMOS, J. (1989) *Race and Racism in Contemporary Britain*, London: Macmillan.

STRATHERN, A.M. (1982) The place of kinship: kin, class and village status in Elmdon, Essex, in COHEN, A. (ed.) (1982), pp. 72–100.

STRATHERN, M. (1982) The village as an idea: constructs of village-ness in Eldom, Essex in COHEN, A (ed.) (1982), pp. 247–77.

SUTTLES, G. (1972) *The Social Construction of Communities*, Chicago: University of Chicago Press.

TAUB, R., TAYLOR, D. & DUNHAM, J. (1984) *Paths of Neighbourhood Change: race and crime in urban America*, Chicago: Chicago University Press.

TAYLOR, I., EVANS, K. & FRASER, P. (1996) *A Tale of Two Cities: Global change, local feeling and everyday life in the North of England. A Study in Manchester and Sheffield*, London: Routledge.

THOMPSON, T. (1995) *Gangland Britain: Inside Britain's Most Dangerous Gangs*, London: Hodder & Stoughton.

TOMPSON, K. (1988) *Under Siege: Racial Violence in Britain Today*, London: Penguin.

URBAN CHANGE GROUP (undated) *Isle of Dogs Survey, Householders Summary*, London: University College London, Department of Geography.

WALFORD, E. (c.1900) *Greater London: A narrative of its history, its people and its places*, London: Cassell & Company Ltd.

WALLMAN, S., BELL, L., DHOOGE, Y., EVANS, C., MILLS, G. & MIRA-SMITH, C. (1987) *The Millwall Survey*, London: University College London, Department of Geography, Urban Change Group.

WATSON, S. (1991) Gilding the smokestacks: the new symbolic representations of deindustrialised regions, *Environment and Planning D: Society and Space*, **9**, 59–70.

WEINREB, B. & HIBBERT, C. (1983) *The London Encyclopaedia*, London: Macmillan.

WHIPP, R. (1990) *Patterns of Labour: Work and Social Change in the Pottery Industry*, London: Routledge.

WHITE, J. (1986) *The Worst Street in North London: Campbell Bunk, Islington, between the wars*, London: RKP.

WHYATT, A. (1996) London East: gateway to regeneration, in BUTLER, T. & RUSTIN, M. (eds) (1996), pp. 265–87.

WIDGERY, D. (1991) *Some Lives! A GP's East End*, London: Sinclair-Stevenson.

WILLIAMS, P. & SMITH, N. (1986) From "renaissance" to restructuring: the dynamics of contemporary urban development, in SMITH, N. & WILLIAMS, P. (eds) (1986).

WILLMOTT, P. & HUTCHISON, R. (1992) *Urban trends 1: a report on Britain's deprived urban areas*, London: Policy Studies Institute.

WILSON, D.F. (1972) *Dockers: The Impact of Industrial Change*, London: Fontana.

WOLKOWITZ, C. (1994) Unemployment, dependency and regeneration strategies, review article, *Work, Employment and Society*, **8**(3), 439–52.

YOUNG, M. & WILLMOTT, P. (1972) *Family and Kinship in East London*, Harmondsworth: Penguin (First published 1957).

ZUKIN, S. (1982) *Loft Living: Culture and Capital in Urban Change*, Baltimore: John Hopkins University Press.

Index